Universitext

Universitext

Universitext is a series of textbooks that presents material from a wide variety of mathematical disciplines at master's level and beyond. The books, often well class-tested by their author, may have an informal, personal even experimental approach to their subject matter. Some of the most successful and established books in the series have evolved through several editions, always following the evolution of teaching curricula, into very polished texts.

Thus as research topics trickle down into graduate-level teaching, first textbooks written for new, cutting-edge courses may make their way into *Universitext*.

More information about this series at http://www.springer.com/series/223

David Borthwick

Introduction to Partial Differential Equations

Springer

David Borthwick
Department of Mathematics
and Computer Science
Emory University
Atlanta, GA
USA

ISSN 0172-5939 ISSN 2191-6675 (electronic)
Universitext
ISBN 978-3-319-84051-2 ISBN 978-3-319-48936-0 (eBook)
DOI 10.1007/978-3-319-48936-0

Printed on acid-free paper

This Springer imprint is published by Springer Nature
The registered company is Springer International Publishing AG
The registered company address is: Gewerbestrasse 11, 6330 Cham, Switzerland

Dedicated to my parents

Preface

Partial differential equations (PDE) first appeared over 300 years ago, and the vast scope of the theory and applications that have since developed makes it challenging to give a reasonable introduction in a single semester. The modern mathematical approach to the subject requires considerable background in analysis, including topics such as metric space topology, measure theory, and functional analysis.

This book is intended for an introductory course for students who do not necessarily have this analysis background. Courses taught at this level traditionally focus on some of the more elementary topics, such as Fourier series and simple boundary value problems. This approach risks giving students a somewhat narrow and outdated view of the subject.

My goal here is to give a balanced presentation that includes modern methods, without requiring prerequisites beyond vector calculus and linear algebra. To allow for some of the more advanced methods to be reached within a single semester, the treatment is necessarily streamlined in certain ways. Concepts and definitions from analysis are introduced only as they will be needed in the text, and the reader is asked to accept certain fundamental results without justification. The emphasis is not on the rigorous development of analysis in its own right, but rather on the role that tools from analysis play in PDE applications.

The text generally focuses on the most important classical PDE, which are the wave, heat, and Laplace equations. Nonlinear equations are discussed to some extent, but this coverage is limited. (Even at a very introductory level, the nonlinear theory merits a full course to itself.)

I have tried to stress the interplay between modeling and mathematical analysis wherever possible. These connections are vital to the subject, both as a source of problems and as an inspiration for the development of methods.

I owe a great debt of gratitude to my colleague Alessandro Veneziani, with whom I collaborated on this project originally. The philosophy of the book and choice of topics were heavily influenced by our discussions, and I am grateful for his support throughout the process. I would also like to thank former student Dallas Albritton, for offering comments and suggestions on an early draft. Thanks also to the series editors at Springer, for comments that helped improve the writing and presentation.

Atlanta, USA
September 2016

David Borthwick

The original version of the book was revised: Belated corrections from author have been incorporated. The erratum to the book is available at https://doi.org/10.1007/978-3-319-48936-0_14

Contents

Notations

$:=$	Equal by definition
$\equiv$	Equal except on a set of measure zero
$\sim$	Asymptotic to (ratio approaches 1)
$\asymp$	Comparable to (ratio bounded above and below)
$\Vert\cdot\Vert$	Norm
$\Vert\cdot\Vert_p$	L^p norm
$\langle\cdot,\cdot\rangle$	Inner product (L^2 by default)
$(\cdot,\cdot)$	Distributional pairing
A_n	Volume of the unit sphere $\mathbb{S}^{n-1} \subset \mathbb{R}^n$
$\mathbb{B}^n$	Open unit ball $\{\lvert\boldsymbol{x}\rvert < 1\} \subset \mathbb{R}^n$
$B(\boldsymbol{x}_0; R)$	Open ball $\{\lvert\boldsymbol{x} - \boldsymbol{x}_0\rvert < R\} \subset \mathbb{R}^n$
$c_k[\cdot]$	Fourier coefficient
$\mathbb{C}$	Complex numbers
$C^m(\Omega)$	m-times continuously differentiable functions $\Omega \to \mathbb{C}$
$C^m(\Omega; \mathbb{R})$	Real valued C^m functions
$C^m(\overline{\Omega})$	C^m functions that admit extension across $\partial\Omega$
$C^m_{cpt}(\Omega)$	C^m functions with compact support in Ω
χ_A	Characteristic (or indicator) function of a set A
$\mathbb{D}$	Unit disk in $\mathbb{R}^2$
D^α	Multivariable derivative
$\delta_{\boldsymbol{x}}$	Dirac delta function
$\partial\Omega$	Boundary of Ω
$\mathcal{D}'(\Omega)$	Distributions on Ω
$\mathcal{D}_f[\cdot]$	Dirichlet energy
dS	Surface integral element
$\mathcal{E}$	Energy of a solution
$\mathcal{F}$	Fourier transform
Γ	Gamma function
∇	Gradient operator
$H^m(\Omega)$	Sobolev space, functions with weak derivatives in L^2 to order m

$H^m_{loc}(\Omega)$	Local Sobolev functions
$H^1_0(\Omega)$	Closure of $C^\infty_{cpt}(\Omega)$ in $H^1(\Omega)$
H_t	Heat kernel
Δ	Laplacian operator on $\mathbb{R}^n$
$L^p(\Omega)$	p-th power integrable functions $\Omega \to \mathbb{C}$
$L^1_{loc}(\Omega)$	Locally integrable functions $\Omega \to \mathbb{C}$
ℓ^p	Discrete L_p space
$\mathbb{N}$	Natural numbers $\{1, 2, 3, \ldots\}$
$\mathbb{N}_0$	Non-negative integers $\{0, 1, 2, 3, \ldots\}$
ν	Outward unit normal
Ω	Domain (open, connected set) in $\mathbb{R}^n$
$\bar{\Omega}$	Closure of Ω ($\Omega \cup \partial\Omega$)
Φ	Fundamental solution
r	$\lvert \boldsymbol{x} \rvert$ in $\mathbb{R}^n$
$\mathbb{R}$	Real numbers
$\mathcal{R}[\cdot]$	Rayleigh quotient
$\mathbb{S}^{n-1}$	Unit sphere $\{\lvert \boldsymbol{x} \rvert = 1\} \subset \mathbb{R}^n$
$\mathcal{S}(\mathbb{R}^n)$	Schwartz functions (smooth functions with rapid decay)
$\mathcal{S}'(\mathbb{R}^n)$	Tempered distributions
$S_n[\cdot]$	Partial sum of Fourier series
$\mathbb{T}$	$\mathbb{R}/2\pi\mathbb{Z}$
W_t	Wave kernel
$\mathbb{Z}$	Integers $\{\ldots, -1, 0, 1, \ldots\}$

Chapter 1
Introduction

1.1 Partial Differential Equations

Continuous phenomena, such as wave propagation or fluid flow, are generally modeled with *partial differential equations* (PDE), which express relationships between rates of change with respect to multiple independent variables. In contrast, phenomena that can be described with a single independent variable, such as the motion of a rigid body in classical physics, are modeled by *ordinary differential equations* (ODE).

A general PDE for a function u has the form

$$F\left(\boldsymbol{x}, u(\boldsymbol{x}), \frac{\partial u}{\partial x_j}(\boldsymbol{x}), \ldots, \frac{\partial^m u}{\partial x_{j_1} \ldots \partial x_{j_m}}(\boldsymbol{x})\right) = 0. \tag{1.1}$$

The *order* of this equation is m, the order of the highest derivative appearing (which is assumed to be finite). A *classical solution* u admits continuous partial derivatives up to order m and satisfies (1.1) at all points $\boldsymbol{x}$ in its domain. In certain situations the differentiability requirements can be relaxed, allowing us to define *weak solutions* that do not solve the equation literally.

A somewhat subtle aspect of the definition (1.1) is the fact that the equation is required to be *local*. This means that functions and derivatives appearing in the equation are all evaluated at the same point.

Although classical physics provided the original impetus for the development of PDE theory, PDE models have since played a crucial role in many other fields, including engineering, chemistry, biology, ecology, medicine, and finance. Many industrial applications of mathematics are based on the numerical analysis of PDE.

Most PDE are not solvable in the explicit sense that a simple calculus problem can be solved. That is, we typically cannot obtain a exact formula for $u(\boldsymbol{x})$. Therefore much of the analysis of PDE is focused on drawing meaningful conclusions from an equation without actually writing down a solution.

D. Borthwick, *Introduction to Partial Differential Equations*,
Universitext, DOI 10.1007/978-3-319-48936-0_1

1.2 Example: d'Alembert's Wave Equation

One of the earliest and most influential PDE models was the *wave equation*, developed by Jean d'Alembert in 1746 to describe the motion of a vibrating string. With physical constants normalized to 1, the equation reads

$$\frac{\partial^2 u}{\partial t^2} - \frac{\partial^2 u}{\partial x^2} = 0, \tag{1.2}$$

where $u(t, x)$ denotes the vertical displacement of the string at position x and time t. If the string has length ℓ and is attached at both ends, then we also require that $u(t, 0) = u(t, \ell) = 0$ for all t. We will discuss the formulation of this model in Sect. 4.1.

D'Alembert also found a general formula for the solution of (1.2), based on the observation that (1.2) is solved by any function of the form $f(x \pm t)$, assuming f is twice-differentiable. Given two such functions on $\mathbb{R}$, we can write a general solution

$$u(t, x) := f_1(x + t) + f_2(x - t). \tag{1.3}$$

A similar formula applies in the case of a string with fixed ends. If f is 2ℓ-periodic on $\mathbb{R}$, meaning $f(x + 2\ell) = f(x)$ for all x, then it is easy to check that

$$u(t, x) := \frac{1}{2}[f(x + t) - f(t - x)] \tag{1.4}$$

satisfies $u(t, 0) = u(t, \ell) = 0$ for any t.

One curious feature of this formula is that it appears to give a sensible solution even in cases where f is not differentiable. For example, to model a plucked string we might take the initial displacement to be a simple piecewise linear function in the form of a triangle from the fixed endpoints, as shown in Fig. 1.1.

If we extend this to an odd, 2ℓ-periodic function on $\mathbb{R}$, then the formula (1.4) yields the result illustrated in Fig. 1.2. The initial kink splits into two kinks which travel in opposite directions on the string and and appear to rebound from the fixed ends.

This is not a classical solution because u is not differentiable at the kinks. However, u does satisfy the requirements for a weak solution, as we will see in Chap. 10.

Although a physical string could not exhibit sharp corners without breaking, the piecewise linear solutions are nevertheless physically reasonable. Direct observations of plucked and bowed strings were first made in the late 19th century by Hermann von Helmholtz, who saw patterns of oscillation quite similar to what is shown in Fig. 1.2. The appearance of kinks propagating along the string is striking, although the corners are not exactly sharp.

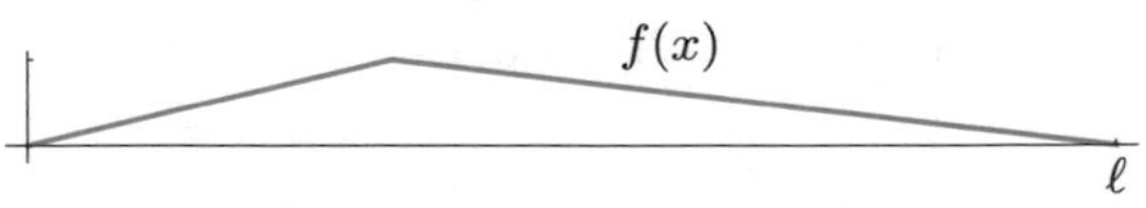

Fig. 1.1 Initial state of a plucked string

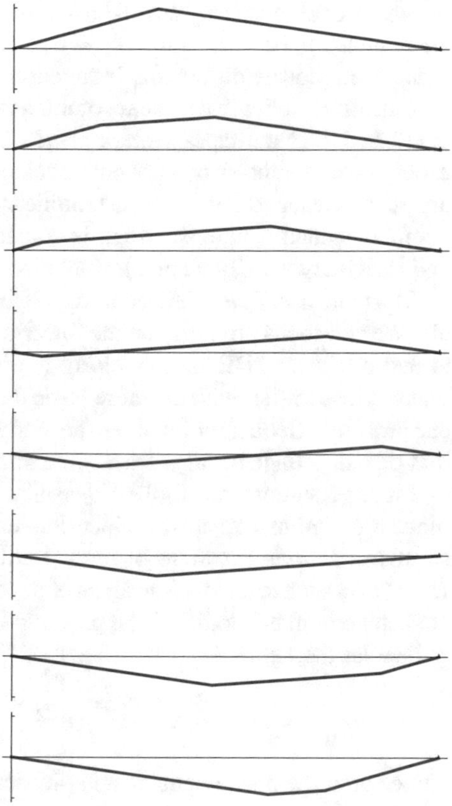

Fig. 1.2 Evolution of the plucked string, starting from $t = 0$ at the *top*

1.3 Types of Equations

There is no general theory of PDE that allows us to analyze all equations of the form (1.1). To make progress it is necessary to restrict our attention to certain classes of equations and develop methods appropriate to those.

The most fundamental distinction between PDE is the property of *linearity*. A PDE is called linear if it can be written in the form

$$Lu = f, \tag{1.5}$$

where f is some function independent of u, and L is a differential operator. Many of the important classical PDE that we will discuss in this book are linear and of first or second order. For such cases L has the general form

$$L = -\sum_{i,j=1}^{n} a_{ij} \frac{\partial^2}{\partial x_i \partial x_j} + \sum_{j=1}^{n} b_j \frac{\partial}{\partial x_j} + c, \tag{1.6}$$

where the coefficients a_{ij}, b_j, and c are functions of $\boldsymbol{x}$. The second-order coefficients are assumed to be symmetric, $a_{ij} = a_{ji}$, because the mixed partials derivatives of a twice continuously differentiable function commute.

Linearity implies that a linear combination of solutions is still a solution, a fact that is referred to as the *superposition principle*. Superposition often lets us decompose problems into simpler components, which is the main reason that linear problems are much easier to handle than nonlinear. It also makes it possible to work with complex-valued solutions, which is sometimes more convenient, because the real and imaginary parts of a complex solution will solve the equation independently.

Most linear PDE are derived as approximations to more realistic, nonlinear models. We will focus primarily on the linear case in this book. The main reason for this is that nonlinear PDE are inherently more complicated, and for an introduction it makes sense to start with the more basic theory. Furthermore, the analysis of nonlinear problems frequently involves the study of associated linear approximations, so that one must understand at least some of the linear theory first.

Linear equations are further classified by the properties of the terms with the highest orders of derivatives, since this determines many qualitative properties of solutions. *Elliptic* equations of second order are associated to an operator L of the form (1.6), such that the eigenvalues of the symmetric matrix $[a_{ij}]$ are strictly positive at each point in the domain. The prototype of an elliptic operator is $L = -\Delta$ where Δ denotes the *Laplacian*,

$$\Delta := \frac{\partial^2}{\partial x_1^2} + \cdots + \frac{\partial^2}{\partial x_n^2}, \tag{1.7}$$

named after the mathematician and physicist Pierre-Simon Laplace.

Equations that include time as an independent variable are called *evolution* equations. The time variable usually plays a very different role from the spatial variables, so in such cases we adapt the form (1.6) by separating out the time derivatives explicitly.

The two classic types of second-order evolution equations are *hyperbolic* and *parabolic*. Hyperbolic equations are exemplified by d'Alembert's wave equation (1.2). The general form is (1.5) with

$$L = \frac{\partial^2 u}{\partial t^2} - \sum_{i,j=1}^{n} a_{ij}(\boldsymbol{x}) \frac{\partial^2 u}{\partial x_i x_j^2} + \text{(lower order terms)}, \tag{1.8}$$

where once again $[a_{ij}]$ is assumed to be a strictly positive matrix. Hyperbolic equations are used to model oscillatory phenomena.

Parabolic evolution equations have the form (1.5) with

$$L = \frac{\partial u}{\partial t} - \sum_{i,j=1}^{n} a_{ij}(\boldsymbol{x}) \frac{\partial^2 u}{\partial x_i x_j^2} + \text{(lower order terms)}, \tag{1.9}$$

where $[a_{ij}]$ is a strictly positive matrix. The heat equation, whose derivation we will discuss in detail in Sect. 6.1, is the prototype for this type of equation. Parabolic equations are generally used to model phenomena of conduction and diffusion.

Note that hyperbolic and parabolic equations revert to elliptic equations in the spatial variables if the solution is independent of time. Elliptic equations thus serve to model the equilibrium states of evolution equations.

Because of their association with phenomenological properties of a system, the terms “elliptic”, “hyperbolic”, and “parabolic” are frequently applied more broadly than this simple classification would suggest. A nonlinear equation is typically described by the category of its linear approximations, which can change depending on the conditions.

For problems on a bounded domain, the application usually dictates some restriction on the solutions at the boundary. Two very common types are *Dirichlet boundary conditions*, specifying the values of u at the boundary, and *Neumann conditions*, specifying the normal derivatives of u at the boundary. These conditions are named for Gustave Lejeune Dirichlet and Carl Neumann, respectively. By default we will use these terms in the homogeneous sense, meaning that the boundary values of the function or derivative are set equal to zero. For evolution equations, we also impose *initial conditions*, specifying the values of u and possibly its time derivatives at some initial time.

1.4 Well Posed Problems

The set of functions used to formulate a PDE, which might include coefficients or terms in the equation itself as well as boundary and initial conditions, is collectively referred to as the input *data*. The most basic question for any PDE is whether a solution exists for a given set of data. However, for most purposes we want to require something more. A PDE problem is said to be *well posed* if, for a given set of data:

1. A solution exists.
2. The solution is uniquely determined by the data.
3. The solution depends continuously on the data.

These criteria were formulated by Jacques Hadamard in 1902. The first two properties hold for ODE under rather general assumptions, but not necessarily for PDE. It is easy to find nonlinear equations that admit no solutions, and even in the linear case there is no guarantee.

The third condition, continuous dependence on the input data, is sometimes called *stability*. One practical justification for this requirement is it is not possible to specify input data with absolute accuracy. Stability implies that the effects of small variations in the data can be controlled.

For certain PDE, especially the classical linear cases, we have a good understanding of the requirements for well-posedness. For other important problems, for

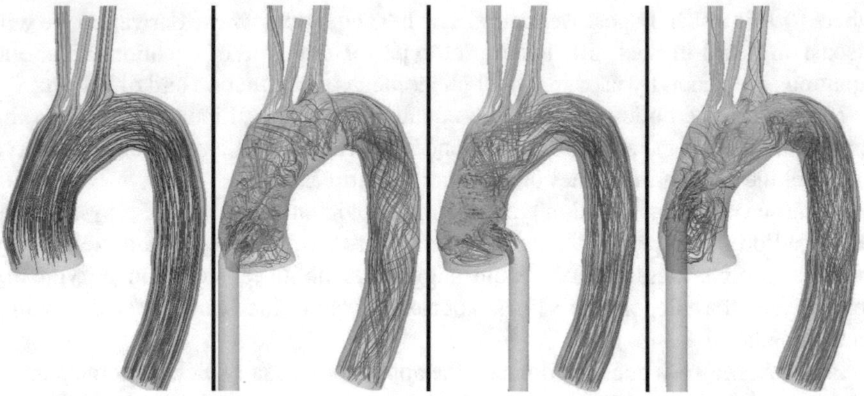

Fig. 1.3 Numerical simulations of blood flow in the aorta. Courtesy of D. Gupta, Emory University Hospital, and T. Passerini, M. Piccinelli and A. Veneziani, Emory Mathematics and Computer Science

example in fluid mechanics, well-posedness remains a difficult unsolved conjecture. Furthermore, many interesting problems are known not to be well posed. For example, problems in image processing are frequently ill posed, because information is lost due to noise or technological limitations.

1.5 Approaches

We can organize the methods for handling PDE problems according to three basic goals:

1. *Solving*: finding explicit formulas for solutions.
2. *Analysis*: understanding general properties of solutions.
3. *Approximation*: calculating solutions numerically.

Solving PDE is certainly worth understanding in those special cases where it is possible. The solution formulas available for certain classical PDE provide insight that is important to the development of the theory.

The goals of theoretical analysis of PDE are extremely broad. We wish to learn as much as we can about the qualitative and quantitative properties of solutions and their relationship to the input data.

Finally, numerical computation is the primary means by which applications of PDE are carried out. Computational methods rely on a foundation of theoretical analysis, but also bring up new considerations such as efficiency of calculation.

Example 1.1 Figure 1.3 shows a set of numerical simulations modeling the insertion in the aorta of a pipe-like device designed to improve blood flow. The leftmost frame shows the aorta before surgery, and the three panes on the right model the insertion at different locations. The PDE model is a complex set of fluid equations called the *Navier-Stokes* equations. These fluid equations are famously difficult to analyze and an exact solution is almost never possible. However, the cylinder is one case that can be handled explicitly. For the numerical simulations, exact solutions for a cylindrical pipe were used to provide boundary data at the point where the pipe meets the aorta.

Theoretical analysis also plays an important role here, in that the regularity theory for the fluid equations is used to predict the accuracy of the simulation. (The complete well-posedness analysis of the Navier-Stokes equations remains a famously unsolved problem, however.)

The simulated flows displayed in Fig. 1.3 were computed numerically by a technique called the *finite element method*. This involves discretizing the problem to reduce the PDE to a system of linear algebraic equations. Modeling a single heartbeat in this simulation require solving a linear system of about 500 million equations.

Chapter 2
Preliminaries

In this chapter we set the stage for the study of PDE with a review of some core background material.

2.1 Real Numbers

The real number system $\mathbb{R}$ is constructed as the "completion" of the field of rational numbers. This means that in addition to the algebraic axioms for addition and multiplication, $\mathbb{R}$ satisfies an additional axiom related to the existence of limits. To state this axiom we use the concept of the *supremum* (or "least upper bound") of subset $A \subset \mathbb{R}$. The supremum is a number $\sup(A) \in \mathbb{R}$ such that (1) all elements of A are less than or equal to $\sup(A)$; and (2) no number strictly less than $\sup(A)$ has this property. The *completeness* axiom says that every nonempty subset of $\mathbb{R}$ that is bounded above has a supremum. An equivalent statement is that a nonempty subset that is bounded below has an *infimum* ("greatest lower bound"), which is denoted $\inf(A)$.

It is convenient to extend these definitions to unbounded sets by defining $\sup(A) := \infty$ when A is not bounded above, and $\inf(A) := -\infty$ when the set is not bounded below. We also set $\sup(\emptyset) = -\infty$ and $\inf(\emptyset) := +\infty$. With these extensions, sup and inf are defined for all subsets of $\mathbb{R}$.

To illustrate the definition, we present a simple result that will prove useful in the construction of approximating sequences for solutions of PDE.

Lemma 2.1 *For a nonempty set $A \subset \mathbb{R}$, there exists a sequence of points $x_k \in A$ such that*

$$\lim_{k\to\infty} x_k = \sup A,$$

and similarly for $\inf A$.

The original version of the book was revised: Belated corrections from author have been incorporated. The erratum to the book is available at https://doi.org/10.1007/978-3-319-48936-0_14

D. Borthwick, *Introduction to Partial Differential Equations*,
Universitext, DOI 10.1007/978-3-319-48936-0_2

Proof If A is not bounded above then there exists a sequence of $x_k \in A$ with $x_k \to \infty$. Therefore the claim holds when $\sup A = \infty$.

Now suppose that $\sup A = \alpha \in \mathbb{R}$. By the definition of the supremum, $\alpha - 1/k$ is not an upper bound of A for $k \in \mathbb{N}$. Therefore, for each k there exists $x_k \in A$ such that $\alpha - 1/k < x_k \leq \alpha$. This yields a sequence such that $x_k \to \alpha$. □

There is an important distinction between supremum and infimum and the related concepts of *maximum* and *minimum*. The latter are required to be elements of the set and thus may not exist. For example, the interval $(0, 1)$ has $\sup = 1$ and $\inf = 0$, but has neither max nor min.

2.2 Complex Numbers

The complex number system $\mathbb{C}$ consists of numbers of the form $z = x + iy$, where $x, y \in \mathbb{R}$ and $i^2 := -1$. The numbers x and y are called the real and imaginary parts of z. The *conjugate* of z is

$$\bar{z} := x - iy,$$

so that

$$\operatorname{Re} z := \frac{z + \bar{z}}{2}, \qquad \operatorname{Im} z := \frac{z - \bar{z}}{2i}.$$

A nonzero complex number has a multiplicative inverse, given by

$$\frac{1}{x + iy} = \frac{x - iy}{x^2 + y^2}.$$

The absolute value on $\mathbb{C}$ is the vector absolute value from Euclidean $\mathbb{R}^2$,

$$|z| := \sqrt{x^2 + y^2}.$$

This can be written in terms of conjugation,

$$|z| = \sqrt{z\bar{z}},$$

which shows in particular that the absolute value is multiplicative,

$$|zw| := |z||w|$$

for $z, w \in \mathbb{C}$.

The basic theory of sequences and series carries over from $\mathbb{R}$ to $\mathbb{C}$ with only minor changes. A sequence $\{z_k\}$ in $\mathbb{C}$ converges to z if

$$\lim_{k\to\infty} |z_k - z| = 0,$$

and a series $\sum a_k$ converges if the sequence of partial sums $\sum_{k=1}^{n} a_k$ is convergent. The series converges *absolutely* if

$$\sum_{k=1}^{\infty} |a_k| < \infty.$$

It follows from the completeness axiom of $\mathbb{R}$ that absolute convergence of a series in $\mathbb{C}$ implies convergence.

The exponential series,

$$e^z := \sum_{k=0}^{\infty} \frac{z^k}{k!}, \tag{2.1}$$

converges absolutely for all $z \in \mathbb{C}$. The special case where z is purely imaginary gives an important relation called *Euler's formula*:

$$\begin{aligned} e^{i\theta} &= \left(1 - \frac{\theta^2}{2!} + \frac{\theta^4}{4!} - \cdots\right) + i\left(\frac{\theta}{1!} - \frac{\theta^3}{3!} + \cdots\right) \\ &= \cos\theta + i\sin\theta. \end{aligned} \tag{2.2}$$

Leonhard Euler, arguably the most influential mathematician of the 18th century, published this identity in 1748. It yields a natural polar-coordinate representation of complex numbers,

$$z = re^{i\theta},$$

where $r = |z|$ and θ is the angle between z and the positive real axis.

The product rule for complex exponentials,

$$e^z e^w = e^{z+w}, \tag{2.3}$$

follows from the power series definition just as in the real case. In combination with (2.2) this allows for a very convenient manipulation of trigonometric functions. For example, setting $z = i\alpha$ and $w = i\beta$ in (2.3) and taking the real and imaginary parts recovers the identities

$$\begin{aligned} \cos(\alpha + \beta) &= \cos\alpha\cos\beta - \sin\alpha\sin\beta, \\ \sin(\alpha + \beta) &= \cos\alpha\sin\beta + \sin\alpha\cos\beta. \end{aligned}$$

The calculus rules for differentiating and integrating exponentials are derived from the power series expansion, and thus extend to the complex case. In particular,

$$\frac{d}{dx} e^{ax} = ae^{ax}$$

for a $\in \mathbb{C}$.

2.3 Domains in $\mathbb{R}^n$

For points in $\mathbb{R}^n$ we will use the vector notation $\boldsymbol{x} = (x_1, \ldots, x_n)$. The Euclidean dot product is denoted $\boldsymbol{x} \cdot \boldsymbol{y}$, and the Euclidean length of a vector is written

$$|\boldsymbol{x}| := \sqrt{\boldsymbol{x} \cdot \boldsymbol{x}}.$$

The Euclidean distance is used to define limits: $\lim_{k\to\infty} \boldsymbol{x}_k = \boldsymbol{w}$ means that

$$\lim_{k\to\infty} |\boldsymbol{x}_k - \boldsymbol{w}| = 0.$$

The *ball* of radius $R > 0$ centered at a point $\boldsymbol{x}_0 \in \mathbb{R}^n$ is

$$B(\boldsymbol{x}_0; R) := \left\{\boldsymbol{x} \in \mathbb{R}^n;\ |\boldsymbol{x} - \boldsymbol{x}_0| < R\right\}.$$

A small ball centered at $\boldsymbol{x}_0$ is called a *neighborhood* of $\boldsymbol{x}_0$. If $\boldsymbol{x} \in A$ has a neighborhood contained in A then $\boldsymbol{x}$ is called an *interior point*.

A subset $U \subset \mathbb{R}^n$ is *open* if all of its points are interior. This generalizes the notion of an open interval in one dimension. The ball $B(\boldsymbol{x}_0; R)$ is open, for example, as is $\mathbb{R}^n$ itself. The empty set is open by default.

A *boundary point* of $A \subset \mathbb{R}^n$ is a point $\boldsymbol{x} \in \mathbb{R}^n$ such that every neighborhood of $\boldsymbol{x}$ intersects both A and its complement. The distinction between interior and boundary points is illustrated in Fig. 2.1. Note that boundary points may or may not be included in the set itself. The boundary of A is denoted

$$\partial A := \{\text{boundary points of } A\}.$$

For example, the boundary of the ball $B(\boldsymbol{x}_0; R)$ is the sphere

$$\partial B(\boldsymbol{x}_0; R) = \left\{\boldsymbol{x} \in \mathbb{R}^n;\ |\boldsymbol{x} - \boldsymbol{x}_0| = R\right\}.$$

A set is open if and only if it contains no boundary points.

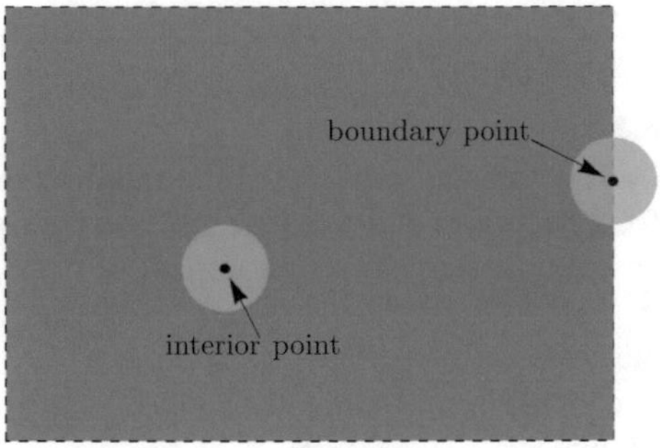

Fig. 2.1 Open rectangle in $\mathbb{R}^2$

A subset of $\mathbb{R}^n$ is *connected* if any two points in the set can be joined by a continuous path within the set. For an open set U this is equivalent to the condition that U cannot be written as the disjoint union of two nonempty open sets.

We will refer to a connected open subset $\Omega \subset \mathbb{R}^n$ as a *domain*, and reserve the notation Ω for this usage. For some problems we assume the domain is *bounded*, meaning that $\Omega \subset B(0; R)$ for sufficiently large R.

The concept of a closed interval can also be generalized to higher dimension. A subset $F \subset \mathbb{R}^n$ is *closed* if it contains all of its boundary points, i.e.,

$$\partial F \subset F.$$

The union of a subset $A \subset \mathbb{R}^n$ with its boundary is called the *closure* of A and denoted

$$\overline{A} := A \cup \partial A.$$

For example,

$$\overline{B(\boldsymbol{x}_0; R)} := \left\{\boldsymbol{x} \in \mathbb{R}^n;\ |\boldsymbol{x} - \boldsymbol{x}_0| \le R\right\}.$$

It is potentially confusing that an overline is used for set closure and complex conjugation, but these notations are standard. Note that closure applies only to sets and not to numbers or functions.

A closed set $F \in \mathbb{R}^n$ contains the limits of all sequences in F that converge in $\mathbb{R}^n$. This is because the limit of a sequence contained in a set must either be a point in the set or a boundary point.

Closed and open sets are related in the sense that the complement of an open set is closed, and vice versa. However, the terms are not mutually exclusive, and a set might not have either property. The interval $(a, b] \subset \mathbb{R}$ is neither open nor closed, for example. The only subsets of $\mathbb{R}^n$ with both properties are $\mathbb{R}^n$ itself and $\emptyset$.

2.4 Differentiability

The space of continuous, complex-valued functions on a domain $\Omega \subset \mathbb{R}^n$ which admit continuous partial derivatives up to order m is denoted by $C^m(\Omega)$. The assumption of continuity for derivatives insures that mixed partials are independent of the order of differentiation. A *smooth* function has continuous derivatives to all orders; the corresponding space is written $C^\infty(\Omega)$.

We use the notation $C^m(\Omega; \mathbb{R})$ to specify real-valued functions, and similarly $C^m(\Omega; \mathbb{R}^n)$ denotes the space of vector-valued functions. It is common to use C^m as an adjective, short for "m-times continuously differentiable".

The definition of $C^m(\Omega)$ makes no conditions on the behavior of functions as the boundary is approached. To impose such restrictions, we use the notation $C^m(\overline{\Omega})$ to denote the space of functions that admit C^m extensions across the boundary. For example, the function $\sqrt{x} \in C^\infty(0, 1)$ is an element of $C^0[0, 1]$, but not $C^1[0, 1]$.

The *support* of $f \in C^0(\Omega)$ is defined as

$$\operatorname{supp} f := \overline{\{\boldsymbol{x} \in \Omega;\ f(\boldsymbol{x}) \neq 0\}}. \tag{2.4}$$

Note that the definition includes a closure. This means that the support does not exclude points where the function merely "passes through" zero. For example, the support of $\sin(x)$ is $\mathbb{R}$ rather than $\mathbb{R}\backslash\pi\mathbb{Z}$.

A closed and bounded subset of $\mathbb{R}^n$ is said to be *compact*. We denote by $C^m_{\text{cpt}}(\Omega)$ the space of functions on Ω that have *compact support*, meaning that $\operatorname{supp} f$ is a compact subset of Ω. Since Ω is open and the support is closed, this requires in particular that $\operatorname{supp} f$ be a strict subset of Ω. For example, $1 - x^2$ vanishes at the boundary of $(-1, 1)$, but does not have compact support in this domain because its support is $[-1, 1]$.

Example 2.2 To demonstrate the existence of compactly supported smooth functions, consider

$$h(x) = \begin{cases} e^{-1/(1-x^2)}, & |x| < 1, \\ 0, & |x| \geq 1, \end{cases}$$

which has support $[-1, 1]$. As illustrated in Fig. 2.2, the function becomes extremely flat as $x \to \pm 1$.

To show that h is in fact smooth, we note that

$$h^{(m)}(x) = \begin{cases} \frac{q_m(x)}{(1-x^2)^m} e^{-1/(1-x^2)}, & |x| < 1, \\ 0, & |x| \geq 1, \end{cases}$$

where q_m denotes a polynomial of degree m. As $x \to \pm 1$, the term $(1-x^2)^{-m}$ blows up while the exponential term tends rapidly to zero. Using l'Hôpital's rule one can check that the exponential dominates this limit, so that all derivatives of h vanish as $x \to \pm 1$ from $|x| \leq 1$. This shows that $h \in C^\infty_{\text{cpt}}(\mathbb{R})$.

The function h can be integrated to produce a smooth function that is constant for $|x| \geq 1$. By translating and rescaling, this construction gives, for $a < b$, a function $\varphi \in C^\infty(\mathbb{R})$ satisfying

$$\varphi(x) = \begin{cases} 0, & x \leq a, \\ 1, & x \geq b. \end{cases}$$

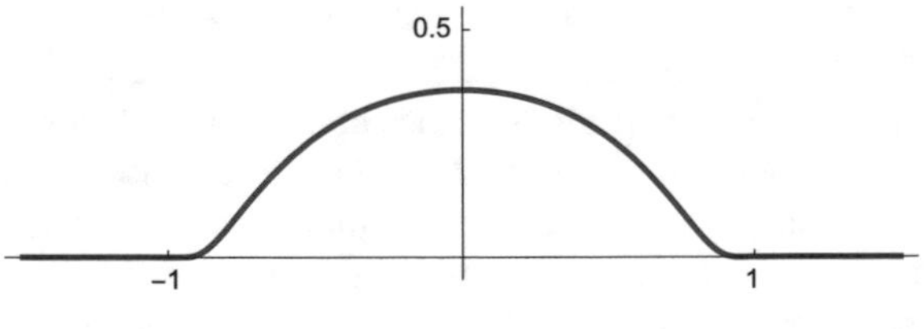

Fig. 2.2 Compactly supported smooth function

These "smooth step functions" are useful as building blocks for pasting together smooth functions on different domains. ◊

Certain problems require regularity of the boundary of Ω. For example, many theorems in vector calculus require the existence of a normal vector on the boundary, which does not exist for a general domain. A standard hypothesis for such theorems is that $\partial\Omega$ is piecewise C^1. This means that $\partial\Omega$ consists of a finite number of components which admit *regular coordinate parametrization.* A coordinate parametrization is a map $\boldsymbol{\sigma} \in C^1(\overline{U}; \mathbb{R}^n)$ where U is a domain in $\mathbb{R}^{n-1}$. To say the parametrization is regular means that the tangent vectors defined by

$$\frac{\partial \boldsymbol{\sigma}}{\partial w_j} := \left(\frac{\partial \sigma_1}{\partial w_j}, \dots, \frac{\partial \sigma_n}{\partial w_j}\right),$$

$j = 1, \dots, n-1$, are linearly independent at each point of $\partial\Omega$.

The piecewise C^1 boundary assumption guarantees that the points of each boundary component have well-defined tangent spaces and normal directions. As an example, the unit cube in $\mathbb{R}^3$ has piecewise C^1 boundary, consisting of 6 planar components with normal directions parallel to the coordinate axes.

In this text we will focus on relatively simple domains with straightforward boundary parametrizations.

2.5 Ordinary Differential Equations

Our development of PDE theory will not rely on any advanced techniques for the solution of ODE, but it will be useful to recall some basic material.

First-order ODE can often be solved directly by methods from calculus. The easiest cases are equations of the form

$$\frac{dy}{dt} = g(y)h(t),$$

where the variables can be separated to yield an integral formula

$$\int \frac{dy}{g(y)} = \int \frac{dt}{h(t)}.$$

Integrating both sides yields a family of solutions with one undetermined constant.

Example 2.3 Consider the equation

$$\frac{dy}{dt} = ay, \quad y(0) = y_0.$$

for $a \neq 0$. This is called the growth or decay equation, depending on the sign of a. Separating the variables gives

$$\int \frac{dy}{y} = \int a\, dt,$$

which integrates to $\ln y = at + C$. Solving for y and using the initial condition gives

$$y(t) = y_0 e^{at}.$$

◊

Higher-order ODE are generally analyzed by reducing to a system of first-order equations. To reduce the nth-order equation

$$y^{(n)}(t) = F\left(t, y, y', \dots, y^{(n-1)}\right), \tag{2.5}$$

we define the vector-valued function $\boldsymbol{w} = (y, y', \dots, y^{(n-1)})$. This satisfies the first-order system

$$\frac{d}{dt}\begin{pmatrix} w_1 \\ \vdots \\ w_{n-1} \\ w_n \end{pmatrix} = \begin{pmatrix} w_2 \\ \vdots \\ w_n \\ F(t, \boldsymbol{w}) \end{pmatrix}.$$

First-order systems can be solved generally by the strategy of Picard iteration, named for mathematician Émile Picard. The first step is to write the vector equation,

$$\frac{d\boldsymbol{w}}{dt} = \boldsymbol{F}(t, \boldsymbol{w}), \qquad \boldsymbol{w}(t_0) = \boldsymbol{w}_0, \tag{2.6}$$

in an equivalent form as a recursive integral equation,

$$\boldsymbol{w}(t) = \boldsymbol{w}_0 + \int_{t_0}^{t} \boldsymbol{F}(s, \boldsymbol{w}(s))\, ds.$$

For the construction, we set $\boldsymbol{u}_0(t) := \boldsymbol{w}_0$ and define a sequence of functions by

$$\boldsymbol{u}_k(t) = \boldsymbol{w}_0 + \int_{t_0}^{t} \boldsymbol{F}(s, \boldsymbol{u}_{k-1}(s))\, ds \tag{2.7}$$

for $k = 1, 2, \ldots$. It can be shown that the limit of this sequence exists and solves (2.6) under some general assumptions on $\boldsymbol{F}$, which leads to a proof of the following result.

Theorem 2.4 (Picard iteration) *Suppose that* $\boldsymbol{F}$ *is a continuous function on* $I \times \Omega$ *where* I *is an open interval containing* t_0 *and* Ω *is a domain in* $\mathbb{R}^n$ *containing* $\boldsymbol{w}_0$, *and that* $\boldsymbol{F}$ *is continuously differentiable with respect to* $\boldsymbol{w}$. *Then* (2.6) *admits a unique solution on some interval* $(t_0 - \varepsilon, t_0 + \varepsilon)$ *with* $\varepsilon > 0$.

Applying Theorem 2.4 to (2.5) shows that an nth order ODE satisfying the regularity assumptions has a unique local solution specified by the initial values of the function and its first $n - 1$ derivatives.

The C^1 hypothesis on $\boldsymbol{F}$ is stronger than necessary, but this version will suffice for our purposes. The point we would like to stress here is the relative ease with which ODE can be analyzed under very general conditions. This is very different from the PDE theory, where no such general results are possible.

Example 2.5 The *harmonic ODE* is the equation

$$\frac{d^2y}{dt^2} = -\kappa^2 y,$$

for $\kappa > 0$. In view of the solution to the growth/decay equation in Example 2.3, it is reasonable to start with an exponential solution as a guess. Substituting $e^{\alpha t}$ into the equation yields $\alpha^2 = -\kappa^2$. From $\alpha = \pm i\kappa$ we obtain the general solution,

$$y(t) = c_1 e^{i\kappa t} + c_2 e^{-i\kappa t}.$$

To see how this relates to the Picard iteration method described above, consider the corresponding system (2.6) for $\boldsymbol{w} = (y, y')$:

$$\frac{d\boldsymbol{w}}{dt} = \begin{pmatrix} 0 & 1 \\ -\kappa^2 & 0 \end{pmatrix} \boldsymbol{w}.$$

With $\boldsymbol{w}_0 = (a, b)$, the recursive formula (2.7) yields the sequence of functions

$$\boldsymbol{u}_k(t) = \sum_{j=0}^{k} \frac{t^j}{j!} \begin{pmatrix} 0 & 1 \\ -\kappa^2 & 0 \end{pmatrix}^j \begin{pmatrix} a \\ b \end{pmatrix}$$

for $k \in \mathbb{N}$. In the limit $k \to \infty$ this gives

$$\begin{aligned} \boldsymbol{w}(t) = &\left[1 - \frac{(\kappa t)^2}{2!} + \frac{(\kappa t)^4}{4!} - \cdots\right] \begin{pmatrix} a \\ b \end{pmatrix} \\ &+ \frac{1}{\kappa}\left[\kappa t - \frac{(\kappa t)^3}{3!} + \frac{(\kappa t)^5}{5!} - \cdots\right] \begin{pmatrix} b \\ -\kappa^2 a \end{pmatrix}. \end{aligned}$$

Reading off y as the first component of $\boldsymbol{w}$ gives the familiar trigonometric solution,

$$y(t) = a\cos(\kappa t) + \frac{b}{\kappa}\sin(\kappa t).$$

The trigonometric and complex exponential solutions are related by Euler's formula (2.2). ◊

2.6 Vector Calculus

The classical theorems of vector calculus were motivated by PDE problems arising in physics. For our purposes the most important of these results is the divergence theorem. We assume that the reader is familiar with the divergence theorem in the context of $\mathbb{R}^2$ or $\mathbb{R}^3$. In this section we will cover the basic definitions needed to state the result in $\mathbb{R}^n$ and develop its corollaries.

As noted in Sect. 2.3, we always take a domain $\Omega \subset \mathbb{R}^n$ to be connected and open. The *gradient* of $f \in C^1(\Omega)$ is the vector-valued function

$$\nabla f := \left(\frac{\partial f}{\partial x_1}, \dots, \frac{\partial f}{\partial x_n}\right).$$

For a vector-valued function $\boldsymbol{v} \in C^1(\Omega; \mathbb{R}^n)$ with components $(v_1, \dots, v_n)$, the *divergence* is

$$\nabla \cdot \boldsymbol{v} := \frac{\partial v_1}{\partial x_1} + \dots + \frac{\partial v_n}{\partial x_n}.$$

The Laplacian operator introduced in (1.7) is the divergence of the gradient

$$\Delta u := \nabla \cdot (\nabla u).$$

For this reason Δ is sometimes written ∇^2.

If Ω is bounded then the Riemannian integral of $f \in C^0(\overline{\Omega})$ exists and is denoted by

$$\int_\Omega f(\boldsymbol{x})\, d^n\boldsymbol{x},$$

where $d^n\boldsymbol{x}$ is a shorthand for $dx_1 \cdots dx_n$. The integral can be extended to unbounded domains if the appropriate limits exist. We will discuss a further generalization of the Riemann definition in Chap. 7.

One issue we will come across frequently is differentiation under the integral. If $\Omega \subset \mathbb{R}^n$ is a bounded domain and u and $\partial u/\partial t$ are continuous functions on $(a, b) \times \overline{\Omega}$, then the Leibniz integral rule says that

$$\frac{d}{dt}\int_{\Omega} u(t,\boldsymbol{x})\, d^n\boldsymbol{x} = \int_{\Omega} \frac{\partial u}{\partial t}(t,\boldsymbol{x})\, d^n\boldsymbol{x}.$$

Differentiation under the integral may still work when the integrals are improper, but this requires greater care.

To set up boundary integrals for a domain with piecewise C^1 boundary, we need to define the surface integral over a regular coordinate patch $\boldsymbol{\sigma} : U \subset \mathbb{R}^{n-1} \to \partial\Omega$. Let $\boldsymbol{\nu} : U \to \mathbb{R}^n$ denote the unit normal vector pointing outwards from the domain. The surface integral for such a patch is defined by

$$\int_{\sigma(U)} f\, dS := \int_U f(\boldsymbol{\sigma}(\boldsymbol{w})) \left| \det\left[\frac{\partial \boldsymbol{\sigma}}{\partial w_1}, \ldots, \frac{\partial \boldsymbol{\sigma}}{\partial w_{n-1}}, \boldsymbol{\nu}\right]\right| d^{n-1}\boldsymbol{w}, \tag{2.8}$$

where det[. . .] denotes the determinant of a matrix of column vectors. The full surface integral over $\partial\Omega$ is defined by summing over the boundary coordinate patches. For simplicity, we notate this as a single integral,

$$\int_{\partial\Omega} f\, dS.$$

In $\mathbb{R}^2$, a boundary parametrization will be a curve $\boldsymbol{\sigma}(t)$ and (2.8) reduces to the arclength integral

$$\int_{\sigma(U)} f\, dS := \int_U f(\boldsymbol{\sigma}(t)) \left|\frac{d\boldsymbol{\sigma}}{dt}\right| dt.$$

In $\mathbb{R}^3$, the unit normal for a surface patch can be computed from the cross product of the tangent vectors. This leads to the surface integral formula

$$\int_{\sigma(U)} f\, dS := \int_U f(\boldsymbol{\sigma}(\boldsymbol{w})) \left|\frac{\partial \boldsymbol{\sigma}}{\partial w_1} \times \frac{\partial \boldsymbol{\sigma}}{\partial w_2}\right| d^2\boldsymbol{w}.$$

Even in low dimensions surface integrals can be rather complicated. We will make explicit use of these formulas only in relatively simple cases, such as rectangular regions and spheres.

We can use (2.8) to decompose integrals into radial and spherical components. This is particularly useful when the domain is a ball. Let $r := |\boldsymbol{x}|$ be the radial coordinate, and define the unit sphere

$$\mathbb{S}^{n-1} := \{r = 1\} \subset \mathbb{R}^n.$$

A point $\boldsymbol{x} \neq 0$ can be written uniquely as $r\boldsymbol{\omega}$ for $\boldsymbol{\omega} \in \mathbb{S}^{n-1}$ and $r > 0$. Let $\boldsymbol{\omega}(\boldsymbol{y})$ be a parametrization of $\mathbb{S}^{n-1}$ by coordinates $\boldsymbol{y} \in U \subset \mathbb{R}^{n-1}$. For the change of variables $(r, \boldsymbol{y}) \mapsto \boldsymbol{x} = r\boldsymbol{\omega}(\boldsymbol{y})$, the Jacobian formula gives

$$d^n \boldsymbol{x} = \left| \det \left[\frac{\partial \boldsymbol{x}}{\partial y_1}, \ldots, \frac{\partial \boldsymbol{x}}{\partial y_{n-1}}, \frac{\partial \boldsymbol{x}}{\partial r} \right] \right| dr\, dy_1 \cdots dy_{n-1} \qquad (2.9)$$
$$= \left| \det \left[\frac{\partial \boldsymbol{\omega}}{\partial y_1}, \ldots, \frac{\partial \boldsymbol{\omega}}{\partial y_{n-1}}, \boldsymbol{\omega} \right] \right| r^{n-1}\, dr\, dy_1 \cdots dy_{n-1}.$$

On the unit sphere, the outward unit normal $\boldsymbol{\nu}$ is equal to $\boldsymbol{\omega}$. Thus (2.9) reduces to

$$d^n \boldsymbol{x} = r^{n-1}\, dr\, dS(\boldsymbol{y}).$$

For an integral over the ball this yields the radial integral formula,

$$\int_{B(0;R)} f(\boldsymbol{x})\, d^n \boldsymbol{x} = \int_{\mathbb{S}^{n-1}} \int_0^R f(r\boldsymbol{\omega}(\boldsymbol{y}))\, r^{n-1} dr\, dS(\boldsymbol{y}). \qquad (2.10)$$

With these definitions in place, we turn to the divergence theorem, which relates the flux of a vector field through a closed surface to the divergence of the field in the interior. This result is generally attributed to Carl Friedrich Gauss, who published a version in 1813 in conjunction with his work on electrostatics.

Theorem 2.6 (Divergence theorem) *Suppose $\Omega \subset \mathbb{R}^n$ is a bounded domain with piecewise C^1 boundary. For a vector field $\boldsymbol{F} \in C^1(\overline{\Omega}; \mathbb{R}^n)$,*

$$\int_\Omega \nabla \cdot \boldsymbol{F}\, d^n \boldsymbol{x} = \int_{\partial\Omega} \boldsymbol{F} \cdot \boldsymbol{\nu}\, dS,$$

where $\boldsymbol{\nu}$ is the outward unit normal to $\partial\Omega$.

A full proof can be found in advanced calculus texts. To illustrate the idea, we will show how the argument works for a spherical domain in $\mathbb{R}^3$.

Example 2.7 Let $\mathbb{B}^3 = \{r < 1\} \subset \mathbb{R}^3$. Because a vector field can be decomposed into components, it suffices to consider a field parallel to one of the coordinate axes, say $\boldsymbol{F} = (0, 0, f)$. The divergence is then

$$\nabla \cdot \boldsymbol{F} = \frac{\partial f}{\partial x_3}.$$

In cylindrical coordinates, $\boldsymbol{x} = (\rho \cos\theta, \rho \sin\theta, z)$, the volume element is

$$d^3 \boldsymbol{x} = \rho\, d\rho\, d\phi\, dz,$$

so the left side of the divergence formula becomes

$$\int_{\mathbb{B}^3} \nabla \cdot \boldsymbol{F}\, d^3\boldsymbol{x} = \int_0^{2\pi} \int_0^1 \int_{-\sqrt{1-\rho^2}}^{\sqrt{1-\rho^2}} \frac{\partial f}{\partial z}\, \rho\, dz\, d\rho\, d\theta$$
$$= \int_0^{2\pi} \int_0^1 \Big[f\left(\rho\cos\theta, \rho\sin\theta, \sqrt{1-\rho^2}\right) \qquad (2.11)$$
$$- f\left(\rho\cos\theta, \rho\sin\theta, -\sqrt{1-\rho^2}\right)\Big]\, \rho\, d\rho\, d\theta.$$

Note that $z = \pm\sqrt{1-\rho^2}$ gives the restriction to the upper and lower hemispheres, respectively.

We denote the two hemispheres $\mathbb{S}^2_\pm \subset \mathbb{S}^2$ and parametrize them as

$$\boldsymbol{\omega}_\pm(\rho, \theta) = \left(\rho\cos\theta, \rho\sin\theta, \pm\sqrt{1-\rho^2}\right).$$

The corresponding surface area elements are given by

$$dS = \left| \frac{\partial \boldsymbol{\omega}_\pm}{\partial \rho} \times \frac{\partial \boldsymbol{\omega}_\pm}{\partial \theta} \right| d\rho\, d\theta$$
$$= \frac{\rho}{\sqrt{1-\rho^2}}\, d\rho\, d\theta.$$

Thus,

$$\int_0^{2\pi} \int_0^1 f\left(\rho\cos\theta, \rho\sin\theta, \pm\sqrt{1-\rho^2}\right)\, \rho\, d\rho\, d\theta = \int_{\mathbb{S}^2_\pm} f\sqrt{1-\rho^2}\, dS.$$

On $\mathbb{S}^2_\pm$ we have $\boldsymbol{F} \cdot \boldsymbol{\nu} = \pm f\sqrt{1-\rho^2}$, so that

$$\int_{\mathbb{S}^2_\pm} f\sqrt{1-\rho^2}\, dS = \pm \int_{\mathbb{S}^2_\pm} \boldsymbol{F} \cdot \boldsymbol{\nu}\, dS.$$

Applying this to (2.11) reduces the equation to

$$\int_{\mathbb{B}^3} \nabla \cdot \boldsymbol{F}\, d^3\boldsymbol{x} = \int_{\mathbb{S}^2} \boldsymbol{F} \cdot \boldsymbol{\nu}\, dS,$$

verifying the divergence theorem in this special case. ◊

Theorem 2.6 can be used to evaluate integrals of the Laplacian of a function by substituting $\boldsymbol{F} = \nabla u$ for the vector field. Inside the volume integral this yields the integrand

$$\nabla \cdot \boldsymbol{F} = \Delta u.$$

On the surface side, the integrand becomes the directional derivative with respect to the outward unit normal, which is denoted

$$\frac{\partial u}{\partial \nu} := \boldsymbol{\nu} \cdot \nabla u\big|_{\partial\Omega}.$$

Corollary 2.8 *If $\Omega \subset \mathbb{R}^n$ is a bounded domain with piecewise C^1 boundary, and $u \in C^2(\overline{\Omega})$, then*

$$\int_\Omega \Delta u \, d^n\boldsymbol{x} = \int_{\partial\Omega} \frac{\partial u}{\partial \nu} \, dS.$$

The application we will encounter most frequently is to the ball $B(0; R) \in \mathbb{R}^n$. The outward unit normal is parallel to the position vector, so that

$$\nu = \frac{\boldsymbol{x}}{R}.$$

It follows from the chain rule that

$$\frac{\partial u}{\partial \nu} = \frac{\partial u}{\partial r}. \tag{2.12}$$

Example 2.9 Consider a radial function $g(r)$ where $r := |\boldsymbol{x}|$ for $\boldsymbol{x} \in \mathbb{R}^n$. For the ball $B(0; a)$, the radial integral formula (2.10) gives

$$\int_{B(0;a)} \Delta g \, d^n\boldsymbol{x} = A_n \int_0^a \Delta g(r) r^{n-1} \, dr,$$

where

$$A_n := \mathrm{vol}(\mathbb{S}^{n-1}). \tag{2.13}$$

By (2.12),

$$\begin{aligned} \int_{\partial B(0;a)} \frac{\partial g}{\partial \nu} \, dS &= \int_{\partial B(0;a)} \frac{\partial g}{\partial r}(a) \, dS \\ &= A_n a^{n-1} \frac{\partial g}{\partial r}(a). \end{aligned}$$

The formula from Corollary 2.8 reduces in this case to

$$\int_0^a \Delta g(r) r^{n-1} \, dr = a^{n-1} \frac{\partial g}{\partial r}(a). \tag{2.14}$$

Differentiating (2.14) with respect to a gives, by the fundamental theorem of calculus,

$$a^{n-1} \Delta g(a) = \frac{\partial}{\partial a}\left[a^{n-1} \frac{\partial g}{\partial r}(a)\right].$$

This holds for all $a > 0$, so evidently the Laplacian of a radial function is given by

$$\Delta g = r^{1-n} \frac{\partial}{\partial r} \left[r^{n-1} \frac{\partial}{\partial r} \right] g. \tag{2.15}$$

In principle one could derive this formula directly from the chain rule, but the direct computation is difficult in high dimensions. ◊

There are two other direct corollaries of the divergence theorem that will be used frequently. These are named for the mathematical physicist George Green, who used them to develop solution formulas for some classical PDE.

The first result is a generalization of Corollary 2.8, obtained from Theorem 2.6 by the substitution $\boldsymbol{F} = v\nabla u$ for a pair of functions u, v. The product rule for differentiation gives

$$\nabla \cdot (v\nabla u) = \nabla v \cdot \nabla u + v\Delta u, \tag{2.16}$$

which can easily be checked by writing out the components of the gradient.

Theorem 2.10 (Green's first identity) *If $\Omega \subset \mathbb{R}^n$ is a bounded domain with piecewise C^1 boundary, then for $u \in C^2(\overline{\Omega})$, and $v \in C^1(\overline{\Omega})$,*

$$\int_\Omega [\nabla v \cdot \nabla u + v\Delta u]\, d^n\boldsymbol{x} = \int_{\partial\Omega} v\frac{\partial u}{\partial \nu}\, dS.$$

The second identity follows from the first by interchanging u with v and then subtracting the result.

Theorem 2.11 (Green's second identity) *If $\Omega \subset \mathbb{R}^n$ is a bounded domain with piecewise C^1 boundary, then for $u, v \in C^2(\overline{\Omega})$,*

$$\int_\Omega [v\Delta u - u\Delta v]\, d^n\boldsymbol{x} = \int_{\partial\Omega} \left(v\frac{\partial u}{\partial \nu} - u\frac{\partial v}{\partial \nu} \right) dS.$$

2.7 Exercises

2.1 For $r := |\boldsymbol{x}|$ in $\mathbb{R}^n$, and $\alpha \in \mathbb{R}$, compute $\nabla(r^\alpha)$ and $\Delta(r^\alpha)$.

2.2 Polar coordinates (r, θ) in $\mathbb{R}^2$ are related to Cartesian coordinates (x_1, x_2) by

$$x_1 = r\cos\theta, \quad x_2 = r\sin\theta.$$

(a) Use the chain rule to compute $\frac{\partial}{\partial r}$ and $\frac{\partial}{\partial \theta}$ in terms of $\frac{\partial}{\partial x_1}$ and $\frac{\partial}{\partial x_2}$.
(b) Find the expression for Δ in the (r, θ) coordinates. (The radial part should agree with (2.15).)

2.3 In $\mathbb{R}^n$ let Ω be the unit cube $(0, 1)^n$. Define

$$\boldsymbol{w}(\boldsymbol{x}) = f(\boldsymbol{x})e_j,$$

where $f \in C^\infty(\overline{\Omega})$ and e_j is the jth coordinate vector, $e_j := (0, \ldots, 1, \ldots, 0)$. Compute both sides of the formula from Theorem 2.6 in this case, and show that the result reduces to an application of the fundamental theorem of calculus in the x_j variable.

2.4 For $f \in C^0(\mathbb{R}^n)$ set

$$h(t) := \int_{B(0;t)} f(\boldsymbol{x})\, d^n\boldsymbol{x}.$$

for $t \geq 0$. Use the radial decomposition formula (2.10) to show that

$$\frac{dh}{dt} = \int_{\partial B(0;t)} f(\boldsymbol{w})\, dS(\boldsymbol{w}).$$

2.5 The gamma function is defined for $z > 0$ by

$$\Gamma(z) := \int_0^\infty t^{z-1} e^{-t}\, dt. \tag{2.17}$$

Note that $\Gamma(1) = 1$ and integration by parts gives the recursion relation $\Gamma(z+1) = z\Gamma(z)$. In this problem we will show that the volume of the unit sphere in $\mathbb{R}^n$ is given by

$$A_n = \frac{2\pi^{\frac{n}{2}}}{\Gamma(\frac{n}{2})}. \tag{2.18}$$

(a) Use the radial formula (2.10) and the substitution $u := r^2$ to compute that

$$\int_{\mathbb{R}^n} e^{-r^2}\, d^n\boldsymbol{x} = \frac{1}{2} A_n \Gamma(\tfrac{n}{2}). \tag{2.19}$$

(b) Observe that we can rewrite

$$\begin{aligned} \int_{\mathbb{R}^n} e^{-r^2}\, d^n\boldsymbol{x} &= \int_{\mathbb{R}^n} e^{-(x_1^2+\cdots+x_n^2)}\, d^n\boldsymbol{x} \\ &= \left[2\int_0^\infty e^{-x^2}\, dx\right]^n. \end{aligned}$$

Substitute $t = x^2$ to evaluate the one-dimensional integral in terms of $\Gamma(\frac{1}{2})$.

(c) Compare (a) to (b) to obtain a formula for A_n.

(d) Use (c) and the fact that $A_2 = 2\pi$ to compute $\Gamma(\frac{1}{2})$ and reduce the formula to (2.18).

Chapter 3
Conservation Equations and Characteristics

A conservation law for a physical system states that a certain quantity (e.g., mass, energy, or momentum) is independent of time. For continuous systems such as fluids or gases, these global quantities can be defined as integrals of density functions. The conservation law then translates into a local form, as a PDE for the density function.

In this section we will study some first-order PDE that arise from conservation laws. We introduce a classic technique, called the method of characteristics, for analyzing these equations.

3.1 Model Problem: Oxygen in the Bloodstream

To derive the conservation equation, we consider a simple model for the concentration of oxygen carried by the bloodstream. For this discussion we ignore any external effects that might break the conservation of mass, such as absorption of oxygen into the walls of a blood vessel. (Some examples of external effects will be considered in the exercises.)

Let us model an artery as a straight tube, as pictured in Fig. 3.1. We assume that the concentration is constant on cross-sections of the tube, so that the problem reduces to one spatial dimension. For the moment, suppose that the artery extends along the real line and is parametrized by $x \in \mathbb{R}$.

Let $u(t, x)$ denote the oxygen concentration, expressed in units of mass per unit length. Within a fixed interval $[a, b]$, as highlighted in Fig. 3.1, the total mass at time t is given by an integral,

$$m(t) := \int_a^b u(t, x)\, dx. \tag{3.1}$$

The original version of the book was revised: Belated corrections from author have been incorporated. The erratum to the book is available at https://doi.org/10.1007/978-3-319-48936-0_14

D. Borthwick, *Introduction to Partial Differential Equations*,
Universitext, DOI 10.1007/978-3-319-48936-0_3

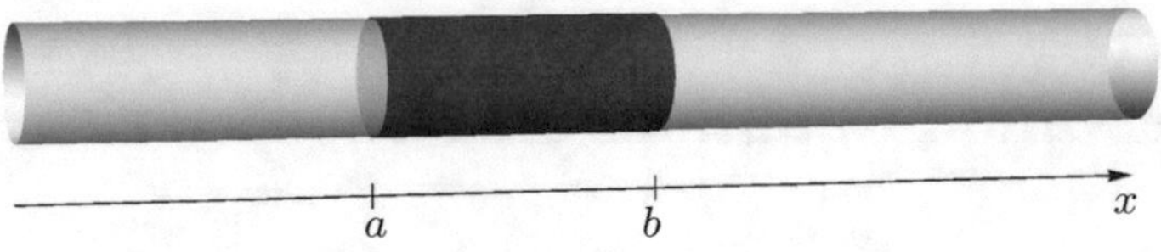

Fig. 3.1 One-dimensional model of an artery

The instantaneous flow rate at a given point x is called the *flux* $q(t, x)$, expressed as mass per unit time. The general relationship between flux and concentration is

$$\text{flux} = (\text{concentration}) \times (\text{velocity}).$$

For the bloodstream model we can reasonably assume that velocity is independent of the oxygen concentration (because oxygen accounts for a relatively small portion of the total density). This assumption implies that q has a linear dependence on u. In other models the velocity might depend on the concentration, making q a nonlinear function of u.

Conservation of mass implies that the total amount of oxygen within the segment changes only as oxygen flows across the boundary points at $x = a$ and $x = b$. Since the flow across these points is given by the flux, the corresponding equation is

$$\frac{dm}{dt}(t) = q(t, a) - q(t, b). \tag{3.2}$$

If q is continuously differentiable with respect to position, then the fundamental theorem of calculus allows us to write the right-hand side of (3.2) as an integral,

$$q(t, a) - q(t, b) = -\int_a^b \frac{\partial q}{\partial x}\, dx.$$

We can also differentiate the integral in (3.1) to obtain

$$\frac{dm}{dt} = \int_a^b \frac{\partial u}{\partial t}\, dx,$$

provided that $u(t, x)$ is continuously differentiable with respect to time. These calculations transform (3.2) into the integral equation

$$\int_a^b \left(\frac{\partial u}{\partial t} + \frac{\partial q}{\partial x} \right) dx = 0. \tag{3.3}$$

Since the segment was arbitrary, (3.3) should hold for all values of a, b. This is only possible if the integrand is identically zero, which gives the local form of the law conservation of mass:

$$\frac{\partial u}{\partial t} + \frac{\partial q}{\partial x} = 0. \tag{3.4}$$

This relationship between concentration and flux is called the *continuity equation* (or *transport equation*). The continuity equation applies generally to the physical process of *advection*, which refers to the motion of particles in a bulk fluid flow.

To adapt (3.4) to a particular model, we need to specify the relationship between q and u. As we remarked above, for the bloodstream model it is reasonable to assume a linear relationship,

$$q = vu, \tag{3.5}$$

where the velocity $v(t, x)$ is part of the input data for the equation. Under this assumption (3.4) reduces to

$$\frac{\partial u}{\partial t} + v\frac{\partial u}{\partial x} + u\frac{\partial v}{\partial x} = 0, \tag{3.6}$$

which is called the *linear conservation equation.*

3.2 Lagrangian Derivative and Characteristics

In this section we will discuss the strategy for solving a first-order PDE such as (3.6). The basic idea is to adopt the perspective of an observer traveling with velocity v. This is like taking measurements in a river from a raft drawn by the current. Once we fix a starting point for the observer, the observed concentration depends only on the time variable, thus reducing the equation to an ODE.

This principle applies to any first-order PDE of the form

$$\frac{\partial u}{\partial t} + v\frac{\partial u}{\partial x} + w = 0, \tag{3.7}$$

where $v = v(t, x)$ is independent of u. The zeroth-order term w could be a general function $w(t, x, u)$. A trajectory $t \mapsto x(t)$ is called a *characteristic* for the equation (3.7) if

$$\frac{dx}{dt}(t) = v(t, x(t)). \tag{3.8}$$

For v and $\partial v/\partial x$ continuous, Theorem 2.4 shows that a unique solution exists in the neighborhood of each starting point (t_0, x_0).

Example 3.1 Suppose $v(t, x) = at + b$, with a and b constant. Integration over t gives

$$x(t) = \frac{a}{2}t^2 + bt + x_0.$$

The characteristics are a family of curves indexed by the parameter x_0, as illustrated in Fig. 3.2. ◊

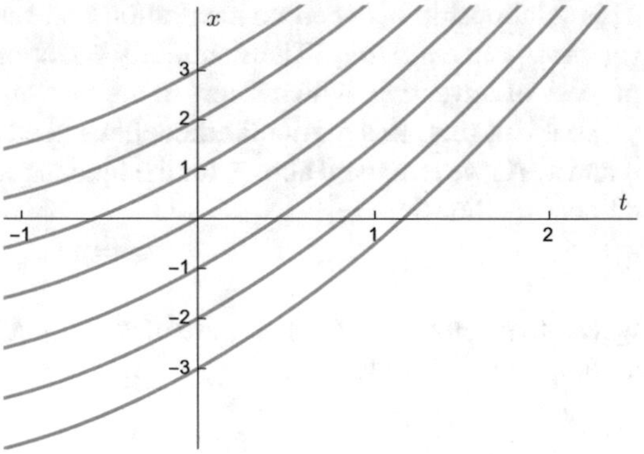

Fig. 3.2 Sample characteristics for the velocity $v(t, x) = 1 + 2t$

From the point of view of an observer carried by the flow, the measured concentration is $u(t, x(t))$. The observed rate of change is the derivative of this quantity,

$$\frac{Du}{Dt}(t) := \frac{d}{dt}u(t, x(t)), \tag{3.9}$$

called the *Lagrangian derivative* (or *material derivative*). This concept was developed by the 18th century mathematician and physicist Joseph-Louis Lagrange. Note that Du/Dt depends also on the initial value (t_0, x_0) that determines the characteristic. For convenience we suppress the initial point from the notation.

Theorem 3.2 *On each characteristic,* (3.7) *reduces to the ODE*

$$\frac{Du}{Dt} + \tilde{w} = 0, \tag{3.10}$$

where $\tilde{w}$ is the restriction of w to the characteristic,

$$\tilde{w}(t) := w\big(t, x(t), u(t, x(t))\big).$$

In particular, if $w = 0$ then u is constant on each characteristic.

Proof Applying the chain rule in (3.9) gives

$$\frac{Du}{Dt} = \frac{\partial u}{\partial t} + \frac{\partial u}{\partial x}\frac{dx}{dt},$$

with the understanding that the partial derivatives on the right are evaluated at the point $(t, x(t))$. Because $x(t)$ solves (3.8), this reduces to

$$\frac{Du}{Dt} = \frac{\partial u}{\partial t} + v\frac{\partial u}{\partial x}, \tag{3.11}$$

If we restrict the variables in (3.7) to $(t, x(t))$, then the first two terms match the right-hand side of (3.11), reducing the equation to (3.10).

If $w = 0$, then (3.10) becomes

$$\frac{Du}{Dt} = 0.$$

This is equivalent to the statement that $u(t, x(t))$ is independent of t. □

With Theorem 3.2 we can effectively reduce the PDE (3.7) to a pair of ODE, namely the characteristic equation (3.8) and the Lagrangian derivative equation (3.10). In many cases, solving these ODE will lead to an explicit formula for $u(t, x)$. This approach is referred to as the *method of characteristics*.

Example 3.3 For constants $a, b \in \mathbb{R}$, assume that $u(t, x)$ satisfies

$$\frac{\partial u}{\partial t} + (at + b)\frac{\partial u}{\partial x} = 0,$$

with the initial condition

$$u(0, x) = g(x),$$

for some function $g \in C^1(\mathbb{R})$. The characteristics for this velocity, $v(t, x) = at + b$, were computed in Example 3.1.

According to Theorem 3.2, u is constant along characteristics, implying that

$$u\left(t, \frac{a}{2}t^2 + bt + x_0\right) = u(0, x_0) = g(x_0), \tag{3.12}$$

for all $t \in \mathbb{R}$. This is not yet a formula for $u(t, x)$, but we can derive the solution formula by identifying

$$x = \frac{a}{2}t^2 + bt + x_0.$$

Solving for x_0 in terms of x and substituting this into (3.12) gives

$$u(t, x) = g\left(x - \frac{a}{2}t^2 - bt\right).$$

◊

Example 3.4 For steady flow through a pipe of changing diameter, the velocity would vary with position rather than time. Let $v(t, x) = a + bx$ for $x \geq 0$, with $a, b > 0$. The resulting characteristic equation (3.8) is

$$\frac{dx}{dt} = a + bx.$$

This can be solved by the standard ODE technique of separating the t and x variables to different sides of the equation:

$$\frac{dx}{a + bx} = dt.$$

Integration of both sides gives the general solution

$$\frac{1}{b} \ln(a + bx) = t + C,$$

with C a constant of integration. (Note that $a + bx > 0$ by our assumptions.) Solving for x gives

$$x(t) = \frac{1}{b} \left[e^{b(t+C)} - a \right].$$

Given the assumption $x \geq 0$, it is natural to index the characteristics by the start time t_0 such that $x(t_0) = 0$. With this convention, the family of solutions is

$$x(t) = \frac{a}{b} \left[e^{b(t-t_0)} - 1 \right]. \tag{3.13}$$

These characteristic curves are illustrated in Fig. 3.3.

With $v = a + bx$ the linear conservation equation (3.6) becomes

$$\frac{\partial u}{\partial t} + (a + bx) \frac{\partial u}{\partial x} + bu = 0.$$

Let us find the solution under the boundary condition

$$u(t, 0) = f(t). \tag{3.14}$$

Since $\partial v/\partial x = b$, (3.10) gives

$$\frac{Du}{Dt} + bu = 0.$$

This is a decay equation, with the family of exponential solutions

$$u(t, x(t)) = Ae^{-bt}.$$

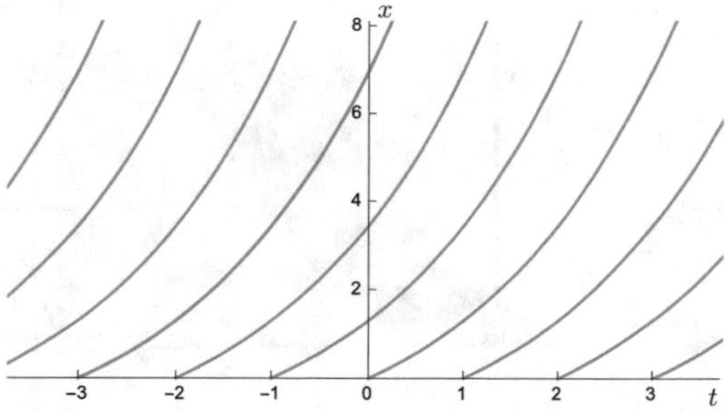

Fig. 3.3 Characteristic lines for the position-dependent velocity function of Example 3.4

To fix A, we substitute the starting point $(t_0, 0)$ into the equation and obtain

$$u(t, x(t)) = f(t_0)e^{-b(t-t_0)}. \tag{3.15}$$

Putting together (3.13) and (3.15) and applying the boundary condition (3.14) gives

$$u\left(t, \frac{a}{b}\left[e^{b(t-t_0)} - 1\right]\right) = f(t_0)e^{-b(t-t_0)}. \tag{3.16}$$

To express this as a function of (t, x), we set

$$x = \frac{a}{b}\left[e^{b(t-t_0)} - 1\right],$$

and solve for t_0 to obtain

$$t_0 = t + \frac{1}{b}\ln\left(\frac{a}{a+bx}\right).$$

Substituting this expression into (3.16) gives the final form of the solution:

$$u(t, x) = \left(\frac{a}{a+bx}\right) f\left(t + \frac{1}{b}\ln\left(\frac{a}{a+bx}\right)\right).$$

A sample solution is illustrated in Fig. 3.4 for $a = 1$, $b = \frac{1}{2}$. For this example the boundary condition $f(t)$ was taken to have support between $t = -1$ and $t = 1$, with a maximum at $t = 0$. The plots of $u(t, x)$ on the right show concentrations at a succession of times. Mass conservation is reflected in the fact that the total area under each of these curves is independent of t. ◊

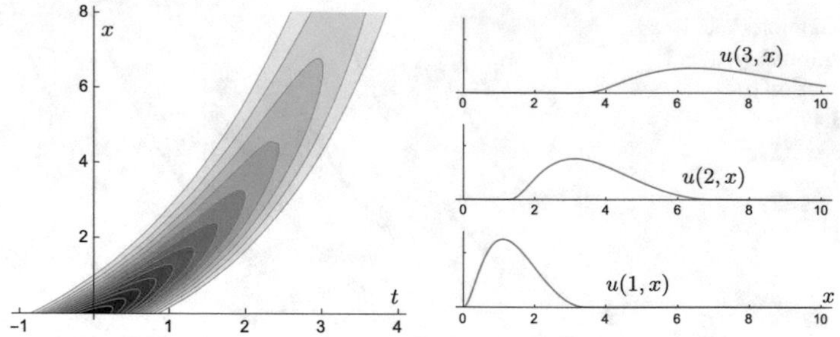

Fig. 3.4 Behavior of solutions for Example 3.4. In the contour plot on the *left*, darker regions correspond to higher concentration. The change in colors corresponds to exponential decay along the characteristics illustrated in Fig. 3.3

3.3 Higher-Dimensional Equations

For flow problems in more than one spatial dimension, we can develop a continuity equation analogous to (3.4) by the same reasoning as in Sect. 3.1. Suppose $u(t, \boldsymbol{x})$ represents a concentration defined for $t \in \mathbb{R}$ and $\boldsymbol{x} \in \mathbb{R}^n$. Let $\mathcal{R} \subset \mathbb{R}^n$ be a bounded region with C^1 boundary. The total mass within this region is given by the volume integral

$$m(t) := \int_{\mathcal{R}} u(t, \boldsymbol{x})\, d^n\boldsymbol{x}.$$

The flow of u is represented by a vector-valued flux density $\boldsymbol{q}(t, \boldsymbol{x})$. The interpretation of the flux density is that the rate at which mass passes through an $(n-1)$-dimensional surface is given by the surface integral of $\boldsymbol{q}$ over this surface. In particular, the rate at which mass exits $\mathcal{R}$ through the boundary is the quantity

$$\int_{\partial\mathcal{R}} \boldsymbol{\nu} \cdot \boldsymbol{q}\, dS,$$

where $\boldsymbol{\nu}$ is the outward unit normal vector defined on $\partial\mathcal{R}$.

Conservation of mass dictates that the mass within $\mathcal{R}$ can change only as mass enters or leaves through the boundary. In other words,

$$\frac{dm}{dt} = -\int_{\partial\mathcal{R}} \boldsymbol{\nu} \cdot \boldsymbol{q}\, dS. \tag{3.17}$$

Assuming that $\boldsymbol{q}$ is C^1 with respect to $\boldsymbol{x}$, the Divergence Theorem (Theorem 2.6) allows us to rewrite the flux integral as

$$\int_{\partial\mathcal{R}} \boldsymbol{q} \cdot \boldsymbol{\nu}\, dS = \int_{\mathcal{R}} \nabla \cdot \boldsymbol{q}\, d^n\boldsymbol{x}. \tag{3.18}$$

Note that since $\boldsymbol{q}$ depends on both t and $\boldsymbol{x}$, the notation $\nabla \cdot \boldsymbol{q}$ is slightly ambiguous. We follow the standard convention that vector calculus operators such as ∇ and Δ act only on spatial variables.

If u is C^1 with respect to t, then we can also differentiate the integral for m to obtain

$$\frac{dm}{dt} = \int_{\mathcal{R}} \frac{\partial u}{\partial t}\, d^n\boldsymbol{x}.$$

Combining this with (3.17) and (3.18) gives

$$\int_{\mathcal{R}} \left(\frac{\partial u}{\partial t} + \nabla \cdot \boldsymbol{q} \right) d^n\boldsymbol{x} = 0. \tag{3.19}$$

As in the one-dimensional case, we now observe that since (3.19) holds for an arbitrary region $\mathcal{R}$, the integrand must vanish. This is the higher-dimensional continuity equation:

$$\frac{\partial u}{\partial t} + \nabla \cdot \boldsymbol{q} = 0. \tag{3.20}$$

Suppose we make the linear assumption that $\boldsymbol{q} = \boldsymbol{v}u$ for a velocity field $\boldsymbol{v}$ which is independent of $\boldsymbol{u}$. The product rule for the divergence of a vector field is

$$\nabla \cdot (\boldsymbol{v}u) = (\nabla \cdot \boldsymbol{v})u + \boldsymbol{v} \cdot \nabla u.$$

Substituting this into (3.20) gives the higher-dimensional form of the linear conservation equation

$$\frac{\partial u}{\partial t} + \boldsymbol{v} \cdot \nabla u + (\nabla \cdot \boldsymbol{v})u = 0. \tag{3.21}$$

In the special case where $\nabla \cdot \boldsymbol{v} = 0$ the velocity field is called *solenoidal* (or *divergence-free*). This situation arises frequently in applications, because incompressible fluids like blood or water have solenoidal velocity fields.

The method of characteristics from Sect. 3.2 can be adapted directly to (3.21). Consider a somewhat more general first-order PDE in the form

$$\frac{\partial u}{\partial t} + \boldsymbol{v} \cdot \nabla u + w = 0, \tag{3.22}$$

with $\boldsymbol{v} = \boldsymbol{v}(t, \boldsymbol{x})$ and $w = w(t, x, u)$. The characteristics associated to this equation are by definition the solutions of

$$\frac{d\boldsymbol{x}}{dt}(t) = \boldsymbol{v}(t, \boldsymbol{x}(t)). \tag{3.23}$$

Theorem 2.4 guarantees that characteristics exist in the neighborhood of each starting point $(t_0, \boldsymbol{x}_0)$ provided $\boldsymbol{v}(t, \boldsymbol{x})$ and its partial derivatives with respect to $\boldsymbol{x}$ are continuous.

The Lagrangian derivative of $\boldsymbol{u}$ along $\boldsymbol{x}(t)$ is defined as before by

$$\frac{Du}{Dt}(t) := \frac{d}{dt}u(t, \boldsymbol{x}(t)).$$

The higher-dimensional version of Theorem 3.2 is the following:

Theorem 3.5 *On each characteristic curve, the PDE* (3.22) *reduces to the ODE*

$$\frac{Du}{Dt} + \tilde{w} = 0, \tag{3.24}$$

where $\tilde{w}$ denotes the restriction of w to the characteristic. In particular, if $w = 0$ then u is constant on each characteristic.

Proof By the chain rule,

$$\frac{Du}{Dt}(t) = \frac{\partial u}{\partial t}(t, \boldsymbol{x}(t)) + \nabla u(t, \boldsymbol{x}(t)) \cdot \frac{d\boldsymbol{x}}{dt}(t).$$

Since $\boldsymbol{x}(t)$ satisfies (3.23), this gives

$$\frac{Du}{Dt} = \frac{\partial u}{\partial t} + \boldsymbol{v} \cdot \nabla u.$$

Substituting this into (3.22) reduces the equation to (3.24).

If $w = 0$ the equation becomes

$$\frac{Du}{Dt} = 0,$$

which means precisely that u is constant along the characteristic curves. □

Example 3.6 Consider a two-dimensional channel modeled as $\Omega = \mathbb{R} \times [-1, 1]$ with coordinates $\boldsymbol{x} = (x_1, x_2)$. The velocity field

$$\boldsymbol{v}(t, \boldsymbol{x}) := (1 - x_2^2, 0). \tag{3.25}$$

is solenoidal and vanishes on the boundary $\{x_2 = \pm 1\}$. The characteristic line originating from $(a, b) \in \Omega$ at $t = 0$ is

$$\boldsymbol{x}(t) = \big(a + (1 - b^2)t, b\big).$$

Let us consider the conservation equation (3.21) for $(t, \boldsymbol{x}) \in \mathbb{R} \times \Omega$, with $\boldsymbol{v}$ given by (3.25), subject to the initial condition

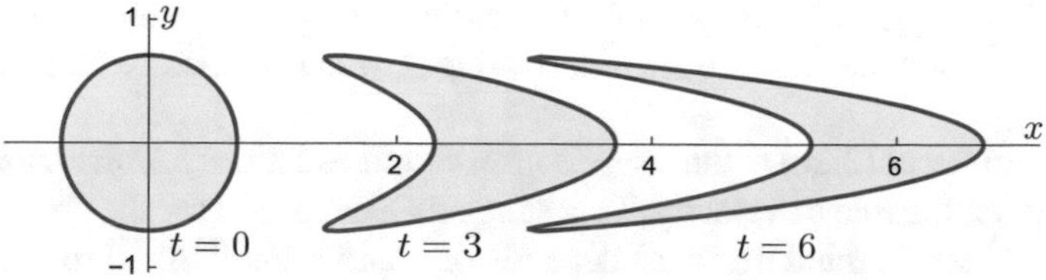

Fig. 3.5 Evolution of a circle according to the two-dimensional flow in Example 3.6

$$u(0, \boldsymbol{x}) = g(\boldsymbol{x}),$$

for $g \in C^1(\Omega)$. Since $\boldsymbol{v}$ is solenoidal, Theorem 3.5 implies that u is constant on characteristics. This gives the relation

$$u\big(t, a + (1 - b^2)t, b\big) := g(a, b).$$

Rewriting this as a function of (t, x, y) gives

$$u(t, x, y) = g\big(x - (1 - b^2)t, y\big).$$

Figure 3.5 illustrates the evolution of a circular "ink spot" distribution under this flow. Conservation of mass is reflected in the fact that the area of the spot is independent of t. ◊

For applications of Theorem 3.5 on a bounded domain $\Omega \in \mathbb{R}^n$, the specification of boundary conditions can be quite a complicated problem, especially if the velocity is time-dependent. (We avoided this problem in Example 3.6 by taking $\boldsymbol{v}$ tangent to $\partial\Omega$.) We will illustrate this issue in the exercises.

3.4 Quasilinear Equations

The method of characteristics remains an important tool for analysis of first-order PDE even in the nonlinear case. In this section we will illustrate the application of this method to the continuity equation (3.20) in the case of a flux term $\boldsymbol{q}$ that depends on the concentration u.

To simplify the analysis, we assume that $\boldsymbol{q} = \boldsymbol{q}(u)$, with no explicit dependence on t and x. By the chain rule, (3.20) then reduces to the form

$$\frac{\partial u}{\partial t} + \mathbf{a}(u) \cdot \nabla u = 0, \tag{3.26}$$

where $\mathbf{a}(u) := d\boldsymbol{q}/du$. This type of PDE is called *quasilinear*, which means that the equation is linear in the highest-order derivatives (which are merely first order in this case).

A comparison of (3.26) to the linear conservation equation (3.21) shows that $a(u)$ is now playing the role of velocity. This suggests a definition for the characteristics, but we must keep in mind that $\mathbf{a}(u)$ depends on t and x implicitly through u.

Theorem 3.7 *Suppose that* $u \in C^1([0, T] \times \Omega)$ *is a solution of* (3.26) *for some region* $\Omega \subset \mathbb{R}^n$, *with* $\mathbf{a} \in C^1(\mathbb{R}; \mathbb{R}^n)$. *Then for each* $\boldsymbol{x}_0 \in \Omega$, *u is constant along the characteristic line defined by*

$$\boldsymbol{x}(t) = \boldsymbol{x}_0 + \mathbf{a}(u(0, \boldsymbol{x}_0))t.$$

Proof Suppose that a solution u exists. Let $\boldsymbol{x}(t)$ be the solution to the ODE

$$\frac{d\boldsymbol{x}}{dt}(t) = \mathbf{a}(u(t, \boldsymbol{x}(t))), \qquad \boldsymbol{x}(0) = \boldsymbol{x}_0,$$

for $t \in [0, T]$. Existence of such a characteristic is guaranteed by Theorem 2.4, at least for t near 0, because the composition $\mathbf{a} \circ u$ is C^1 as a function of (t, x) by the assumptions on $\mathbf{a}$ and u.

To establish the claim that $u(t, \boldsymbol{x}(t))$ is independent of t, we use the chain rule to differentiate

$$\begin{aligned} \frac{d}{dt}u(t, \boldsymbol{x}(t)) &= \frac{\partial u}{\partial t}(t, \boldsymbol{x}(t)) + \nabla u(t, \boldsymbol{x}(t)) \cdot \frac{d\boldsymbol{x}}{dt}(t) \\ &= \frac{\partial u}{\partial t}(t, \boldsymbol{x}(t)) + \mathbf{a}(u(t, \boldsymbol{x}(t))) \cdot \nabla u(t, \boldsymbol{x}(t)). \end{aligned}$$

The right-hand side vanishes by (3.26), so that

$$\frac{d}{dt}u(t, \boldsymbol{x}(t)) = 0.$$

This implies that

$$u(t, \boldsymbol{x}(t)) = u(0, \boldsymbol{x}_0),$$

which means that $\mathbf{a}(u(t, \boldsymbol{x}(t)))$ is also constant. The characteristic equation reduces to

$$\frac{d\boldsymbol{x}}{dt}(t) = \mathbf{a}(u(0, \boldsymbol{x}_0)),$$

and we can integrate over t to compute $\boldsymbol{x}(t)$. □

In contrast to the characteristic equation (3.8) in the linear case, the equation for $\boldsymbol{x}(t)$ here depends on the initial condition $u(0, \boldsymbol{x}_0)$. Furthermore, it is important to keep in mind that Theorem 3.7 does not imply that a solution to (3.26) exists; this is assumed as a hypothesis. As we will see below, it is possible that the conclusion of the theorem will lead to a contradiction, in the form of multiple values for the solution at the same point. The implication in such a case is that a classical solution does not exist.

To illustrate the application of Theorem 3.7, let us consider a simple model for traffic on a single-lane road of infinite length, parametrized by $x \in \mathbb{R}$. Let $u(t, x)$ denote the linear density of cars at a given point and time. Cars are discrete objects, of course, but for modeling purposes we can assume that u is a C^1 function that describes the density in an aggregate sense.

In traffic flow, the density of cars affects the flow velocity, with traffic slowing down and possibly stopping as the density increases. A standard way to model this effect is to set a maximum value for the velocity v_m (presumably the speed limit). The velocity is assumed to take its maximum value at $u = 0$ and decrease linearly as u increases, up to some maximum value u_m for which $v = 0$. In other words, for this model $u \in [0, u_m]$ and

$$v(u) := v_m \left(1 - \frac{u}{u_m}\right).$$

Since $v \geq 0$, the model always assumes that traffic moves to the right.

To eliminate the constants and focus on the equation itself, let us set $v_m = 1$ and $u_m = 1$, reducing the velocity equation to

$$v(u) = 1 - u$$

for $u \in [0, 1]$. The corresponding flux is

$$q(u) = u - u^2.$$

Substituting these assumptions into (3.26), we obtain a quasilinear equation called the *traffic equation*:

$$\frac{\partial u}{\partial t} + (1 - 2u)\frac{\partial u}{\partial x} = 0. \tag{3.27}$$

Suppose we impose a general initial condition of the form

$$u(0, x) = h(x),$$

for some $h : \mathbb{R} \to [0, 1]$. Assuming a solution exists, Theorem 3.7 gives the family of characteristics

$$x(t) = x_0 + (1 - 2h(x_0))t. \tag{3.28}$$

Therefore, the solution u must satisfy

$$u\big(t, x_0 + (1 - 2h(x_0))t\big) = h(x_0). \tag{3.29}$$

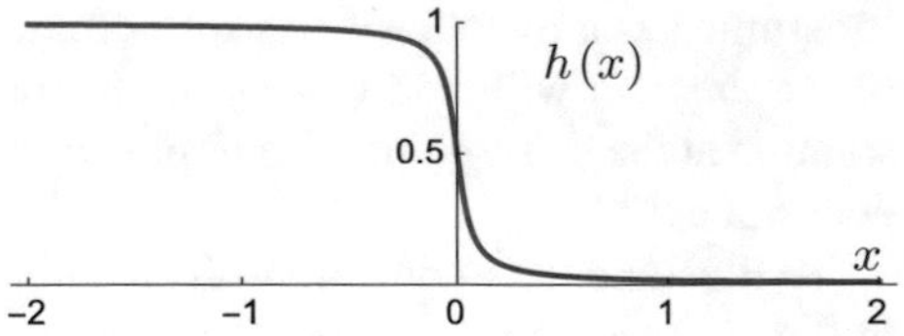

Fig. 3.6 Initial traffic density modeling a line of cars stopped at a traffic light

As we will demonstrate in the examples below, (3.29) leads to a solution formula for some choices of h, while for others it leads to a contradiction.

Example 3.8 Figure 3.6 shows a plot of the initial condition

$$h(x) = \frac{1}{2} - \frac{1}{\pi}\arctan(20x),$$

which could represent a line of cars stopped at a traffic light at the point $x = 0$. The corresponding characteristic lines as given by (3.28) are plotted in Fig. 3.7.

To derive a formula for $u(t, x)$ from (3.29), we need to invert the equation

$$x = x_0 + (1 - 2h(x_0))t,$$

to express x_0 as a function of t and x. For the function h given above it is not possible to do this explicitly. However, there is a unique solution for each (t, x), which can easily be calculated numerically. The resulting solutions are shown in Fig. 3.8.

□

Example 3.9 In order to solve the traffic equation explicitly, let us simplify the initial condition to the piecewise linear function

$$h(x) = \begin{cases} 1, & x \le 0, \\ 1 - x, & 0 < x < 1, \\ 0, & x \ge 1. \end{cases}$$

This is not C^1, but the resulting solution could be interpreted as a weak solution in the sense described in Sect. 1.2. We will discuss the precise definition in Chap. 10.

By the formula from Theorem 3.7, the characteristic lines are

$$x(t) = \begin{cases} x_0 - t, & x_0 \le 0, \\ x_0 + (2x_0 - 1)t, & 0 < x_0 < 1, \\ x_0 + t, & x_0 \ge 1. \end{cases} \tag{3.30}$$

Solving these equations for x_0 gives

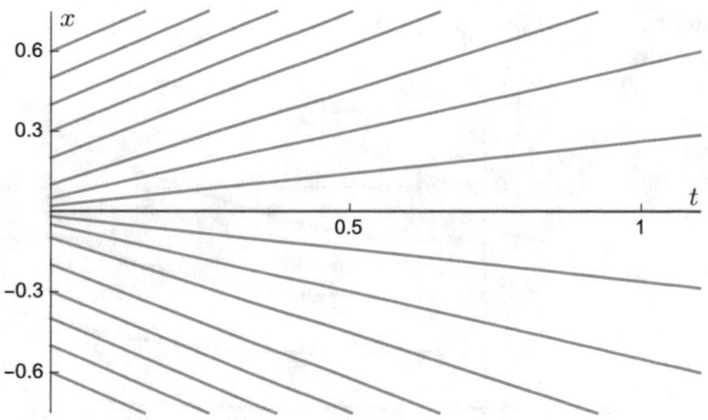

Fig. 3.7 Characteristic lines for the initial density shown in Fig. 3.6

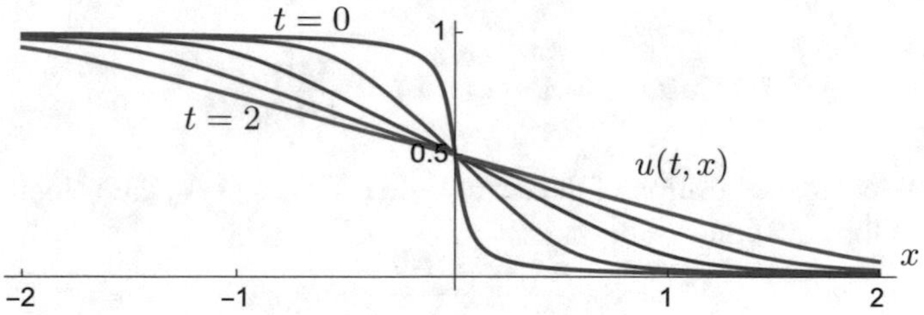

Fig. 3.8 Solutions for the traffic light problem

$$x_0 = \begin{cases} x+t, & x \le -t, \\ \frac{x+t}{1+2t}, & -t < x < 1+t, \\ x-t. & x \ge 1+t. \end{cases}$$

Therefore, by the solution formula (3.29), the solution is

$$u(t,x) = \begin{cases} 1, & x \le -t, \\ 1 - \frac{x+t}{1+2t}, & -t < x < 1+t, \\ 0. & x \ge 1+t. \end{cases} \tag{3.31}$$

This is a continuous function, but differentiability fails on the lines $x = -t$ and $x = 1+t$. Away from these lines it is easy to check that u solves (3.27).

Despite the lack of smoothness, this solution is quite reasonable. To illustrate this, let us trace the motion of a particular car starting from the position $x_0 \le 0$. The velocity of the car is given by the flow rate $v(u) = 1 - u$. The initial density at x_0 is $u = 1$, so the car is stationary for a time. According to (3.31), at $t = -x_0$ the value of (t, x) enters the region where $-t < x < 1+t$ and so at this time the density starts to decrease and the car starts to move. For (t, x) in this range, (3.31) gives

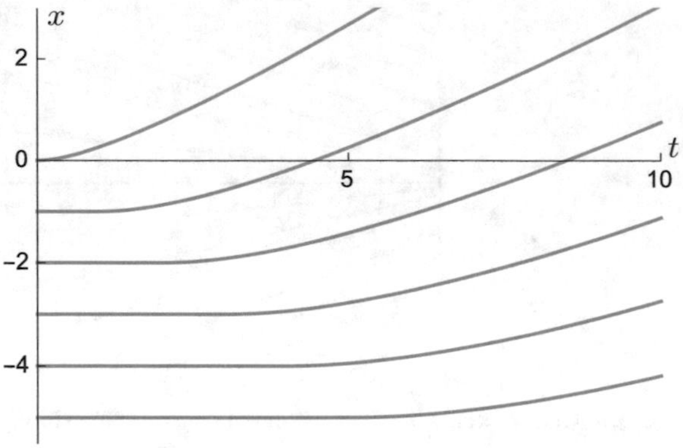

Fig. 3.9 Trajectories of individual cars according to the model of Example 3.9

$$v(t,x) = 1 - u(t,x) = \frac{x+t}{1+2t}. \tag{3.32}$$

Let $s(t)$ denote the position of the car at time t. For $t \geq -x_0$ the velocity formula (3.32) gives the equation

$$\frac{ds}{dt} = \frac{s+t}{1+2t}, \tag{3.33}$$

The initial condition at $t = -x_0$ is the original starting point $s(-x_0) = x_0$. The standard ODE method of integrating factors can be used to solve (3.33), yielding

$$s(t) = \begin{cases} x_0, & 0 \leq t \leq -x_0, \\ 1 + t - \sqrt{(1-2x_0)(1+2t)}, & t \geq -x_0. \end{cases}$$

These trajectories are illustrated in Fig. 3.9. As we might expect, the cars further back in the line wait longer before moving, but each car eventually moves forward and gradually accelerates. ◊

Example 3.10 Consider the initial condition

$$h(x) = \frac{1}{2} + \frac{1}{\pi}\arctan(20x),$$

as shown in Fig. 3.10. This is the reverse of the initial condition of Example 3.8. The characteristics specified in Theorem 3.7 now cross each other, as illustrated in Fig. 3.11. The existence of crossings implies that a classical solution with this initial condition cannot exist beyond the time of the first crossing.

If we were to trace the trajectories of individual cars, as we did in Example 3.9, we would see that these also intersect each other at the points where characteristics cross. In effect, the model predicts the formation of a traffic jam. ◊

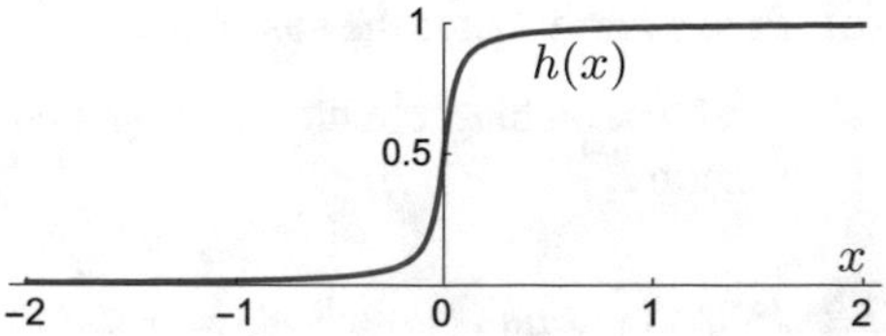

Fig. 3.10 Initial traffic density with a near-maximum density of cars to the right

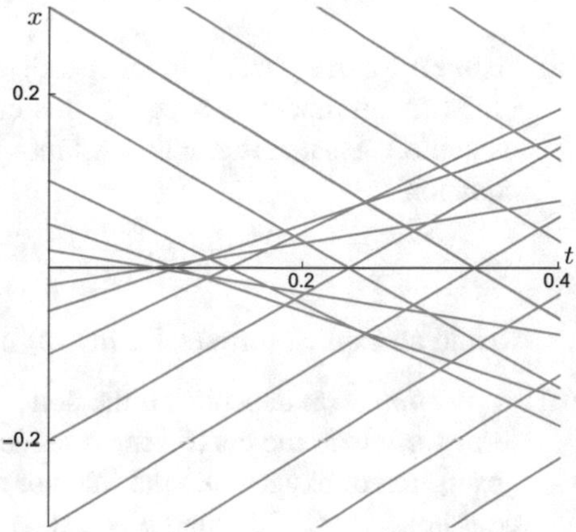

Fig. 3.11 Conflicting characteristic lines for the initial density shown in Fig. 3.10

A crossing of characteristics as observed in Example 3.10 is called a *shock*. After the shock, the solution is forced to have discontinuities. The proper interpretation of this situation requires weak solutions, for which discontinuities are allowed. We will return to this issue in Chap. 10.

3.5 Exercises

3.1 Consider the conservation equation with a constant velocity $c > 0$,

$$\frac{\partial u}{\partial t} + c\frac{\partial u}{\partial x} = 0,$$

on the quadrant $t \geq 0$, $x \geq 0$. Suppose the boundary and initial conditions are

$$\begin{cases} u(0, x) = g(x), & x \geq 0, \\ u(t, 0) = h(t), & t \geq 0, \end{cases}$$

for $g, h \in C^1[0, \infty)$.

(a) Find a formula for the solution $u(t, x)$ in terms of g and h.

(b) Find a matching condition for g and h that will ensure that $u(t, x)$ is a C^1 function.

3.2 In the continuity equation (3.4), external factors that break the conservation of mass are accounted for by adding terms to the right-hand side.

(a) A *forcing term* $f(t, x)$ is independent of the existing concentration. (In the bloodstream model of Sect. 3.1, this could represent intravenous injection, for example.) Assume that c is constant, $f \in C^1(\mathbb{R}^2)$, and $g \in C^1(\mathbb{R})$. Solve the equation

$$\frac{\partial u}{\partial t} + c\frac{\partial u}{\partial x} = f, \qquad u(0, x) = g(x),$$

to find an explicit formula for $u(t, x)$ in terms of f and g.

(b) A *reaction term* depends on the concentration u. The simplest case is a linear term γu where the coefficient is some function $\gamma(t, x)$. (This could represent absorption of oxygen into the walls of the artery, for example.) Assume that c is constant, $\gamma \in C^1(\mathbb{R}^2)$, and $g \in C^1(\mathbb{R})$. Solve the equation

$$\frac{\partial u}{\partial t} + c\frac{\partial u}{\partial x} = \gamma u, \qquad u(0, x) = g(x),$$

to find an explicit formula for $u(t, x)$ in terms of γ and g.

3.3 Assume that u satisfies the linear conservation equation

$$\frac{\partial u}{\partial t} + 2t\frac{\partial u}{\partial x} = 0,$$

for $t \in \mathbb{R}$ and $x \in [0, 1]$. Suppose the boundary conditions are given by

$$u(t, 0) = h_0(t), \quad u(t, 1) = h_1(t).$$

Find a relation between h_0 and h_1. (This shows that we can only impose a boundary condition at one side of the interval $[0, 1]$.)

3.4 If the spatial domain in the linear conservation equation (3.21) is a bounded region $\Omega \subset \mathbb{R}^n$, then for a given velocity field $\boldsymbol{v}$, the *inflow boundary* $\partial\Omega_{\text{in}} \in \partial\Omega$ is defined as the set of boundary points where $\boldsymbol{v}$ points into Ω. Fixing boundary conditions on the inflow boundary will generally determine the solution in the interior. Suppose $\Omega = (-1, 1) \times (-1, 1) \in \mathbb{R}^2$ with coordinates (x_1, x_2). For the velocity fields below, determine the characteristics and specify the inflow boundary. Draw a sketch of Ω for each case, indicating these features.

(a) $\boldsymbol{v}(x_1, x_2) = (x_2, 1)$.

(b) $\boldsymbol{v}(x_1, x_2) = (1, -x_2)$.

3.5 Suppose that a section of of a river is modeled as a rectangle $\Omega = (0, \ell) \times (0, 1) \subset \mathbb{R}^2$, parametrized by (x_1, x_2). Assume the flow is parallel to the x_1-axis, with velocity

$$\boldsymbol{v}(x_1, x_2) = (f(x_2), 0),$$

for some positive function f on $(0, 1)$. Assume also that the concentration on the left boundary $\{x_1 = 0\}$ is given by

$$u(t, 0, x_2) = h(t, x_2).$$

Find a formula for $u(t, x_1, x_2)$ in terms of the functions h and f.

3.6 *Burgers' equation* is a simple quasilinear equation that appears in models of gas dynamics,

$$\frac{\partial u}{\partial t} + u\frac{\partial u}{\partial x} = 0.$$

(a) Use the method of characteristics as described in Sect. 3.4 to find a formula for the solution $u(t, x)$ given the initial condition

$$u(0, x) = \begin{cases} 0, & x \le 0, \\ \frac{x}{a}, & 0 < x < a, \\ 1, & x \ge a. \end{cases}$$

(b) Suppose $a > b$ and

$$u(0, x) = \begin{cases} a, & x \le 0, \\ a(1 - x) + bx, & 0 < x < 1, \\ b, & x \ge 1. \end{cases}$$

Show that all of the characteristics originating from $x_0 \in [0, 1]$ meet at the same point (thus creating a shock).

3.7 In the mid-19th century, William Hamilton and Carl Jacobi developed a formulation of classical mechanics based on ideas from geometric optics. In this approach the dynamics of a free particle in $\mathbb{R}$ are described by a generating function $u(t, x)$ satisfying the *Hamilton-Jacobi equation*:

$$\frac{\partial u}{\partial t} + \frac{1}{2}\left(\frac{\partial u}{\partial x}\right)^2 = 0. \tag{3.34}$$

Assume that $u \in C^1([0,\infty) \times \mathbb{R}^n)$ is a solution of (3.34). By analogy with Theorem 3.7, a characteristic of (3.34) is defined as a solution of

$$\frac{dx}{dt}(t) = \frac{\partial u}{\partial x}(t, x(t)), \quad x(0) = x_0. \tag{3.35}$$

(a) Assuming that $x(t)$ solves (3.35), use the chain rule to compute d^2x/dt^2.

(b) Differentiate (3.34) with respect to x and then restrict the result to $(t, x(t))$, where $x(t)$ solves (3.35). Conclude from (a) that to

$$\frac{d^2x}{dt^2} = 0.$$

Hence, for some constant v_0 (which depends on the characteristic),

$$x(t) = x_0 + v_0 t.$$

(c) Show that the Lagrangian derivative of u along $x(t)$ satisfies

$$\frac{Du}{Dt} = \frac{1}{2}v_0^2,$$

implying that

$$u(t, x_0 + v_0 t) = u(0, x_0) + \frac{1}{2}v_0^2 t.$$

(d) Use this approach to find the solution $u(t, x)$ under the initial condition

$$u(0, x) = x^2.$$

(For the characteristic starting at $(0, x_0)$, note that you can compute v_0 by evaluating (3.35) at $t = 0$.)

Chapter 4
The Wave Equation

As we noted in Sect. 1.2, d'Alembert's derivation of the wave equation in the 18th century was an early milestone in the development of PDE theory. In this chapter we will develop this equation as a model for the vibrating string problem, and derive d'Alembert's explicit solution in one dimension using the method of characteristics introduced in Chap. 3.

In higher dimensions the wave equation is used to model electromagnetic or acoustic waves. We will discussion the derivation of the acoustic model later in Sect. 4.5. A clever reduction trick allows the solution formula for $\mathbb{R}^n$ to be deduced from the one-dimensional case. The resulting integral formula yields insight into the propagation of waves in different dimensions.

The chapter concludes with a discussion of the energy of a solution, based on the physical principles of kinetic and potential energy.

4.1 Model Problem: Vibrating String

Consider a flexible string that is stretched tight between two points, like the strings on a violin or guitar. The stretching of the string creates a tension force T that pulls in both directions at each point along its length. For simplicity, let us assume that any other forces acting on the string, including gravity, are negligible compared to the tension. The linear density of mass ρ is taken to be constant along the string.

For a violin string it is also reasonable to assume that the displacement of the string is extremely small relative to its length. This assumption justifies taking T to be a fixed constant, ignoring the additional stretching that occurs when the string is displaced. It also allows us to treat horizontal and vertical components of the displacement independently, so we can restrict our attention to the vertical.

The original version of the book was revised: Belated corrections from author have been incorporated. The erratum to the book is available at https://doi.org/10.1007/978-3-319-48936-0_14

D. Borthwick, *Introduction to Partial Differential Equations*,
Universitext, DOI 10.1007/978-3-319-48936-0_4

Let the string be parametrized by $x \in [0, \ell]$. The vertical displacement as a function of time is denoted by $u(t, x)$. To develop an equation for u, we first discretize the model by subdividing the total length ℓ into segments of length $\Delta x = \ell/n$ for some large n. Each segment has a mass $\rho\Delta x$ and is subject to the tension forces pulling in the direction of its neighbors on either side.

For $j = 0, \ldots n$, let $x_j := j\Delta x$ be the position of the jth segment along the string. The segments $j = 0$ and $j = n$ represent the fixed endpoints, with $j = 1, \ldots, n-1$ in the interior. Let $u(t, x_j)$ denote the vertical displacement of the jth segment as a function of time. Figure 4.1 illustrates this discretization (with displacements greatly exaggerated).

To develop an equation for the string, we apply Newton's laws of motion to the segments of the discretization, as if they were single particles. The jth particle is being pulled by its neighbors with a force T on each side. Unless the string is straight, these forces are not quite aligned.

In terms of the angles labeled in Fig. 4.2, the net vertical force on a single segment is

$$\Delta F(t, x_j) = T \sin \alpha_j + T \sin \beta_j.$$

We have assumed that the relative displacements are extremely small, so the angles α_j, β_j will be very small also. To leading order, we can replace the sines by tangents, which are linear in u,

$$\sin \alpha_j \approx \frac{u(t, x_{j-1}) - u(t, x_j)}{\Delta x}, \quad \sin \beta_j \approx \frac{u(t, x_{j+1}) - u(t, x_j)}{\Delta x}.$$

With this linear approximation, the net vertical force at the point x_j becomes

$$\Delta F(t, x_j) = \frac{T}{\Delta x}\big[u(t, x_{j+1}) + u(t, x_{j-1}) - 2u(t, x_j)\big]. \tag{4.1}$$

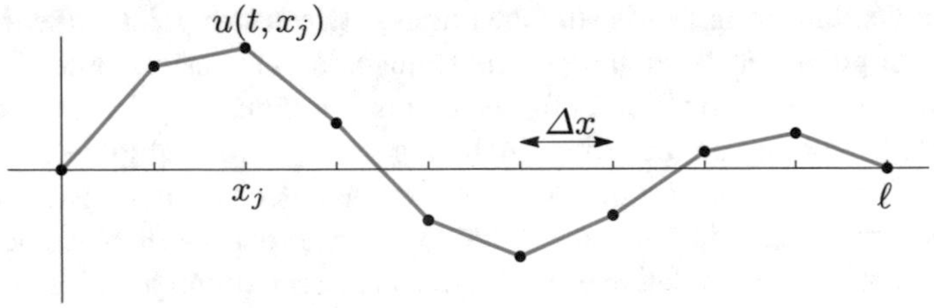

Fig. 4.1 Discrete model for the displacement of the string

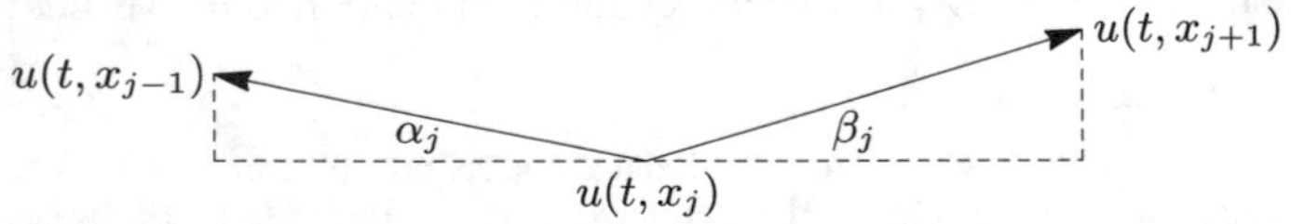

Fig. 4.2 Discrete model for the displacement of the string

The equation of motion for the jth segment now comes from Newton's law: mass times acceleration equals force. At the point x_j this translates to

$$\rho \Delta x \frac{\partial^2 u}{\partial t^2}(t, x_j) = \Delta F(t, x_j). \tag{4.2}$$

Using (4.1) on the right then gives

$$\frac{\partial^2 u}{\partial t^2}(t, x_j) = \frac{T}{\rho} \frac{u(t, x_{j+1}) + u(t, x_{j-1}) - 2u(t, x_j)}{(\Delta x)^2}. \tag{4.3}$$

The final step is to take the continuum limit $n \to \infty$ and $\Delta x \to 0$. Assuming that u is twice continuously differentiable as a function of x, we can deduce from the quadratic Taylor approximation of $u(t, x)$ that

$$\lim_{\Delta x \to 0} \frac{u(t, x + \Delta x) + u(t, x - \Delta x) - 2u(t, x)}{(\Delta x)^2} = \frac{\partial^2 u}{\partial x^2}(t, x).$$

Hence, taking $\Delta x \to 0$ in (4.3) gives

$$\frac{\partial^2 u}{\partial t^2} - \frac{T}{\rho} \frac{\partial^2 u}{\partial x^2} = 0. \tag{4.4}$$

This is the one-dimensional *wave equation*. The fixed ends of the string correspond to Dirichlet boundary conditions,

$$u(t, 0) = u(t, \ell) = 0.$$

4.2 Characteristics

For convenience, set $c^2 := T/\rho$ in (4.4), assuming $c > 0$, and rewrite the equation as

$$\frac{\partial^2 u}{\partial t^2} - c^2 \frac{\partial^2 u}{\partial x^2} = 0. \tag{4.5}$$

The constant c is called the *propagation speed*, for reasons that will become apparent as we analyze the equation.

Let the physical domain be $x \in \mathbb{R}$ for the moment; we will discuss boundary conditions later. The key to applying the method of characteristics to (4.5) is that the differential operator appearing in the equation factors as a product of two first-order operators, i.e.,

$$\frac{\partial^2}{\partial t^2} - c^2 \frac{\partial^2}{\partial x^2} = \left(\frac{\partial}{\partial t} + c\frac{\partial}{\partial x}\right)\left(\frac{\partial}{\partial t} - c\frac{\partial}{\partial x}\right). \tag{4.6}$$

Individually, these operators have characteristic lines $t \mapsto x_0 \pm ct$. Both sets of characteristics will play an important role here.

Theorem 4.1 *Under the initial conditions*

$$u(0, x) = g(x), \qquad \frac{\partial u}{\partial t}(0, x) = h(x), \tag{4.7}$$

for $g \in C^2(\mathbb{R})$ and $h \in C^1(\mathbb{R})$, the wave equation (4.5) *admits a unique solution*

$$u(t, x) = \frac{1}{2}\Big[g(x + ct) + g(x - ct)\Big] + \frac{1}{2c}\int_{x-ct}^{x+ct} h(\tau)\, d\tau. \tag{4.8}$$

Proof Consider the auxiliary function $w(t, x)$ defined by

$$w := \frac{\partial u}{\partial t} - c\frac{\partial u}{\partial x}. \tag{4.9}$$

By (4.6), w satisfies the linear conservation equation

$$\frac{\partial w}{\partial t} + c\frac{\partial w}{\partial x} = 0.$$

The characteristics for this equation are given by $x_+(t) = x_0 + ct$. By Theorem 3.2 the unique solution with an initial condition $w(0, x) = w_0(x)$ is

$$w(t, x) = w_0(x - ct). \tag{4.10}$$

We will relate w_0 back to the initial conditions g and h in a moment.

With w given by (4.10), the definition (4.9) can be regarded as a linear conservation equation for u,

$$\frac{\partial u}{\partial t} - c\frac{\partial u}{\partial x} = w, \tag{4.11}$$

where w acts as a forcing term as described in Exercise 3.2. The characteristics of (4.11) are $x_-(t) = x_0 - ct$. By Theorem 3.2, we can thus reduce the equation to the form

$$\frac{d}{dt}u(t, x_0 - ct) = w(t, x_0 - ct). \tag{4.12}$$

The unique solution to (4.12) under the initial condition $u(0, x) = g(x)$ is given by direct integration with respect to time:

$$u(t, x_0 - ct) = g(x_0) + \int_0^t w(s, x_0 - cs)\, ds.$$

Setting $x = x_0 - ct$ then gives

$$u(t,x) = g(x+ct) + \int_0^t w(s, x - c(s-t))\, ds.$$

Using the formula (4.10) for the solution w, we obtain

$$u(t,x) = g(x+ct) + \int_0^t w_0(x - 2cs + ct)\, ds.$$

As a final step, the substitution $\tau := x + ct - 2cs$ gives

$$u(t,x) = g(x+ct) + \frac{1}{2c}\int_{x-ct}^{x+ct} w_0(\tau)\, d\tau. \tag{4.13}$$

The function w_0 can be computed from the initial conditions (4.7),

$$\begin{aligned} w_0(x) &:= \frac{\partial u}{\partial t}(0,x) - c\frac{\partial u}{\partial x}(0,x) \\ &= h(x) - c\frac{\partial g}{\partial x}(x). \end{aligned}$$

The w_0 contribution to (4.13) is then given by

$$\begin{aligned} \frac{1}{2c}\int_{x-ct}^{x+ct} w_0(\tau)\, d\tau &= \frac{1}{2c}\int_{x-ct}^{x+ct} h(\tau)\, d\tau - \frac{1}{2}\int_{x-ct}^{x+ct} \frac{\partial g}{\partial x}(\tau)\, d\tau \\ &= \frac{1}{2c}\int_{x-ct}^{x+ct} h(\tau)\, d\tau - \frac{1}{2}\big[g(x+ct) - g(x-ct)\big]. \end{aligned}$$

Substituting back into (4.13) now gives the formula (4.8). □

To highlight the role played by the characteristic lines in the solution of Theorem 4.1, consider the functions

$$u_\pm(x) := \frac{1}{2}g(x) \mp \frac{1}{2c}\int_0^x h(\tau)\, d\tau.$$

In terms of $u_\pm$, the solution (4.8) simplifies to

$$u(x,t) = u_+(x-ct) + u_-(x+ct), \tag{4.14}$$

matching the form of the solution stated in (1.3). The subscripts in $u_\pm$ indicate the propagation direction, i.e., u_+ propagates to the right and u_- to the left. In either direction the speed of propagation is the parameter c.

Example 4.2 Consider the wave equation (4.5) with the initial conditions $h(x) = 0$ and

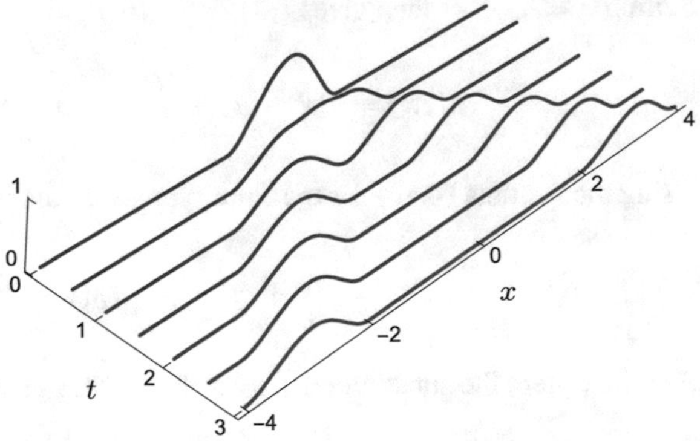

Fig. 4.3 Evolution of a solution to the wave equation

$$g(x) = \begin{cases} (1-x^2)^2, & |x| \leq 1, \\ 0, & |x| > 1. \end{cases}$$

By (4.8) the solution is the superposition of two localized bumps, which propagate in opposite directions as illustrated in Fig. 4.3. ◊

In Example 4.2 the initial condition was supported in $[-1, 1]$, and we can see in Fig. 4.3 that the resulting solution has support in a V-shaped region. This region could be identified as the span of the characteristic lines emerging from the initial support interval.

This restriction of the support of a solution is closely related to *Huygens' principle*, an empirical law for propagation of light waves published by Christiaan Huygens in 1678. The one-dimensional wave equation exhibits a special, strict form of this principle:

Theorem 4.3 **(Huygens' principle in dimension one)** *Suppose u solves the wave equation* (4.5) *for $t \geq 0$, $x \in \mathbb{R}$, with initial data given by* (4.7). *If the functions g, h are supported in a bounded interval $[a, b]$, then*

$$\operatorname{supp} u \subset \Big\{(t, x) \in \mathbb{R}^+ \times \mathbb{R};\ x \in [a - ct, b + ct]\Big\}.$$

Proof Consider the components of the solution (4.8). The g term will vanish unless $x \pm ct \in [a, b]$. The support of this term is thus restricted to $x \in [a - ct, b - ct]$ or $x \in [a+ct, b+ct]$. As for the h term, the integral over τ will vanish unless the interval $[x - ct, x + ct]$ intersects $[a, b]$, which occurs only when $x \in [a - ct, b + ct]$. □

The restriction of support described in Theorem 4.3 is illustrated in Fig. 4.4. The term g contributes only in the regions shown in blue, but the h term may contribute throughout the full support region. However, the solution is constant (equal to $\int_a^b h(\tau)d\tau$ when $[a, b]$ is contained in $[x - ct, x + ct]$. This constant region shown in purple in Fig. 4.4.

Example 4.4 Suppose the initial data from Example 4.2 are altered to include a singularity at $x = 0$. For example,

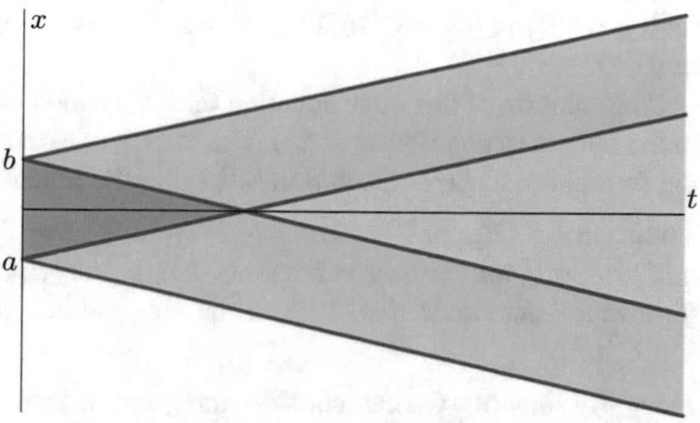

Fig. 4.4 Support of a wave solution with initial data in a bounded interval

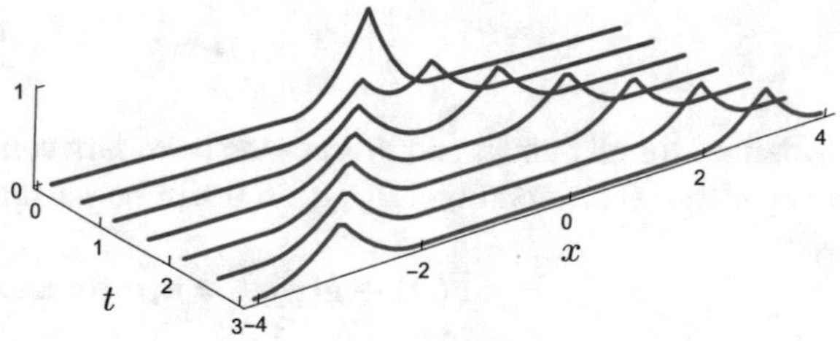

Fig. 4.5 Propagation of singularities of the wave equation along characteristic lines

$$g(x) = \begin{cases} (1 - |x|)^2, & |x| \leq 1, \\ 0, & |x| \geq 1. \end{cases}$$

Then (4.8) still gives a formula for the solution even though g is not differentiable. (This is a weak solution in the sense we will describe in Chap. 10). A set of solutions at different points in time is plotted in Fig. 4.5. Observe that the original singularity splits into two singularities, which propagate outward along the two characteristic lines emanating from $x = 0$. ◊

4.3 Boundary Problems

In the string model of Sect. 4.1 the domain of the wave equation (4.5) was restricted to $x \in [0, \ell]$, with Dirichlet boundary conditions

$$u(t, 0) = u(t, \ell) = 0, \quad \text{for all } t \geq 0. \tag{4.15}$$

Suppose the initial data are given for $x \in [0, \ell]$ by

$$u(0, x) = g(x), \qquad \frac{\partial u}{\partial t}(0, x) = h(x), \tag{4.16}$$

with $g \in C^2[0, \ell]$, $h \in C^1[0, \ell]$. Both g and h are assumed to vanish at the endpoints of $[0, \ell]$.

The solution of the wave equation on $\mathbb{R}$ provided in Theorem 4.1 can be adapted to the boundary conditions (4.15). The idea is to extend g, h to $\mathbb{R}$ in such a way that the formula (4.8) gives a solution satisfying the boundary conditions for all t.

Theorem 4.5 *The wave equation* (4.5) *on* $[0, \ell]$*, with Dirichlet boundary conditions and satisfying the initial conditions* (4.16)*, admits a solution of the form* (4.8)*, only if the initial data extensions to* $\mathbb{R}$ *as odd,* 2ℓ*-periodic functions, with* $g \in C^2(\mathbb{R})$ *and* $h \in C^1(\mathbb{R})$.

Proof By linearity we can consider the g and h terms independently. Assume that the g term,

$$\frac{1}{2}\big[g(x + ct) + g(x - ct)\big], \tag{4.17}$$

is defined for all t and x and satisfies the boundary conditions on $[0, \ell]$ for all values of t. At $x = 0$ the condition $u(t, 0) = 0$ will be satisfied if and only if

$$g(ct) + g(-ct) = 0, \quad \text{for all } t \geq 0.$$

In other words, $u(t, 0) = 0$ if and only if g is odd. At $x = \ell$ the condition is

$$g(\ell + ct) + g(\ell - ct) = 0, \quad \text{for all } t \geq 0.$$

This is equivalent to the condition that g is odd with respect to reflection at the point $x = \ell$.

The composition of the reflections about 0 and ℓ gives translation by 2ℓ. Hence the expression (4.17) satisfies the boundary conditions if and only if g is odd and 2ℓ-periodic.

A similar argument works for the h term,

$$u(t, x) = \frac{1}{2c} \int_{x-ct}^{x+ct} h(\tau)\, d\tau. \tag{4.18}$$

The requirement at $x = 0$ is

$$\int_{-ct}^{ct} h(\tau)\, d\tau = 0, \quad \text{for all } t \geq 0. \tag{4.19}$$

Differentiation with respect to t, using the fundamental theorem of calculus, shows that (4.19) is satisfied if and only if h is odd with respect to reflection at 0. Similarly, the condition

$$\int_{\ell-ct}^{\ell+ct} h(\tau)\, d\tau = 0, \quad \text{for all } t \geq 0$$

requires odd symmetry with respect to reflection at $x = \ell$. □

Example 4.6 Consider the vibrating string problem with $c = 1$ and $\ell = 1$. Suppose that the solution initially has the form (4.14) with $u_+ = 0$ and the left-propagating solution given by the function u_- shown in Fig. 4.6. For small $t > 0$ the solution is

$$u(t, x) = u_-(x + t), \tag{4.20}$$

but eventually the bump hits the boundary at $x = 0$, and we would like to understand what happens then.

To apply Theorem 4.5, we must first solve for g and h in terms of u_+. By (4.20) we set $g(x) = u_+(x)$ and

$$h(x) = \frac{\partial}{\partial t} u_-(x + t)\Big|_{t=0} = \frac{du_-}{dx}(x).$$

The resulting functions g and h, extended to odd functions on $\mathbb{R}$, are shown in Fig. 4.7.

According to Theorem 4.5 we can compute the solution from (4.8) using these odd periodic extensions of g and h. The results are shown in Fig. 4.8. The bump temporarily disappears at $t = 0.3$ and then reemerges as an inverted bump traveling in the opposite direction. ◊

4.4 Forcing Terms

The derivation of the string model in Sect. 4.1 assumed that no external forces act on the string. Additional forces could be incorporated by adding extra terms to the expression (4.1) for the force on a segment. In the continuum limit this yields a

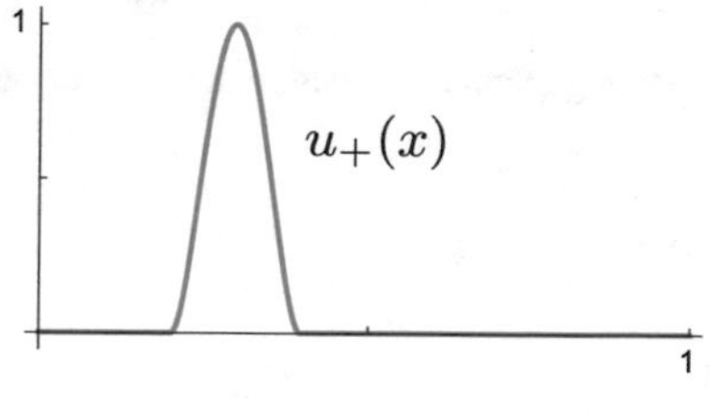

Fig. 4.6 The initial waveform u_+

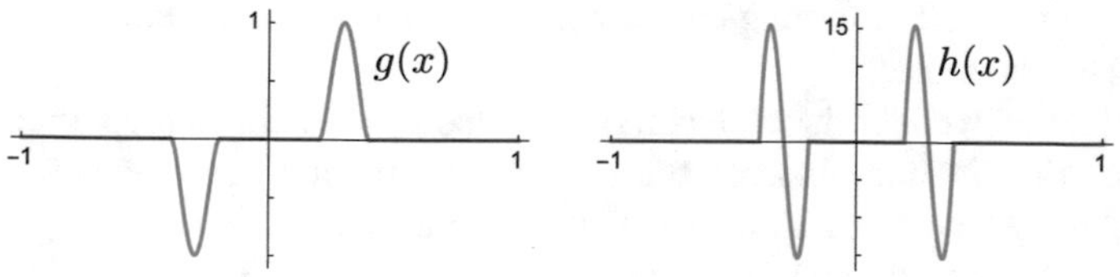

Fig. 4.7 The odd extensions of the initial conditions g and h

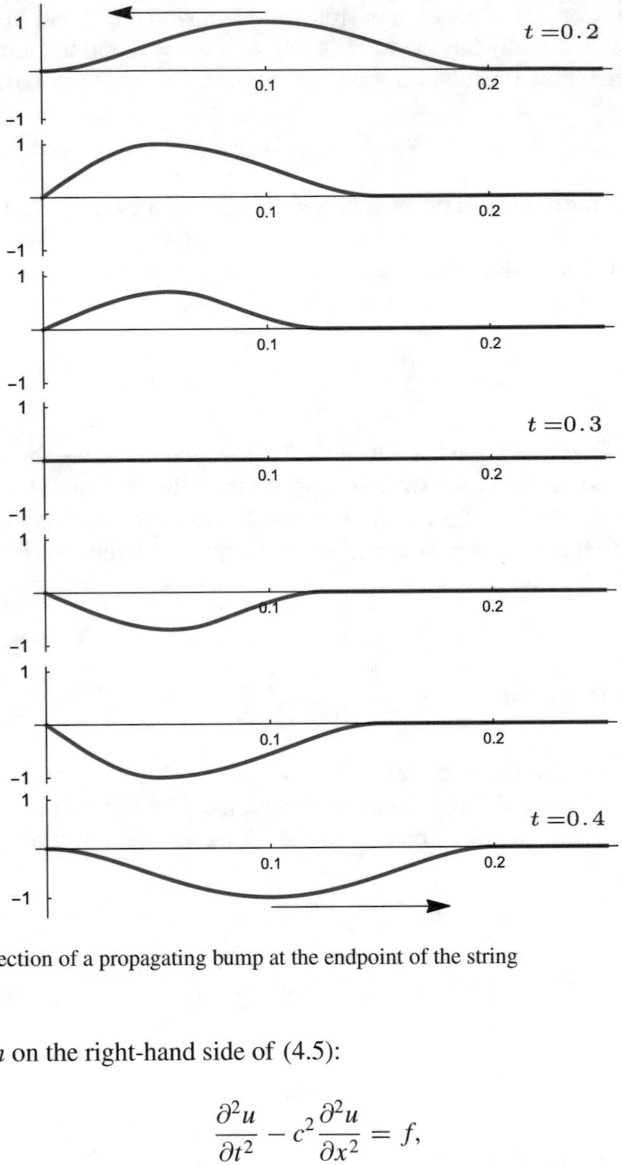

Fig. 4.8 Reflection of a propagating bump at the endpoint of the string

forcing term on the right-hand side of (4.5):

$$\frac{\partial^2 u}{\partial t^2} - c^2 \frac{\partial^2 u}{\partial x^2} = f, \tag{4.21}$$

where $f = f(t, x)$. The forcing term could be used to model plucking or bowing of the string, for example.

In this section we introduce a technique, called *Duhamel's method*, that allows us to adapt solution methods for evolution equations to include a forcing term. The idea, which is closely related to a standard ODE technique called *variation of parameters*, is to reformulate the forcing term as an initial condition. This technique is named for the 19th century French mathematician and physicist Jean-Marie Duhamel, who developed the idea in a study of the heat equation.

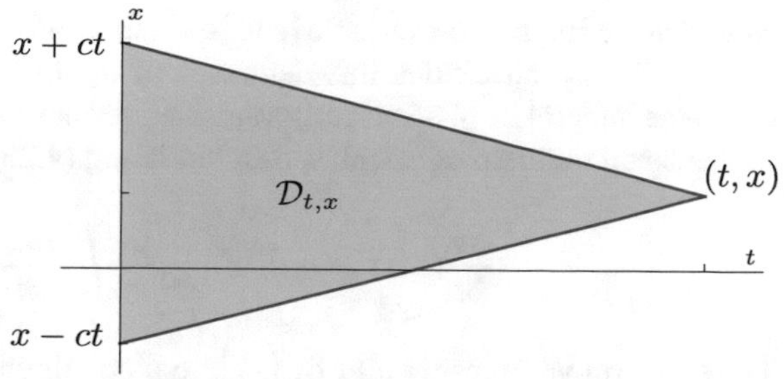

Fig. 4.9 Domain of dependence for the point (t, x)

To focus our attention on the driving term, let us consider (4.21) on the domain $x \in \mathbb{R}$ with the initial conditions set to zero. For a given c, define the *domain of dependence* of a point (t, x) with $t > 0$ and $x \in \mathbb{R}$ by

$$\mathcal{D}_{t,x} := \left\{(s, x') \in \mathbb{R}_+ \times \mathbb{R} : \ x - c(t - s) \le x' \le x + c(t - s)\right\}.$$

This is a triangular region, as pictured in Fig. 4.9. The terminology refers to the fact that the solution $u(t, x)$ is influenced only by the values of f within $\mathcal{D}_{t,x}$, as the next result shows.

Theorem 4.7 *For $f \in C^1(\mathbb{R})$, the unique solution of* (4.21) *satisfying the initial conditions*

$$u(0, x) = 0, \quad \frac{\partial u}{\partial t}(0, x) = 0,$$

is given by

$$u(t, x) = \frac{1}{2c} \int_{\mathcal{D}_{t,x}} f(s, x')\, dx'\, ds. \tag{4.22}$$

Proof For each $s \ge 0$, let $\eta_s(t, x)$ be the solution of the homogeneous wave equation (4.5) for $t \ge s$, subject to the initial conditions

$$\eta_s(t, x)\Big|_{t=s} = 0, \qquad \frac{\partial \eta_s}{\partial t}(t, x)\Big|_{t=s} = f(s, x). \tag{4.23}$$

This function can be written explicitly by shifting t to $t - s$ in (4.8),

$$\eta_s(t, x) = \frac{1}{2c} \int_{x-c(t-s)}^{x+c(t-s)} f(s, x')\, dx'. \tag{4.24}$$

We claim that the solution of (4.21) is given by the integral

$$u(t, x) := \int_0^t \eta_s(t, x)\, ds. \tag{4.25}$$

Note that the integration variable here is s rather than t.

We will first check that this definition of u satisfies the initial conditions. For $t = 0$ the integral in (4.25) clearly vanishes, so that $u(0, x) = 0$ is satisfied. By the fundamental theorem of calculus, differentiating (4.25) with respect to t gives

$$\frac{\partial u}{\partial t}(t, x) = \eta_s(t, x)\big|_{s=t} + \int_0^t \frac{\partial \eta_s}{\partial t}(t, x)\, ds.$$

The first term vanishes for all t by the initial condition (4.23), leaving

$$\frac{\partial u}{\partial t}(t, x) = \int_0^t \frac{\partial \eta_s}{\partial t}(t, x)\, ds, \tag{4.26}$$

for all $t \geq 0$. Setting $t = 0$ gives

$$\frac{\partial u}{\partial t}(0, x) = 0.$$

Now let us check that the u defined in (4.25) solves (4.21). Differentiating (4.26) once more gives

$$\frac{\partial^2 u}{\partial t^2}(t, x) = \frac{\partial \eta_s}{\partial t}(t, x)\Big|_{s=t} + \int_0^t \frac{\partial^2 \eta_s}{\partial t^2}(t, x)\, ds. \tag{4.27}$$

By (4.23) the first term on the right is equal to $f(t, x)$. To simplify the second term, we use the fact that η_s solves (4.5) and the definition of u to compute

$$\begin{aligned}\int_0^t \frac{\partial^2 \eta_s}{\partial t^2}(t, x)\, ds &= c^2 \int_0^t \frac{\partial^2 \eta_s}{\partial x^2}(t, x)\, ds \\ &= c^2 \frac{\partial^2 u}{\partial x^2}.\end{aligned}$$

Therefore, (4.27) reduces to

$$\frac{\partial^2 u}{\partial t^2} = f + c^2 \frac{\partial^2 u}{\partial x^2},$$

proving that u solves (4.21).

Combining (4.24) and (4.25), we can write the formula for u as

$$u(t, x) = \frac{1}{2c} \int_0^t \left(\int_{x-c(t-s)}^{x+c(t-s)} f(s, x')\, dx' \right) ds,$$

which is equivalent to (4.22).

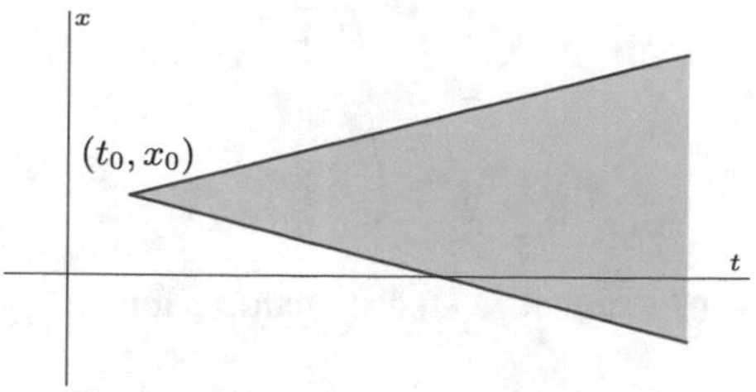

Fig. 4.10 Range of influence of the point (t_0, x_0)

To prove uniqueness, suppose u_1 and u_2 are solutions of (4.21). Then $u_2 - u_1$ is a solution of (4.5). Since $u_2 - u_1$ also has vanishing initial conditions, Theorem 4.1 implies that $u_2 - u_1 = 0$. Hence the solution is unique. □

By the superposition principle, Theorem 4.7 is easily extended to the case of nonzero initial conditions, by setting $u = v + w$ where v is a solution of the form (4.22) and w is a solution of the form (4.8).

The concept of domain of dependence still applies when the initial conditions are nonzero. In the solution formula (4.8), $u(t, x)$ depends only on the values of g and h at the base of the triangle, $\mathcal{D}_{t,x} \cap \{t = 0\}$. Thus it is still the case that the solution $u(t, x)$ depends only on the data within $\mathcal{D}_{t,x}$.

The existence of the domain of dependence is a limitation imposed by the propagation speed c. For systems governed by the wave equations (4.5) or (4.21), no information can travel at a speed faster than c.

The region of the space-time plane in which solutions can be affected by the data at a particular point (t_0, x_0) is called the *range of influence* of this point. By the definition of the domain of dependence, the range of influence consists of the points (t, x) such that $(t_0, x_0) \in \mathcal{D}_{t,x}$. This region is a triangle with vertex (t_0, x_0) and sides given by the characteristics $(t, x_0 \pm c(t - t_0))$, as shown in Fig. 4.10.

Duhamel's method applies also to the case of a vibrating string with fixed ends. Assuming that $f(t, x)$ satisfies the boundary conditions at $x = 0$ and $x = \ell$, we extend f to an odd 2ℓ-periodic function on $\mathbb{R}$, just as in Theorem 4.5. This extension guarantees that the intermediate solution η_s defined by (4.24) will satisfy the boundary conditions also. And then so will the solution $u(t, x)$ defined by (4.25).

Example 4.8 Consider a string of length ℓ with propagation speed $c = 1$. Suppose the forcing term is given by

$$f(t, x) = \cos(\omega t) \sin(\omega_0 x), \tag{4.28}$$

where $\omega_0 := \pi/\ell$ and $\omega > 0$ is the driving frequency. Since $\sin(\omega_0 x)$ is odd and 2π-periodic, the extension required by Theorem 4.5 is automatic. As in Theorem 4.7, let us set the initial conditions $g = h = 0$ to focus on the forcing term.

Substituting (4.28) into (4.22) gives

$$u(t,x) = \frac{1}{2}\int_0^t \int_{x-t+s}^{x+t-s} \cos(\omega s)\sin(\omega_0 x')\,dx'\,ds$$
$$= \frac{1}{2\omega_0}\int_0^t [\cos(\omega_0(x-t+s)) - \cos(\omega_0(x+t-s))]\cos(\omega s)\,ds$$

A trigonometric identity reduces this to

$$u(t,x) = \frac{\sin(\omega_0 x)}{\omega_0}\int_0^t \sin(\omega_0(t-s))\cos(\omega s)\,ds. \tag{4.29}$$

For $\omega \neq \omega_0$ we obtain

$$u(t,x) = \frac{\sin(\omega_0 x)}{\omega_0^2 - \omega^2}\,[\cos(\omega t) - \cos(\omega_0 t)]\,.$$

Note that the x dependence of the solution matches that of the forcing term. The interesting part of this solution is the oscillation, which includes both frequencies ω and ω_0. Figure 4.11 illustrates the behavior of the amplitude as a function of time, in a case where $\omega \ll \omega_0$. The large-scale oscillation has a period $1/\omega$, corresponding to the low driving frequency. The solution also exhibits fast oscillations at the frequency ω_0 which depends only on ℓ.

For $\omega = \omega_0$ the formula (4.29) gives the solution

$$u(t,x) = \frac{t}{2\omega_0}\sin(\omega_0 x)\sin(\omega_0 t).$$

The resulting amplitude grows linearly, as shown in Fig. 4.12. ◊

The physical phenomenon illustrated by Example 4.8 is called *resonance*. If the string is driven at its natural frequency ω_0 then it will continually absorb energy from the driving force. Of course, there is a limit to how much energy a physical string could absorb before it breaks. Once the displacement amplitude becomes sufficiently large, the linear wave equation (4.5) no longer serves as an appropriate model.

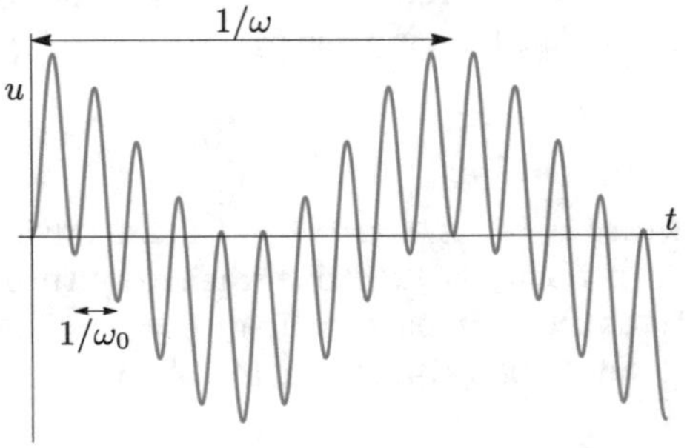

Fig. 4.11 Oscillation pattern with a driving frequency $\omega = \omega_0/10$

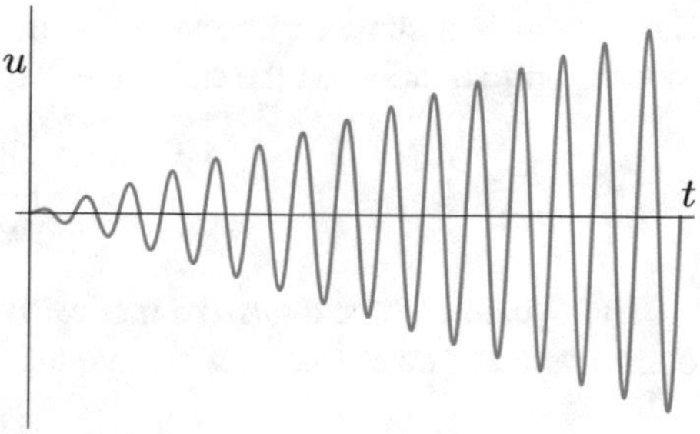

Fig. 4.12 Growth of the amplitude at the resonance frequency $\omega = \omega_0$

4.5 Model Problem: Acoustic Waves

The vibration of a drumhead can be modeled on a bounded domain $\Omega \subset \mathbb{R}^2$, with a function $u(t, \boldsymbol{x})$ representing the vertical displacement of the membrane at time t and position $\boldsymbol{x} \in \Omega$. With arguments similar to those in Sect. 4.1, one can derive the equation

$$\frac{\partial^2 u}{\partial t^2} - c^2 \Delta u = 0, \tag{4.30}$$

where Δ is the Laplacian operator (1.7). The wave equation (4.30) appears in many other contexts as well, including the propagation of light and all other forms of electromagnetic radiation. In all these cases the constant c represents the speed of propagation.

In this section we will derive the three-dimensional wave equation as a model for acoustic waves traveling through the air. Acoustic waves consist of fluctuations of pressure which propagate through a gas. To analyze them, we must consider the relationships between the pressure P, the velocity field $\boldsymbol{v}$, and the density ρ. For a gas in motion these are all functions of both time and position.

Because acoustic waves involve minute pressure fluctuations with very little heat transfer, the relationship between pressure and density is given by the *adiabatic gas law*

$$P = C\rho^\gamma, \tag{4.31}$$

where C and γ are physical constants. We will fix background atmospheric values of the pressure P_0 and density ρ_0 and focus on the deviations

$$u := P - P_0, \qquad \sigma := \rho - \rho_0.$$

Applying (4.31) to P/P_0 gives the equation

$$1 + \frac{u}{P_0} = \left(1 + \frac{\sigma}{\rho_0}\right)^\gamma.$$

Since σ/ρ_0 is assumed to be very small, we can linearize by taking a first-order Taylor approximation on the right side. This yields

$$u = \frac{\gamma P_0}{\rho_0}\sigma. \tag{4.32}$$

The dynamics of the gas are modeled with two conservation laws. The first is conservation of mass (3.20), which yields

$$\frac{\partial \rho}{\partial t} + \nabla \cdot (\rho \boldsymbol{v}) = 0.$$

Since σ and $\boldsymbol{v}$ are both assumed to be very small, for the leading approximation we can replace ρ by ρ_0 to obtain

$$\frac{\partial \sigma}{\partial t} + \rho_0 \nabla \cdot \boldsymbol{v} = 0. \tag{4.33}$$

The second dynamical law is conservation of momentum. This is encapsulated in a fluid equation derived by Euler in 1757, called *Euler's force equation*:

$$-\nabla P = \rho \left(\frac{\partial}{\partial t} + \boldsymbol{v} \cdot \nabla \right) \boldsymbol{v}.$$

Euler's equation is an aggregate form of Newton's second law (force equals mass times acceleration). Note that the "acceleration" term on the right is the Lagrangian derivative of the velocity field $\boldsymbol{v}$. As above, we substitute $P = P_0 + u$ and $\rho = \rho_0 + \sigma$ and keep only the first order terms to derive the linearization

$$-\nabla u = \rho_0 \frac{\partial \boldsymbol{v}}{\partial t}. \tag{4.34}$$

The final step is to eliminate the velocity field from the equation. Substituting (4.32) into (4.33) and differentiating with respect to time gives

$$\frac{\partial^2 u}{\partial t^2} = -\gamma P_0 \frac{\partial}{\partial t}(\nabla \cdot \boldsymbol{v}). \tag{4.35}$$

On the other hand, by (4.34),

$$\begin{aligned}
\frac{\partial}{\partial t}(\nabla \cdot \boldsymbol{v}) &= \nabla \cdot \frac{\partial \boldsymbol{v}}{\partial t} \\
&= \nabla \cdot \left(-\frac{\nabla u}{\rho_0} \right) \\
&= -\frac{1}{\rho_0} \Delta u.
\end{aligned}$$

Substituting this in (4.35) yields the *acoustic wave equation*,

$$\frac{\partial^2 u}{\partial t^2} - \frac{\gamma P_0}{\rho_0}\Delta u = 0.$$

As with our previous derivations, many approximations are required to produce a linear equation. These simplifications are well justified for sound waves at ordinary volume levels, but more dramatic pressure fluctuations would require a nonlinear equation.

4.6 Integral Solution Formulas

Let us consider the wave equation (4.30) on $\mathbb{R}^3$ with $c = 1$,

$$\frac{\partial^2 u}{\partial t^2} - \Delta u = 0.$$

This problem can be reduced to the one-dimensional case by a clever averaging trick.

For $f \in C^0(\mathbb{R}^3)$, define

$$\tilde{f}(\boldsymbol{x};\rho) := \frac{1}{4\pi\rho}\int_{\partial B(\boldsymbol{x};\rho)} f(\boldsymbol{w})\,dS(\boldsymbol{w}), \tag{4.36}$$

where $\boldsymbol{x} \in \mathbb{R}^3$ and $\rho > 0$. The surface area of $\partial B(\boldsymbol{x};\rho)$ is $4\pi\rho^2$, so $\tilde{f}$ is ρ times the spherical average of f. By continuity the spherical average approaches the value of the function at the center point as $\rho \to 0$, so that

$$\lim_{\rho\to 0}\frac{\tilde{f}(\boldsymbol{x};\rho)}{\rho} = f(\boldsymbol{x}). \tag{4.37}$$

The dimensional reduction of the wave equation is based on the following formula of Jean-Gaston Darboux.

Lemma 4.9 (Darboux's formula) *For $f \in C^2(\mathbb{R}^3)$,*

$$\frac{\partial^2}{\partial\rho^2}\tilde{f}(\boldsymbol{x};\rho) = \Delta_{\boldsymbol{x}}\tilde{f}(\boldsymbol{x};\rho).$$

Proof To compute the radial derivative of the spherical average, it is helpful to change coordinates by setting $\boldsymbol{w} = \boldsymbol{x} + \rho\boldsymbol{y}$, so that the domain of $\boldsymbol{y}$ is the unit sphere $\mathbb{S}^2 \subset \mathbb{R}^3$,

$$\frac{1}{4\pi\rho^2}\int_{\partial B(\boldsymbol{x};\rho)} f(\boldsymbol{w})\,dS(\boldsymbol{w}) = \frac{1}{4\pi}\int_{\mathbb{S}^2} f(\boldsymbol{x}+\rho\boldsymbol{y})\,dS(\boldsymbol{y}).$$

Differentiation under the integral gives,

$$\frac{\partial}{\partial \rho}\int_{\mathbb{S}^2} f(\boldsymbol{x}+\rho \boldsymbol{y})\, dS(\boldsymbol{y})\frac{1}{4\pi} = \int_{\mathbb{S}^2} \nabla f(\boldsymbol{x}+\rho \boldsymbol{y})\cdot \boldsymbol{y}\, dS(\boldsymbol{y}).$$

In the original coordinates this implies

$$\begin{aligned}\frac{\partial}{\partial \rho}&\left[\frac{1}{4\pi\rho^2}\int_{\partial B(\boldsymbol{x};\rho)} f(\boldsymbol{w})\, dS(\boldsymbol{w})\right]\\ &= \frac{1}{4\pi\rho^2}\int_{\partial B(\boldsymbol{x};\rho)} \nabla f(\boldsymbol{w})\cdot\left(\frac{\boldsymbol{w}-\boldsymbol{x}}{\rho}\right) dS(\boldsymbol{w}).\end{aligned} \tag{4.38}$$

Since $(\boldsymbol{w}-\boldsymbol{x})/\rho$ is the outward unit normal to $\partial B(\boldsymbol{x};\rho)$,

$$\nabla f(\boldsymbol{w})\cdot\left(\frac{\boldsymbol{w}-\boldsymbol{x}}{\rho}\right) = \frac{\partial f}{\partial \nu}(\boldsymbol{w}).$$

Furthermore, by Corollary 2.8,

$$\int_{\partial B(\boldsymbol{x};\rho)} \frac{\partial f}{\partial \nu}(\boldsymbol{w})\, dS(\boldsymbol{w}) = \int_{B(\boldsymbol{x};\rho)} \Delta f(\boldsymbol{w})\, d^3\boldsymbol{w}.$$

Applying this to the right-hand side of (4.38) gives

$$\frac{\partial}{\partial \rho}\left[\frac{1}{4\pi\rho^2}\int_{\partial B(\boldsymbol{x};\rho)} f(\boldsymbol{w})\, dS(\boldsymbol{w})\right] = \frac{1}{4\pi\rho^2}\int_{B(\boldsymbol{x};\rho)} \Delta f(\boldsymbol{w})\, d^3\boldsymbol{w}. \tag{4.39}$$

Substituting the definition of $\tilde{f}$ in (4.39) yields

$$\frac{\partial}{\partial \rho}\tilde{f}(\boldsymbol{x};\rho) = \frac{1}{4\pi\rho^2}\int_{\partial B(\boldsymbol{x};\rho)} f(\boldsymbol{w})\, dS(\boldsymbol{w}) + \frac{1}{4\pi\rho}\int_{B(\boldsymbol{x};\rho)} \Delta f(\boldsymbol{w})\, d^3\boldsymbol{w}$$

A further differentiation using (4.39) and the radial derivative formula from Exercise 2.4 then gives

$$\begin{aligned}\frac{\partial^2}{\partial \rho^2}\tilde{f}(\boldsymbol{x};\rho) &= \frac{1}{4\pi\rho}\frac{\partial}{\partial \rho}\int_{B(\boldsymbol{x};\rho)} \Delta f(\boldsymbol{w})\, d^3\boldsymbol{w}\\ &= \frac{1}{4\pi\rho}\int_{\partial B(\boldsymbol{x};\rho)} \Delta f(\boldsymbol{w})\, dS(\boldsymbol{w}).\end{aligned} \tag{4.40}$$

On the other hand

$$
\begin{aligned}
\Delta_x \tilde{f}(\boldsymbol{x};\rho) &= \Delta_x \left[\frac{1}{4\pi\rho}\int_{\mathbb{S}^2} f(\boldsymbol{x}+\rho\boldsymbol{y})\,dS(\boldsymbol{y})\right] \\
&= \frac{1}{4\pi\rho}\int_{\mathbb{S}^2} \Delta f(\boldsymbol{x}+\rho\boldsymbol{y})\,dS(\boldsymbol{y}) \\
&= \frac{1}{4\pi\rho}\int_{\partial B(\boldsymbol{x};\rho)} \Delta f(\boldsymbol{w})\,dS(\boldsymbol{w}).
\end{aligned}
\tag{4.41}
$$

The claim thus follows from (4.40). □

Lemma 4.9 allows us to relate the three-dimensional wave equation in variables $(t, \boldsymbol{x})$ to a one-dimensional equation in variables (t, ρ). The result is a solution formula first derived in 1883 by the physicist Gustav Kirchhoff.

Theorem 4.10 (Kirchhoff's integral formula) *For* $u \in C^2([0,\infty)\times\mathbb{R}^3)$*, suppose that*

$$\frac{\partial^2 u}{\partial t^2} - \Delta u = 0$$

under the initial conditions

$$u|_{t=0} = g, \quad \left.\frac{\partial u}{\partial t}\right|_{t=0} = h.$$

Then

$$u(t,\boldsymbol{x}) = \frac{\partial}{\partial t}\tilde{g}(\boldsymbol{x};t) + \tilde{h}(\boldsymbol{x};t),$$

with $\tilde{g}$ *and* $\tilde{h}$ *defined as in* (4.36).

Proof Define

$$\tilde{u}(t,\boldsymbol{x};\rho) := \frac{1}{4\pi\rho}\int_{\partial B(\boldsymbol{x};\rho)} u(t,\boldsymbol{w})\,dS(\boldsymbol{w}).$$

Since u satisfies the wave equation, differentiating under the integral gives

$$\frac{\partial^2}{\partial t^2}\tilde{u}(t,\boldsymbol{x};\rho) = \frac{1}{4\pi\rho}\int_{\partial B(\boldsymbol{x};\rho)} \Delta u(t,\boldsymbol{w})\,dS(\boldsymbol{w}).$$

By the calculation (4.41) this is equivalent to

$$\frac{\partial^2}{\partial t^2}\tilde{u}(t,\boldsymbol{x};\rho) = \Delta_x\tilde{u}(t,\boldsymbol{x};\rho).$$

Lemma 4.9 then shows that

$$\left(\frac{\partial^2}{\partial t^2} - \frac{\partial^2}{\partial\rho^2}\right)\tilde{u}(t,\boldsymbol{x};\rho) = 0. \tag{4.42}$$

The initial conditions for $\tilde{u}$ follow from the initial conditions for u,

$$\tilde{u}(0, \boldsymbol{x}; \rho) = \tilde{g}(\boldsymbol{x}; \rho), \quad \frac{\partial}{\partial t}\tilde{u}(0, \boldsymbol{x}; \rho) = \tilde{h}(\boldsymbol{x}; \rho).$$

By (4.37) we also have a boundary condition at $\rho = 0$,

$$\tilde{u}(t, \boldsymbol{x}; 0) = 0.$$

Using Theorem 4.1 and the reflection argument from Theorem 4.5, we conclude that the unique solution of (4.42) under these conditions is given by extending $\tilde{g}(\boldsymbol{x}; \rho)$ and $\tilde{h}(\boldsymbol{x}; \rho)$ to $\rho \in \mathbb{R}$ with odd symmetry and then using the d'Alembert formula,

$$\tilde{u}(t, \boldsymbol{x}; \rho) = \frac{1}{2}\left[\tilde{g}(\boldsymbol{x}; \rho + t) + \tilde{g}(\boldsymbol{x}; \rho - t)\right] + \frac{1}{2}\int_{\rho - t}^{\rho + t} \tilde{h}(\boldsymbol{x}; \tau)\, d\tau. \tag{4.43}$$

By (4.37), we can recover u from this formula by setting

$$u(t, \boldsymbol{x}) = \lim_{\rho \to 0} \frac{\tilde{u}(t, \boldsymbol{x}; \rho)}{\rho}. \tag{4.44}$$

To evaluate this limit, first note that for $0 \le \rho \le t$ the odd symmetry of $\tilde{g}$ and $\tilde{h}$ with respect to ρ can be used to rewrite (4.43) as

$$\tilde{u}(t, \boldsymbol{x}; \rho) = \frac{1}{2}\left[\tilde{g}(\boldsymbol{x}; t + \rho) - \tilde{g}(\boldsymbol{x}; t - \rho)\right] + \frac{1}{2}\int_{t - \rho}^{t + \rho} \tilde{h}(\boldsymbol{x}; \tau)\, d\tau.$$

The computations are now straightforward:

$$\lim_{\rho \to 0} \frac{1}{2\rho}\left[\tilde{g}(\boldsymbol{x}; t + \rho) - \tilde{g}(\boldsymbol{x}; t - \rho)\right] = \frac{\partial}{\partial t}\tilde{g}(\boldsymbol{x}; t),$$

and

$$\lim_{\rho \to 0} \frac{1}{2\rho}\int_{t - \rho}^{t + \rho} \tilde{h}(\boldsymbol{x}; \tau)\, d\tau = \tilde{h}(\boldsymbol{x}; t).$$

The claimed solution formula thus follows from (4.44). □

One interesting consequence of the Kirchhoff formula is the fact that three-dimensional wave propagation exhibits a strict form of the Huygens' principle. Theorem 4.10 shows that the range of influence of the point $(t_0, \boldsymbol{x}_0)$ is the *forward light cone*,

$$\Gamma_+(t_0, \boldsymbol{x}_0) := \{(t, \boldsymbol{x});\ t > t_0,\ |\boldsymbol{x} - \boldsymbol{x}_0| = t - t_0\}.$$

This matches the result of Theorem 4.3 for the one-dimensional wave equation. The strict Huygens phenomenon is readily observable for acoustic waves, in the fact that

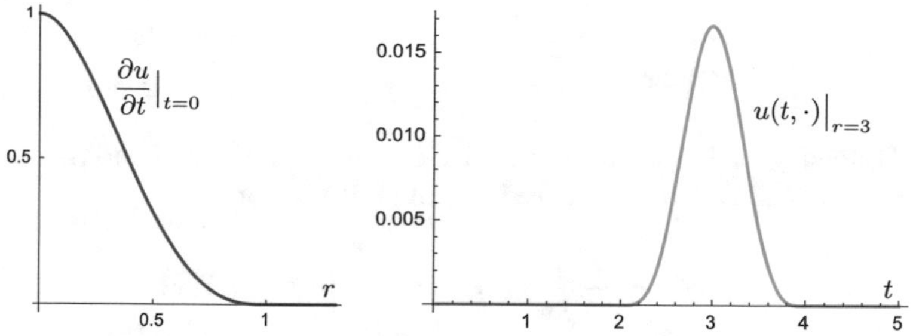

Fig. 4.13 The plot on the right shows the observed waveform at a distance 3 from the origin, caused by the initial radial impulse shown on the left

a sudden sound like a clap propagates as a sharp wavefront that is heard as a single discrete event, without aftereffects unless there are reflective surfaces to cause an echo. Figure 4.13 illustrates this effect; an observer located away from the origin experiences a waveform of duration equal to the diameter of the initial impulse. The strict Huygens' principle holds in every odd dimension greater than 1, but fails in even dimensions, as we will illustrate below.

The spherical averaging trick used for Theorem 4.10 also works in higher odd dimensions, although the solution formulas become more complicated. For even dimensions, solution formulas can be derived from the odd-dimensional case by a technique called the *method of descent*.

We will work this out for the two-dimensional case. Suppose $u \in C^2([0,\infty)\times\mathbb{R}^2)$ solves the wave equation with initial conditions

$$u|_{t=0} = g, \quad \left.\frac{\partial u}{\partial t}\right|_{t=0} = h,$$

with g, h functions on $\mathbb{R}^2$. If we extend g and h to $\mathbb{R}^3$ as functions that are independent of x_3, then Kirchhoff's formula gives a solution to the three dimensional problem. Since this solution is also independent of x_3, it "descends" to a solution in $\mathbb{R}^2$. The resulting formula was first worked out by Siméon Poisson in the early 19th century (well before Kirchhoff's three-dimensional formula).

Corollary 4.11 (Poisson's integral formula) *For* $u \in C^2([0,\infty)\times\mathbb{R}^2)$*, suppose that*

$$\frac{\partial^2 u}{\partial t^2} - \Delta u = 0$$

under the initial conditions

$$u|_{t=0} = g, \quad \left.\frac{\partial u}{\partial t}\right|_{t=0} = h.$$

Then

$$u(t,\boldsymbol{x}) = \frac{\partial}{\partial t}\left(\frac{t}{2\pi}\int_{\mathbb{D}} \frac{g(\boldsymbol{x}-t\boldsymbol{y})}{\sqrt{1-|\boldsymbol{y}|^2}}\,d^2\boldsymbol{y}\right) + \frac{t}{2\pi}\int_{\mathbb{D}} \frac{h(\boldsymbol{x}-t\boldsymbol{y})}{\sqrt{1-|\boldsymbol{y}|^2}}\,d^2\boldsymbol{y}.$$

Proof Following the procedure described above, we extend g and h to functions on $\mathbb{R}^3$ independent of x_3. In this case the integral (4.36) becomes

$$\tilde{g}(\boldsymbol{x};\rho) = \frac{\rho}{4\pi}\int_{\mathbb{S}^2} g(x_1+\rho y_1, x_2+\rho y_2)\,dS(\boldsymbol{y}) \tag{4.45}$$

for $\boldsymbol{x} \in \mathbb{R}^2$. By symmetry we can restrict our attention to the upper hemisphere, parametrized in polar coordinates by

$$\boldsymbol{y} = \left(r\cos\theta, r\sin\theta, \sqrt{1-r^2}\right).$$

The surface area element is

$$dS = \frac{r}{\sqrt{1-r^2}}\,dr\,d\theta,$$

so that (4.45) becomes

$$\begin{aligned}\tilde{g}(\boldsymbol{x};\rho) &= \frac{\rho}{2\pi}\int_0^{2\pi}\int_0^r \frac{g(x_1+\rho r\cos\theta, x_2+\rho r\sin\theta)}{\sqrt{1-r^2}}\,r\,dr\,d\theta \\ &= \frac{\rho}{2\pi}\int_{\mathbb{D}} \frac{g(\boldsymbol{x}+\rho\boldsymbol{y})}{\sqrt{1-|\boldsymbol{y}|^2}}\,d^2\boldsymbol{y}.\end{aligned}$$

The claimed two-dimensional solution follows by substituting this formula for $\tilde{g}$ and the corresponding result for $\tilde{h}$ in the Kirchhoff formula from Theorem 4.10. □

Corollary 4.11 shows that the range of influence of $(t_0, \boldsymbol{x}_0)$ for the two-dimensional wave equation is the solid region bounded by the forward light cone $\Gamma_+(t_0, \boldsymbol{x}_0)$, not just the surface. Thus, in $\mathbb{R}^2$ the wave caused by a sudden disturbance has a lingering "tail" after the initial wavefront has passed, as illustrated in Fig. 4.14.

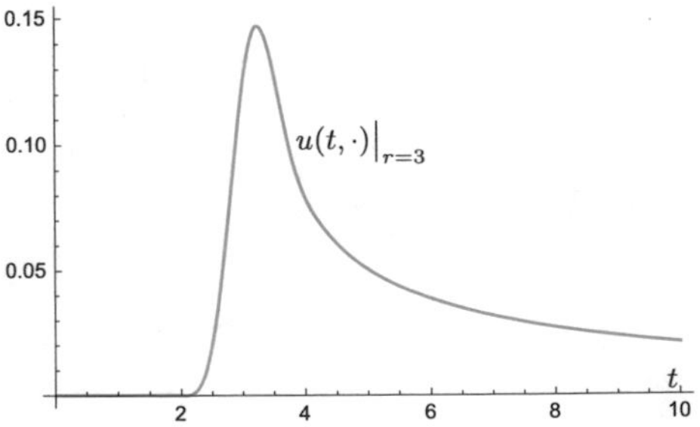

Fig. 4.14 Two-dimensional waveform observed at a distance 3 from the origin, corresponding to the radial impulse shown on the left in Fig. 4.13

In all dimensions, solutions of the wave equation exhibit a phenomenon known as *finite propagation speed*. If the constant c is reinstated as in (4.30), then the range of influence is restricted to spacetime points that are reachable at a speed less than or equal to c.

4.7 Energy and Uniqueness

In Theorem 4.1, uniqueness of solutions was a consequence of the method of characteristics. In this section we will present an alternative approach, which allows us to deduce uniqueness directly from the equation without requiring any knowledge of the solution other than differentiability. This argument is based on the concept of *energy* of a solution, which proves to be a powerful tool for analyzing many different types of PDE.

To motivate the definition, let us specialize again to the case of a string of length ℓ with fixed ends. Assume that $u \in C^2([0,\infty] \times [0,\ell])$ satisfies the string wave equation (4.4) with Dirichlet boundary conditions. In the discrete model of the string used for the derivation of the equation, the segment of length Δx located at x_j had mass $\rho\Delta x$ and velocity $\frac{\partial u}{\partial t}(x_j)$. By the standard expression for the kinetic energy of a moving particle, $\frac{1}{2}(\text{mass})\times(\text{velocity})^2$, the kinetic energy of this segment is therefore

$$\frac{1}{2}\rho\Delta x\left[\frac{\partial u}{\partial t}(x_j)\right]^2.$$

Summing over the segments and passing the continuum limit gives a formula for the total kinetic energy of the string:

$$\mathcal{E}_K := \frac{\rho}{2}\int_0^\ell \left(\frac{\partial u}{\partial t}\right)^2 dx.$$

The *potential energy* of the solution can be calculated as the energy required to move the string from zero displacement into the configuration described by $u(t,\cdot)$. Let us represent this process by scaling the displacement to $su(t,\cdot)$ for $s \in [0,1]$. By (4.1) the opposing force generated by the tension also scales proportionally to s. The work required to shift the segment at x_j from s to $s+\Delta s$ is therefore $s\Delta F(t,x_j)u(t,x_j)\Delta s$. The potential energy associated with this segment is

$$\begin{aligned}\Delta\mathcal{E}_P(t,x_j) &:= -\int_0^1 s\Delta F(t,x_j)u(t,x_j)\,ds \\ &= -\frac{1}{2}u(t,x_j)\Delta F(t,x_j) \\ &\approx -\frac{T}{2}u(t,x_j)\frac{\partial^2 u}{\partial x^2}(t,x_j)\Delta x,\end{aligned}$$

with a minus sign because the displacement and force are in opposing directions.

Summing over the segments and taking the continuum limit gives the total potential energy,

$$\mathcal{E}_P(t) := -\frac{T}{2}\int_0^\ell u\,\frac{\partial^2 u}{\partial x^2}\,dx.$$

For comparison to the kinetic term, it is convenient to integrate by parts and rewrite this in the form

$$\mathcal{E}_P(t) = \frac{T}{2}\int_0^\ell \left(\frac{\partial u}{\partial x}\right)^2 dx.$$

The total energy of the one-dimensional string at time t is given by

$$\mathcal{E} = \mathcal{E}_K + \mathcal{E}_P.$$

For the higher-dimensional wave equation (4.30) on a domain $\Omega \subset \mathbb{R}^n$ the corresponding definition is

$$\mathcal{E}[u](t) := \frac{1}{2}\int_\Omega \left[\left(\frac{\partial u}{\partial t}\right)^2 + c^2|\nabla u|^2\right] d^n\boldsymbol{x} \tag{4.46}$$

This is well-defined for $u \in C^2([0,\infty)\times\overline{\Omega})$, provided Ω is bounded.

Theorem 4.12 *Suppose $\Omega \subset \mathbb{R}^n$ is a bounded domain with piecewise C^1 boundary. If $u \in C^2([0,\infty)\times\overline{\Omega})$ is a solution of* (4.30) *with $u|_{\partial\Omega} = 0$, then the energy $\mathcal{E}[u]$ defined by* (4.46) *is independent of t.*

Proof The assumptions on u justify differentiating under the integral, so that

$$\frac{d}{dt}\mathcal{E}[u] = \int_\Omega \left[\frac{\partial u}{\partial t}\frac{\partial^2 u}{\partial t^2} + c^2\nabla\left(\frac{\partial u}{\partial t}\right)\cdot\nabla u\right] d^n\boldsymbol{x}.$$

Under the condition $u|_{\partial\Omega} = 0$, Green's first identity (Theorem 2.10) applies to the second term to give

$$\int_\Omega \nabla\left(\frac{\partial u}{\partial t}\right)\cdot\nabla u\, d^n\boldsymbol{x} = -\int_\Omega \frac{\partial u}{\partial t}\Delta u\, d^n\boldsymbol{x}. \tag{4.47}$$

Thus

$$\frac{d}{dt}\mathcal{E}[u] = \int_\Omega \left[\frac{\partial u}{\partial t}\left(\frac{\partial^2 u}{\partial t^2} - c^2\Delta u\right)\right] d^n\boldsymbol{x},$$

and (4.30) implies that $\mathcal{E}$ is constant. □

Corollary 4.13 *Suppose $\Omega \subset \mathbb{R}^n$ is a bounded domain with piecewise C^1 boundary. A solution $u \in C^2(\mathbb{R}_+\times\overline{\Omega})$ of the equation*

$$\frac{\partial^2 u}{\partial t^2} - c^2 \Delta u = f, \quad u|_{\partial \Omega} = 0,$$

$$u|_{t=0} = g, \quad \left.\frac{\partial u}{\partial t}\right|_{t=0} = h,$$

is uniquely determined by the functions f, g, h.

Proof If u_1 and u_2 are solutions of the equation with the same initial conditions, then $w := u_1 - u_2$ satisfies (4.30) with the initial conditions

$$w(0, x) = 0, \quad \frac{\partial w}{\partial t}(0, x) = 0.$$

At time $t = 0$ this gives $\mathcal{E}[w] = 0$, and Theorem 4.12 then implies that $\mathcal{E}[w] = 0$ for all t. Since the terms in the integrand of $\mathcal{E}[w]$ are non-negative, they must each vanish. This shows that w is constant, and hence $w = 0$ by the initial conditions. Therefore $u_1 = u_2$. □

4.8 Exercises

4.1 Suppose $u(t, x)$ is a solution of the wave equation (4.5) for $x \in \mathbb{R}$. Let $\mathcal{P}$ be a parallelogram in the (t, x) plane whose sides are characteristic lines. Show that the value of u at each vertex of $\mathcal{P}$ is determined by the values at the other three vertices.

4.2 The wave equation (4.5) is an appropriate model for the longitudinal vibrations of a spring. In this application $u(t, x)$ represents displacement parallel to the spring. Suppose that spring has length ℓ and is free at the ends. This corresponds to the Neumann boundary conditions

$$\frac{\partial u}{\partial x}(t, 0) = \frac{\partial u}{\partial x}(t, \ell) = 0, \quad \text{for all } t \geq 0.$$

Assume the initial conditions are g and h as in (4.16), which also satisfy Neumann boundary conditions on $[0, \ell]$. Determine the appropriate extensions of g and h from $[0, \ell]$ to $\mathbb{R}$ so that the solution $u(t, x)$ given by (4.8) will satisfy the Neumann boundary problem for all t.

4.3 In the derivation in Sect. 4.1, suppose we include the effect of gravity by adding a term $-\rho g \Delta x$ to the discrete equation of motion (4.2), where $g > 0$ is the constant of gravitational acceleration. The wave equation is then modified to

$$\frac{\partial^2 u}{\partial t^2} - c^2 \frac{\partial^2 u}{\partial x^2} = -g. \tag{4.48}$$

Assume that $x \in [0, \ell]$, with u satisfying Dirichlet boundary conditions at the endpoints.

(a) Find an *equilibrium* solution $u_0(x)$ for (4.48), that satisfies the boundary conditions but does not depend on time.

(b) Show that if u_1 is a solution of the original wave equation (4.5), also with Dirichlet boundary conditions, then $u = u_0 + u_1$ solves (4.48).

(c) Given the initial conditions $u(x, 0) = 0$, $\frac{\partial u}{\partial t}(x, 0) = 0$, find the corresponding initial conditions for u_1. Then apply Theorem 4.5 to find u_1 and hence solve for u.

4.4 In Example 4.8, let the forcing term be

$$f(t, x) = \cos(\omega t) \sin(\omega_k x),$$

with $\omega > 0$ and

$$\omega_k := \frac{k\pi}{\ell}.$$

Find the solution $u(t, x)$ given initial conditions $g = h = 0$. Include both cases $\omega \neq \omega_k$ and $\omega = \omega_k$.

4.5 The *telegraph equation* is a variant of the wave equation that describes the propagation of electrical signals in a one-dimensional cable:

$$\frac{\partial^2 u}{\partial t^2} + a\frac{\partial u}{\partial t} + bu - c^2\frac{\partial^2 u}{\partial x^2} = 0,$$

where $u(t, x)$ is the line voltage, c is the propagation speed, and $a, b > 0$ are determined by electrical properties of the cable (resistance, inductance, etc.). Show that the substitution

$$u(t, x) = e^{-at/2} w(t, x)$$

reduces the telegraph equation to an ordinary wave equation for w, provided a and b satisfy a certain condition. Find the general solution in this case. (This result has important practical applications, in that the electrical properties of long cables can be "tuned" to eliminate distortion.)

4.6 An alternative approach to the one-dimensional wave equation is to recast the PDE as a pair of ODE. Consider the wave equation with forcing term,

$$\frac{\partial^2 u}{\partial t^2} - c^2\frac{\partial^2 u}{\partial x^2} = f.$$

(a) Define a vector-valued function $\boldsymbol{v} = (v_1, v_2)$ with components

$$v_1 := \frac{\partial u}{\partial t}, \qquad v_2 := \frac{\partial u}{\partial x}.$$

Show that $\boldsymbol{v}$ satisfies a vector equation

$$\frac{\partial \boldsymbol{v}}{\partial t} - A \cdot \frac{\partial \boldsymbol{v}}{\partial x} = \boldsymbol{b}. \tag{4.49}$$

where $\boldsymbol{b} := (f, 0)$ and A is the matrix

$$A := \begin{pmatrix} 0 & c^2 \\ 1 & 0 \end{pmatrix}.$$

(b) The vector equation (4.49) can be solved by diagonalizing A. Check that if we set

$$T := \begin{pmatrix} 1 & c \\ 1 & -c \end{pmatrix},$$

then

$$TAT^{-1} = \begin{pmatrix} c & 0 \\ 0 & -c \end{pmatrix}.$$

Then show under that the substitution

$$\boldsymbol{w} := T\boldsymbol{v},$$

(4.49) reduces to a pair of linear conservation equations for the components of $\boldsymbol{w}$:

$$\begin{cases} \frac{\partial w_1}{\partial t} - c\frac{\partial w_1}{\partial x} = f, \\ \frac{\partial w_2}{\partial t} + c\frac{\partial w_2}{\partial x} = f. \end{cases} \tag{4.50}$$

(c) Translate the initial conditions

$$u(0, x) = g(x), \qquad \frac{\partial u}{\partial t}(0, x) = h(x),$$

into initial conditions for w_1 and w_2, and then solve (4.50) using the method of characteristics.

(d) Combine the solutions for w_1 and w_2 to compute $v_1 = \partial u/\partial t$, and then integrate to solve for u. Your answer should be a combination of of the d'Alembert formula (4.8) and the Duhamel formula (4.22).

4.7 The evolution of a quantum-mechanical wave function $u(t, \boldsymbol{x})$ is governed by the *Schrödinger equation*:

$$\frac{\partial u}{\partial t} - i\Delta u = 0 \tag{4.51}$$

(ignoring the physical constants). Suppose that $u(t, \boldsymbol{x})$ is a solution of (4.51) for $t \in [0, \infty)$ and $\boldsymbol{x} \in \mathbb{R}^n$, with initial condition

$$u(0, x) = g(x).$$

Assume that

$$\int_{\mathbb{R}^n} |g|^2\, d^n\boldsymbol{x} < \infty.$$

(a) Show that for all $t \geq 0$,

$$\int_{\mathbb{R}^n} |u(t, \boldsymbol{x})|^2\, d^n\boldsymbol{x} = \int_{\mathbb{R}^n} |g|^2\, d^n\boldsymbol{x}.$$

(In quantum mechanics $|u|^2$ is interpreted as a probability density, so this identity is conservation of total probability.)

(b) Show that a solution of Schrödinger's equation is uniquely determined by the initial condition g.

4.8 In $\mathbb{R}^n$ consider the wave equation

$$\frac{\partial^2 u}{\partial t^2} - c^2\Delta u = 0. \tag{4.52}$$

The *plane wave* solutions have the form

$$u(t, x) = e^{i(\boldsymbol{k}\cdot\boldsymbol{x} - \omega t)}, \tag{4.53}$$

where $\omega \in \mathbb{R}$ and $\boldsymbol{k} \in \mathbb{R}^n$ are constants.

(a) Find the condition on $\omega = \omega(\boldsymbol{k})$ for which u solves (4.52).

(b) For fixed $t, \theta \in \mathbb{R}$, show that $\left\{\boldsymbol{x} \in \mathbb{R}^n;\ u(t, \boldsymbol{x}) = e^{i\theta}\right\}$ is a set of planes perpendicular to $\boldsymbol{k}$. Show that these planes propagate, as t increases, in a direction parallel to $\boldsymbol{k}$ with speed given by c. (Hence the term "plane" wave.)

4.9 The *Klein-Gordon equation* in $\mathbb{R}^n$ is a variant of the wave equation that appears in relativistic quantum mechanics,

$$\frac{\partial^2 u}{\partial t^2} - \Delta u + m^2 u = 0, \tag{4.54}$$

where m is the mass of a particle.

(a) Find a formula for $\omega = \omega(\boldsymbol{k}, m)$ under which this equation admits plane wave solutions of the form (4.53).

(b) Show that we can define a conserved energy $\mathcal{E}$ for this equation by adding a term proportional to u^2 to the integrand in (4.46).

Chapter 5
Separation of Variables

Some PDE can be split into pieces that involve distinct variables. For example, the equation

$$\frac{\partial u}{\partial t} - a(t)b(\boldsymbol{x})\Delta u = 0$$

could be written as

$$\frac{1}{a(t)}\frac{\partial u}{\partial t} = b(\boldsymbol{x})\Delta u,$$

provided $a(t) \neq 0$. This puts all of the t derivatives and t-dependent coefficients on the left and all of the terms involving $\boldsymbol{x}$ on the right.

Splitting an equation this way is called *separation of variables*. For PDE that admit separation, it is natural to look for product solutions whose factors depend on the separate variables, e.g., $u(t, \boldsymbol{x}) = v(t)\phi(\boldsymbol{x})$. The full PDE then reduces to a pair of equations for the factors. In some cases, one or both of the reduced equations is an ODE that can be solved explicitly.

This idea is most commonly applied to evolution equations such as the heat or wave equations. The classical versions of these PDE have constant coefficients, and separation of variables can thus be used to split the time variable from the spatial variables. This reduces the evolution equation to a simple temporal ODE and a spatial PDE problem.

Separation among the spatial variables is sometimes possible as well, but this requires symmetry in the equation that is also shared by the domain. For example, we can separate variables for the Laplacian on rectangular or circular domains in $\mathbb{R}^2$. But if the domain is irregular or the differential operator has variable coefficients, then separation is generally not possible.

Despite these limitations, separation of variables plays a significant role the development of PDE theory. Explicit solutions can still yield valuable information even if they are very special cases.

D. Borthwick, *Introduction to Partial Differential Equations*,
Universitext, DOI 10.1007/978-3-319-48936-0_5

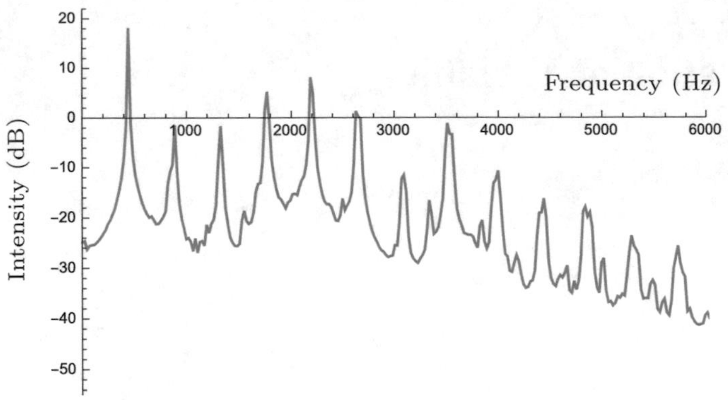

Fig. 5.1 Frequency decomposition for the sound of a violin string

5.1 Model Problem: Overtones

In 1636 the mathematician Marin Mersenne published his observation that a vibrating string produces multiple pitches simultaneously. The most audible pitch corresponds to the lowest frequency of vibration, called the fundamental tone of the string. Mersenne also detected higher pitches, at integer multiples of the fundamental frequency. (The relationship between frequency and pitch is logarithmic; doubling the frequency raises the pitch by one octave.)

The higher multiples of the fundamental frequency are called *overtones* of the string. Figure 5.1 shows the frequency decomposition for a sound sample of a bowed violin string, with a fundamental frequency of 440 Hz. The overtones appear as peaks in the intensity plot at multiples of 440.

At the time of Mersenne's observations, there was no theoretical model for string vibration that would explain the overtones. The wave equation that d'Alembert subsequently developed (a century later) gave the first theoretical justification. However, this connection is not apparent in the explicit solution formula developed in Sect. 4.3. To understand how the overtones are predicted by the wave equation, we need to organize the solutions in terms of frequency.

5.2 Helmholtz Equation

The classical evolution equations on $\mathbb{R}^n$ have the form

$$P_t u - \Delta u = 0, \tag{5.1}$$

where P_t is a first- or second-order differential operator involving only the time variable. Examples include the wave equation ($P_t = \partial^2/\partial t^2$), heat equation ($P_t = \partial/\partial t$), and Schrödinger equation ($P_t = -i\partial/\partial t$).

Lemma 5.1 *If u is a classical solution of* (5.1) *of the form*

$$u(t, \boldsymbol{x}) = v(t)\phi(\boldsymbol{x}),$$

for $t \in \mathbb{R}$ and $\boldsymbol{x} \in \Omega \subset \mathbb{R}^n$, then in any region where u is nonzero there is a constant κ such that the components solve the equations

$$P_t v = \kappa v, \qquad \Delta\phi = \kappa\phi. \tag{5.2}$$

Proof Substituting $u = v\phi$ into (5.1) gives

$$\phi P_t v - v\Delta\phi = 0.$$

Assuming that u is nonzero, we can divide by u to obtain

$$\frac{1}{v}P_t v = \frac{1}{\phi}\Delta\phi.$$

The left hand-side is independent of $\boldsymbol{x}$ and the right is independent of t. We conclude that both sides must be equal to some constant κ. □

The two differential equations in (5.2) are analogous to eigenvalue equations from linear algebra, with the role of the linear operator or matrix taken by the differential operators P_t or Δ.

Let us first focus on the spatial problem, which is usually written in the form

$$-\Delta\phi = \lambda\phi. \tag{5.3}$$

This is called the *Helmholtz equation*, after the 19th century physicist Hermann von Helmholtz. The minus sign is included so that $\lambda \geq 0$ for the most common types of boundary conditions. Adapting the linear algebra terminology, we refer to the number λ in (5.3) as an *eigenvalue* and the corresponding solution ϕ as an *eigenfunction*. The Helmholtz equation is sometimes called the Laplacian eigenvalue equation.

We will present a general analysis of the Helmholtz problem on any bounded domain in $\mathbb{R}^n$ in Chap. 11, and later in this chapter we will consider some two- or three-dimensional cases for which further spatial separation is possible. For the remainder of this section we restrict our attention to problems in one spatial dimensional, for which (5.3) is an ODE.

Theorem 5.2 *For $\phi \in C^2[0, \ell]$ the equation*

$$-\frac{d^2\phi}{dx^2} = \lambda\phi, \quad \phi(0) = \phi(\ell) = 0, \tag{5.4}$$

has nonzero solutions only if

$$\lambda_n := \frac{\pi^2 n^2}{\ell^2}$$

for n $\in \mathbb{N}$*. Up to a constant multiple, the corresponding solutions are*

$$\phi_n(x) := \sin(\sqrt{\lambda_n} x). \tag{5.5}$$

Proof Note that (5.4) implies

$$\lambda \int_0^\ell |\phi|^2 \, dx = -\int_0^\ell \frac{d^2\phi}{dx^2} \bar{\phi} \, dx.$$

Using the Dirichlet boundary conditions we can integrate by parts on the right without any boundary term, yielding

$$\lambda \int_0^\ell |\phi|^2 \, dx = \int_0^\ell \left| \frac{d\phi}{dx} \right|^2 dx. \tag{5.6}$$

Assuming that ϕ is not identically zero, this shows that $\lambda \geq 0$. Furthermore, $\lambda = 0$ implies $d\phi/dx = 0$, which gives a constant solution. The only constant solution is the trivial case $\phi = 0$, because of the boundary conditions.

It therefore suffices to consider the case $\lambda > 0$, for which the ODE in (5.4) reduces to the *harmonic oscillator* equation, with independent solutions given by $\sin(\sqrt{\lambda} x)$ and $\cos(\sqrt{\lambda} x)$. Only sine satisfies the condition $\phi(0) = 0$, so the possible solutions have the form

$$\phi(x) = \sin(\sqrt{\lambda} x).$$

To satisfy the condition $\phi(\ell) = 0$ we must have

$$\sin(\sqrt{\lambda} \ell) = 0.$$

For a nonzero solution this imposes the restriction that $\sqrt{\lambda} \ell \in \pi\mathbb{N}$, which gives the claimed set of solutions. □

Some of the eigenfunctions obtained in Theorem 5.2 are illustrated in Fig. 5.2. For the sake of application to our original string model, let us reinstate the propagation speed $c := \sqrt{T/\rho}$ and write the string equation as

$$\frac{\partial^2 u}{\partial t^2} - c^2 \Delta u = 0, \quad u(t, 0) = u(0, \ell) = 0. \tag{5.7}$$

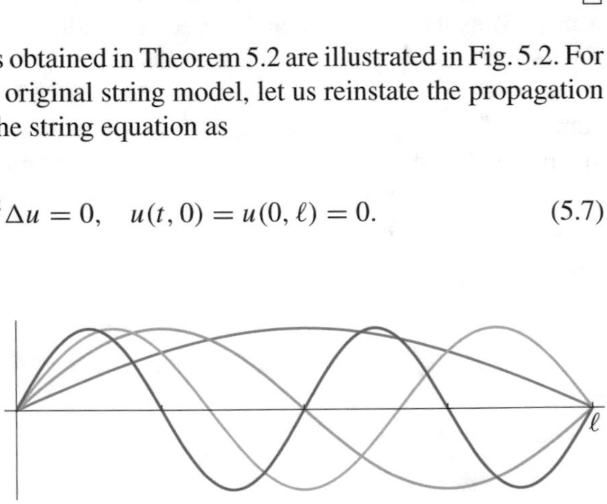

Fig. 5.2 The first four eigenfunctions for a vibrating string with fixed ends

With the spatial solution given by the eigenfunction associated to λ_n, the corresponding temporal eigenvalue equation is also a harmonic oscillator ODE,

$$-\frac{d^2v}{dt^2} = c^2\lambda_n v.$$

The solutions could be written in terms of sines and cosines, but for the temporal component it is usually more convenient to use the complex exponential form. The general complex-valued solution is

$$v_n(t) = a_n e^{i\omega_n t} + b_n e^{-i\omega_n t},$$

with $a_n, b_n \in \mathbb{C}$ and

$$\omega_n := c\sqrt{\lambda_n} = \frac{c\pi n}{\ell}. \tag{5.8}$$

For real-valued solutions, the coefficients are restricted by $b_n = \overline{a_n}$.

Combining the temporal and spatial components gives a set of solutions for the vibrating string problem:

$$u_n(t,x) = \left[a_n e^{i\omega_n t} + b_n e^{-i\omega_n t}\right]\sin(\sqrt{\lambda_n}x), \tag{5.9}$$

for $n \in \mathbb{N}$.

The functions (5.9) are referred to as "pure-tone" solutions, because they model oscillation at a single frequency ω_n. In the case of visible light waves, the frequency corresponds directly to color. For this reason the set of frequencies $\{\omega_n\}$ is called the *spectrum*. By association, the term spectrum is also used for sets of eigenvalues appearing in more general problems. For example, the set $\{\lambda_n\}$ of eigenvalues for which the Helmholtz problem has a nontrivial solution is called the spectrum of the Laplacian, even though λ_n is proportional to the square of the frequency ω_n.

From (5.8) we can deduce the fundamental tone of the string, as predicted by d'Alembert's wave equation model. To convert frequency to the standard unit of Hz (cycles per second), we divide ω_1 by 2π to obtain the formula

$$\frac{\omega_1}{2\pi} = \frac{1}{2\ell}\sqrt{\frac{T}{\rho}}. \tag{5.10}$$

This is known as *Mersenne's law*, published in 1637.

The wave equation model also predicts the higher frequencies $\omega_n = n\omega_1$, corresponding to the sequence of overtones noted illustrated in Fig. 5.1. The fact that each overtone is associated with a particular spatial eigenfunction is significant. The waveforms for higher overtones have *nodes*, meaning points where the string is stationary. As we can see in Fig. 5.2, the nodes associated to the frequency ω_n subdivide the string into n equal segments. Touching the string lightly at one of these nodes will knock out the lower frequencies, a practice string players refer to as playing a "harmonic".

As this discussion illustrates, the "spectral analysis" of the wave equation is more directly connected to experimental observation than the explicit solution formula (4.8). The displacement of a vibrating string is technically difficult to observe directly because the motion is both rapid and of small amplitude. Such observations were first achieved by Hermann von Helmholtz in the mid-19th century.

Example 5.3 The one-dimensional wave equation can be used to model for the fluctuations of air pressure inside a clarinet. The interior of a clarinet is essentially a cylindrical column, and for simplicity we can assume that the pressure is constant on cross-sections of the cylinder, so that the variations in pressure are described by a function $u(t, x)$ with $x \in [0, \ell]$, where ℓ is the length of the instrument. Pressure fluctuations are measured relative to the fixed atmospheric background, with $u = 0$ for atmospheric pressure.

The maximum pressure fluctuation occurs at the mouthpiece at $x = 0$, where a reed vibrates as the player blows air into the instrument. Since a local maximum of the pressure corresponds to a critical point of $u(t, \cdot)$, the appropriate boundary condition is

$$\frac{\partial u}{\partial x}(t, 0) = 0. \tag{5.11}$$

At the opposite end the air column is open to the atmosphere, so the pressure does not fluctuate,

$$u(t, \ell) = 0. \tag{5.12}$$

The evolution of u as a function of t is governed by the wave equation (4.5), with c equal to the speed of sound. The corresponding Helmholtz problem is

$$-\frac{d^2\phi}{dx^2} = \lambda\phi, \quad \phi'(0) = 0, \quad \phi(\ell) = 0. \tag{5.13}$$

The boundary condition at $x = 0$ implies that

$$\phi(x) = \cos(\sqrt{\lambda}x),$$

and the condition at $x = \ell$ then requires

$$\cos(\sqrt{\lambda}\ell) = 0.$$

This means that the eigenvalues are given by

$$\lambda_n := \frac{\pi^2}{\ell^2}(n - \tfrac{1}{2})^2,$$

for $n \in \mathbb{N}$. Some of the resulting eigenfunctions are shown in Fig. 5.3.

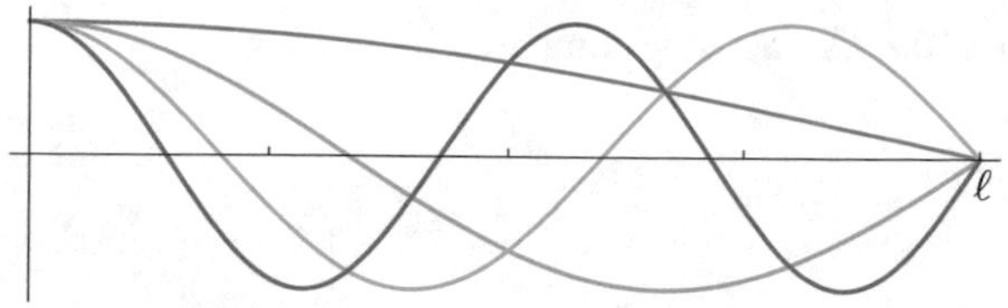

Fig. 5.3 The first four eigenfunctions for pressure fluctuations in a clarinet

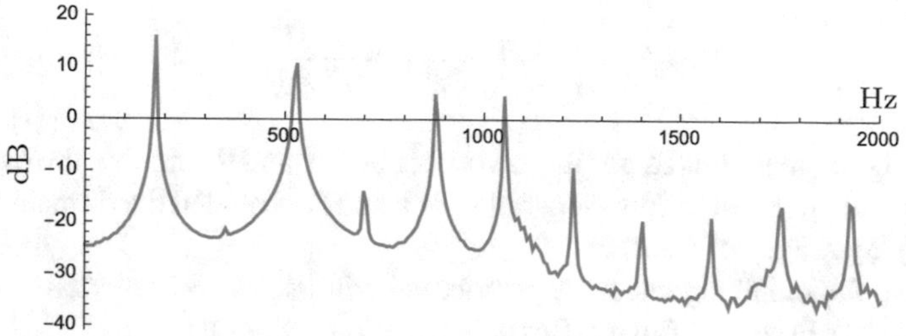

Fig. 5.4 Clarinet frequency decomposition

The corresponding oscillation frequencies are given by

$$\omega_n = \frac{c\pi}{\ell}(n - \tfrac{1}{2}).$$

In contrast to the string, the model predicts that the clarinet's spectrum will contain only odd multiples of the fundamental frequency ω_1. Figure 5.4 shows the frequency decomposition for a clarinet sound sample. The prediction holds true for the first few modes, but the simple model appears to break down at higher frequencies. ◊

5.3 Circular Symmetry

In dimension greater than one, spatial separation of variables is essentially the only way to compute explicit solutions of the Helmholtz equation (5.3), and this only works for very special cases. The most straightforward example is a rectangular domain in $\mathbb{R}^n$, which we will discuss in the exercises.

In this section we consider the simplest non-rectangular case, based on polar coordinates (r, θ) in $\mathbb{R}^2$. Separation in polar coordinates allows us to compute eigenfunctions and eigenvalues on a disk in $\mathbb{R}^2$, for example.

With $\boldsymbol{x} = (x_1, x_2)$ in $\mathbb{R}^2$, polar coordinates are defined by

$$(x_1, x_2) = (r\cos\theta, r\sin\theta).$$

The polar form of the Laplacian is computed by writing

$$\Delta = \frac{\partial^2}{\partial x_1^2} + \frac{\partial^2}{\partial x_2^2}$$

and then converting the partials with respect to x_1 and x_2 into r and θ derivatives using the chain rule. The result is

$$\Delta = \frac{1}{r}\frac{\partial}{\partial r}\left(r\frac{\partial}{\partial r}\right) + \frac{1}{r^2}\frac{\partial^2}{\partial \theta^2}. \tag{5.14}$$

Note that there are no mixed partials involving both r and θ, and that the coefficients do not depend on θ. This allows separation of r and θ, provided the domain is defined by specifying ranges of r and θ.

To solve the radial eigenvalue equation, we will use *Bessel functions*, named for the astronomer Friedrich Bessel. Bessel's equation is the ODE:

$$z^2 f''(z) + zf'(z) + (z^2 - k^2)f(z) = 0, \tag{5.15}$$

with $k \in \mathbb{C}$ in general. For our application k will be an integer. The standard pair of linearly independent solutions is given by the Bessel functions $J_k(z)$ and $Y_k(z)$.

The Bessel J-functions, a few of which are pictured in Fig. 5.5, satisfy

$$J_{-k}(z) = (-1)^k J_k(z), \tag{5.16}$$

for all $k \in \mathbb{Z}$. Bessel represented these solutions as integrals:

$$J_k(z) := \frac{1}{\pi}\int_0^{\pi} \cos\left(z \sin\theta - k\theta\right) d\theta.$$

One can also write J_k as a power series $k \in \mathbb{N}_0$,

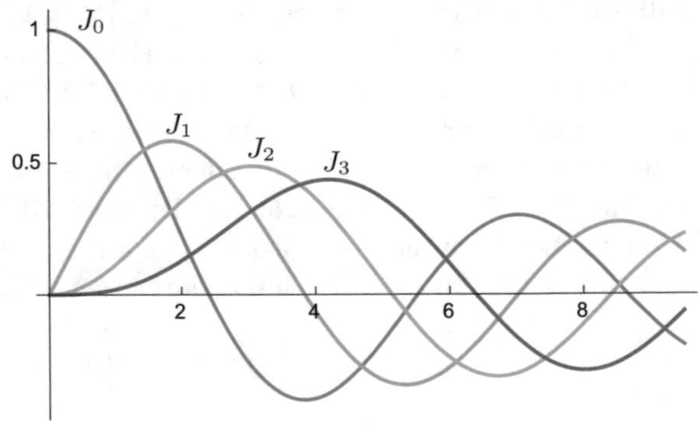

Fig. 5.5 The first four Bessel J-functions

$$J_k(z) = \left(\frac{z}{2}\right)^k \sum_{l=0}^{\infty} \frac{1}{l!(k+l)!} \left(-\frac{z^2}{4}\right)^l. \tag{5.17}$$

Together with (5.16), this shows that $J_k(z) \sim c_k z^{|k|}$ as $z \to 0$ for any $k \in \mathbb{Z}$. In contrast, the Bessel Y-function satisfies $Y_k(z) \sim c_k z^{-|k|}$ as $z \to 0$.

A change of sign in (5.15) gives the equation

$$z^2 f''(z) + z f'(z) + (z^2 + k^2) f(z) = 0. \tag{5.18}$$

Its standard solutions are the *modified Bessel functions* $I_k(z)$ and $K_k(z)$. As $z \to 0$ these satisfy the asymptotics $I_k(z) \sim c_k z^{|k|}$, as illustrated in Fig. 5.6, and $K_k(z) \sim c_k z^{-|k|}$.

Lemma 5.4 *Suppose $\phi \in C^2(\mathbb{R}^2)$ is a solution of*

$$-\Delta\phi = \lambda\phi,$$

that factors as a product $h(r)w(\theta)$. Then, up to a multiplicative constant, ϕ has the form

$$\phi_{\lambda,k}(r, \theta) := h_k(r) e^{ik\theta}, \tag{5.19}$$

for some $k \in \mathbb{Z}$, with

$$h_k(r) := \begin{cases} r^{|k|}, & \lambda = 0, \\ J_k(\sqrt{\lambda} r), & \lambda > 0, \\ I_k(\sqrt{-\lambda} r), & \lambda < 0. \end{cases}$$

Proof Under the assumption $\phi = hw$, the Helmholtz equation reduces by (5.14) to

$$\frac{w}{r} \frac{\partial}{\partial r}\left(r \frac{\partial h}{\partial r}\right) + \frac{h}{r^2} \frac{\partial^2 w}{\partial \theta^2} + \lambda h w = 0.$$

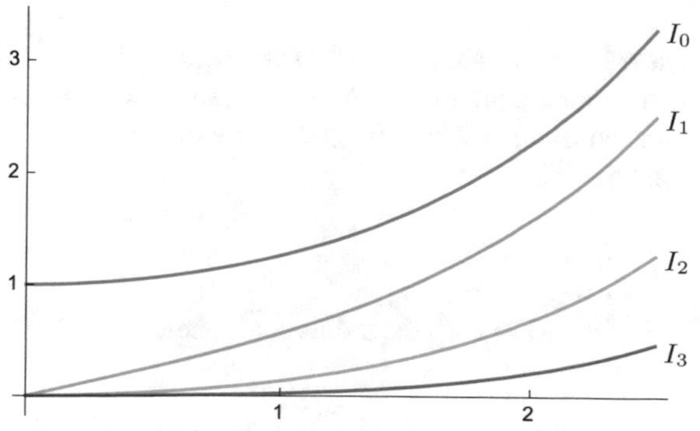

Fig. 5.6 The first four modified Bessel I-functions

With some rearranging, we can separate the r and θ variables,

$$\frac{1}{h}\left(r\frac{\partial}{\partial r}\right)^2 h + \lambda^2 r^2 = -\frac{1}{w}\frac{\partial^2 w}{\partial \theta^2}, \tag{5.20}$$

provided h and w are nonzero.

As in Lemma 5.1, we conclude that both sides must be equal to some constant κ. The θ equation is

$$-\frac{\partial^2 w}{\partial \theta^2} = \kappa w. \tag{5.21}$$

The function $w(\theta)$ is assumed to be 2π-periodic. By the arguments used in Theorem 5.2, a nontrivial solution is possible only if $\kappa = k^2$ where k is an integer. A full set of 2π-periodic solutions of (5.21) is given by

$$w_k(\theta) := e^{ik\theta}, \quad k \in \mathbb{Z}.$$

Before examining the radial equation, let us note that the assumption that ϕ is C^2 imposes a boundary condition at $r = 0$. To see this, first note that the function $r = \sqrt{x_1^2 + x_2^2}$ is continuous at $(0, 0)$ but not differentiable. For $r > 0$,

$$\frac{\partial r}{\partial x_j} = \frac{x_j}{r},$$

which does not have a limit as $r \to 0$. On the other hand, the functions

$$re^{\pm i\theta} = x_1 \pm i x_2$$

are C^∞. Similarly, for $k \in \mathbb{Z}$ we have

$$r^{|k|}e^{ik\theta} = \begin{cases} (x_1 + i x_2)^k, & k \in \mathbb{N}_0, \\ (x_1 - i x_2)^{-k}, & -k \in \mathbb{N}. \end{cases} \tag{5.22}$$

These functions are polynomial and hence C^∞. We will see below that the solutions of the radial equation corresponding to $\kappa = k^2$ satisfy $h(r) \sim ar^{\pm k}$ as $r \to 0$, for some constant a. The differentiability of ϕ at the origin will require the asymptotic condition

$$h_k(r) \sim ar^{|k|} \tag{5.23}$$

as $r \to 0$.

For $w_k(\theta) = e^{ik\theta}$, the radial component of (5.20) is

$$\left(r\frac{\partial}{\partial r}\right)^2 h_k + (\lambda r^2 - k^2)h_k = 0. \tag{5.24}$$

The case $\lambda = 0$ is relatively straightforward to analyze. In this case (5.24) is homogeneous in the r variable (meaning invariant under scaling). Such equations are solved by monomials of the form $h_k(r) = r^\alpha$ with $\alpha \in \mathbb{R}$. If we substitute this guess into (5.24) with $\lambda = 0$, the equation reduces to

$$\alpha^2 - k^2 = 0,$$

with solutions $\alpha = \pm k$. Since a second order ODE has exactly two independent solutions, the functions $r^{\pm k}$ give a full set of solutions for $k \neq 0$. For $k = 0$ the two possibilities are 1 and $\ln r$. By the condition (5.23), the solutions $\ln r$ and $r^{-|k|}$ must be ruled out. The only possible solutions for $\lambda = 0$ are thus

$$h_k(r) = r^{|k|}.$$

Note that the resulting solutions,

$$\phi_{0,k}(r, \theta) := r^{|k|} e^{ik\theta},$$

are precisely the polynomials (5.22).

For $\lambda > 0$ (5.24) can be reduced to the Bessel form (5.15) by the change of variables $z = \sqrt{\lambda} r$. The possible solutions $Y_k(\sqrt{\lambda} r)$ are ruled out because they diverge at $r = 0$. On the other hand, the power series (5.17) shows that the function $h_k(r) = J_k(\sqrt{\lambda} r)$ satisfies the condition (5.23). Thus for $\lambda > 0$ the possible eigenfunction with $k \in \mathbb{Z}$ is

$$\phi_{\lambda,k}(r, \theta) := J_k(\sqrt{\lambda} r) e^{ik\theta}.$$

We should check that this function is at least C^2 at the origin. In fact, it follows from the power series expansion (5.17) that $\phi_{\lambda,k}$ is C^∞ on $\mathbb{R}^2$.

Similar considerations apply for $\lambda < 0$, except that this time the substitution $z = \sqrt{-\lambda} r$ reduces (5.24) to (5.18). The condition (5.23) is satisfied only for the solution $I_k(\sqrt{-\lambda} r)$. □

Example 5.5 The linear model for the vibration of a drumhead is the wave equation (4.30). For a circular drum we can take the spatial domain to be the unit disk $\mathbb{D} := \{r < 1\} \subset \mathbb{R}^2$. Lemma 5.1 reduces the problem of determining the frequencies of the drum to the Helmholtz equation,

$$-\Delta\phi = \lambda\phi, \qquad \phi|_{\partial\mathbb{D}} = 0. \tag{5.25}$$

The possible product solutions are given by Lemma 5.4, subject to the boundary condition $h_k(1) = 0$. This rules out $\lambda \leq 0$, because in that case $h_k(r)$ has no zeros for $r > 0$.

For $\lambda > 0$, we have $h_k(r) = J_k(\sqrt{\lambda} r)$, and the boundary condition takes the form

$$J_k(\sqrt{\lambda}) = 0.$$

Table 5.1 Zeros of the Bessel function J_k. For each k, the spacing between zeros approaches π as $m \to \infty$

k	$j_{k,1}$	$j_{k,2}$	$j_{k,3}$	$j_{k,4}$
0	2.405	5.520	8.654	11.792
1	3.832	7.016	10.174	13.324
2	5.136	8.417	11.620	14.796
3	6.380	9.761	13.015	16.223
4	7.588	11.065	14.373	17.616

(This is analogous to the condition $\sin(\sqrt{\lambda}\ell) = 0$ from the one-dimensional string problem.) Although J_k is not a periodic function, it does have an infinite sequence of positive zeros with roughly evenly spacing. It is customary to write these zeros in increasing order as

$$0 < j_{k,1} < j_{k,2} < \dots.$$

By the symmetry (5.16),

$$j_{-k,m} = j_{k,m}.$$

Table 5.1 lists some of these zeros.

Restricting $\sqrt{\lambda}$ to the set of Bessel zeros gives the set of eigenvalues

$$\lambda_{k,m} = j_{k,m}^2,$$

indexed by $k \in \mathbb{Z}$, $m \in \mathbb{N}$. The corresponding eigenfunctions are

$$\phi_{k,m}(r, \theta) := J_k(j_{k,m}r)e^{ik\theta}. \tag{5.26}$$

The first set of these are illustrated in Fig. 5.7.

The collection of functions (5.26) yields a complete list of eigenfunctions and eigenvalues for $\mathbb{D}$, although that is not something we can prove here. ◊

The eigenvalues calculated in Example 5.5 correspond to vibrational frequencies

$$\omega_{k,m} := cj_{k,m},$$

for $k \in \mathbb{Z}$ and $m \in \mathbb{N}$. The propagation speed c depends on physical properties such as tension and density. The relative size of the frequencies helps to explain the lack of definite pitch in the sound of a drum. The ratios of overtones above the fundamental $\omega_{0,1}$ are shown in Table 5.2. In contrast to the vibrating string case, where the corresponding ratios were integers $1, 2, 3, \dots$, or the clarinet model of Example 5.3 with ratios $1, 3, 5, \dots$, the frequencies of the drum are closely spaced with no evident pattern.

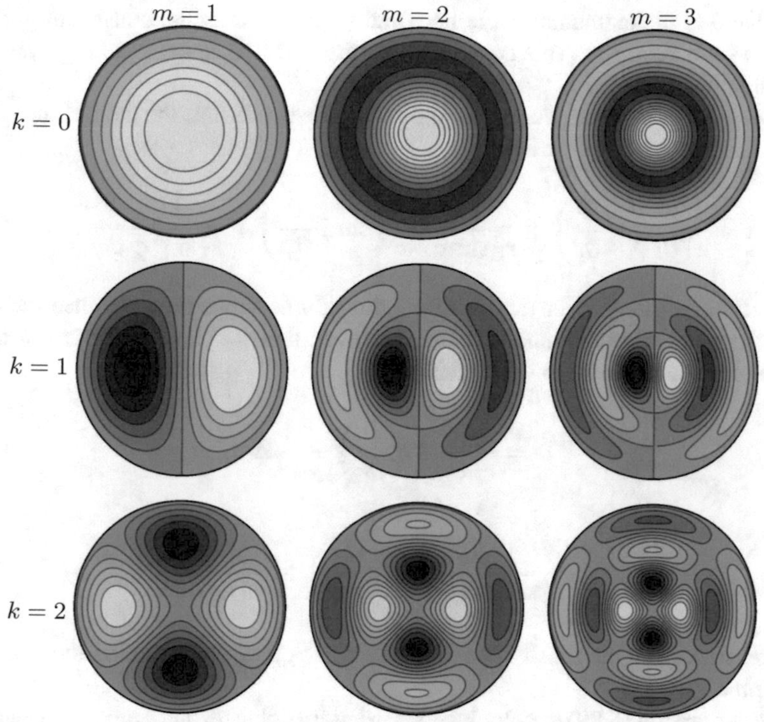

Fig. 5.7 Contour plots of the spatial component of the eigenfunctions of $\mathbb{D}$

Table 5.2 Frequency ratios for a circular drumhead

k	m	$\omega_{k,m}/\omega_{0,1}$
0	1	1
1	1	1.593
2	1	2.136
0	2	2.295
3	1	2.653
1	2	2.917
4	1	3.155

5.4 Spherical Symmetry

Another special case that allows separation of spatial variables is spherical symmetry in $\mathbb{R}^3$. Spherical coordinates (r, φ, θ) are defined through the relation

$$(x_1, x_2, x_3) = (r \sin\varphi \cos\theta, r \sin\varphi \sin\theta, r \cos\varphi).$$

Note that θ is the azimuthal angle here and φ the polar angle, consistent with the notation from Sect. 5.3. (This convention is standard in mathematics; in physics the roles are often reversed.)

As in the circular case, we can use the chain rule to translate the three-dimensional Laplacian into spherical variables:

$$\Delta = \frac{1}{r^2}\frac{\partial}{\partial r}\left(r^2\frac{\partial}{\partial r}\right) + \frac{1}{r^2 \sin\varphi}\frac{\partial}{\partial\varphi}\left(\sin\varphi\frac{\partial}{\partial\varphi}\right) + \frac{1}{r^2\sin^2\varphi}\frac{\partial^2}{\partial\theta^2}. \tag{5.27}$$

It is not immediately clear that this operator admits separation, because the coefficients depend on both r and φ. Note, however, that we can factor r^{-2} out of the angular derivative terms, to write (5.27) as

$$\Delta = \frac{1}{r^2}\frac{\partial}{\partial r}\left(r^2\frac{\partial}{\partial r}\right) + \frac{1}{r^2}\Delta_{\mathbb{S}^2}, \tag{5.28}$$

where

$$\Delta_{\mathbb{S}^2} := \frac{1}{\sin\varphi}\frac{\partial}{\partial\varphi}\left(\sin\varphi\frac{\partial}{\partial\varphi}\right) + \frac{1}{\sin^2\varphi}\frac{\partial^2}{\partial\theta^2}. \tag{5.29}$$

Here $\mathbb{S}^2$ stands for the unit sphere $\{r = 1\} \subset \mathbb{R}^3$, and $\Delta_{\mathbb{S}^2}$ is called the *spherical Laplacian*.

The expression (5.29) may look awkward at first glance, but $\Delta_{\mathbb{S}^2}$ is a very natural operator geometrically. From the fact that Δ is invariant under rotations of $\mathbb{R}^3$ about the origin, we can deduce that $\Delta_{\mathbb{S}^2}$ is also invariant under rotations of the sphere. It is possible to show that $\Delta_{\mathbb{S}^2}$ is the only second-order operator with this property, up to a multiplicative constant. The operator $\Delta_{\mathbb{S}^2}$ is thus as symmetric as possible, and the reason that (5.29) looks so complicated is that the standard coordinate system (θ, φ) does not reflect the full symmetry of the sphere.

We will discuss the radial component of (5.28) in an example below. For now let us focus on the Helmholtz problem on the sphere, which allows further separation of the θ and φ variables.

The classical ODE that arises from separation of the angle variables is the *associated Legendre equation*:

$$(1 - z^2)f''(z) - 2zf'(z) + \left(\nu(\nu+1) - \frac{\mu^2}{1-z^2}\right) f(z) = 0, \tag{5.30}$$

with parameters $\mu, \nu \in \mathbb{C}$. A pair of linearly independent solutions is given by the *Legendre functions* $P_\nu^\mu(z)$ and $Q_\nu^\mu(z)$.

In the special case where ν is replaced by $l \in \mathbb{N}_0$ and μ by a number $m \in \{-l, \ldots, l\}$, respectively, the Legendre P-functions are given by a relatively simple formula:

$$P_l^m(z) = \frac{(-1)^m}{2^l l!}(1 - z^2)^{m/2}\frac{d^{l+m}}{dz^{l+m}}(z^2 - 1)^l. \tag{5.31}$$

Associated to this set of Legendre functions are functions of the angle variables called *spherical harmonics*. These are defined by

$$Y_l^m(\varphi, \theta) := c_{l,m} e^{im\theta} P_l^m(\cos\varphi), \tag{5.32}$$

where $c_{l,m}$ is a normalization constant whose value is not important for us.

From (5.31), using $z = \cos\varphi$ and $1 - z^2 = \sin^2\varphi$, we can see that Y_l^m is a polynomial of degree l in $\sin\varphi$ and $\cos\varphi$. This makes it relatively straightforward to check that each $Y_l^m(\varphi, \theta)$ is a smooth function on $\mathbb{S}^2$.

Lemma 5.6 *Suppose $u \in C^2(\mathbb{S}^2)$ is a solution of the equation*

$$-\Delta_{\mathbb{S}^2} u = \lambda u \tag{5.33}$$

that factors as $u(\varphi, \theta) = v(\varphi)w(\theta)$. Then up to a multiplicative constant, u is equal to a spherical harmonic Y_l^m for $l \in \mathbb{N}_0$ and $m \in \{-l, \dots, l\}$. The corresponding eigenvalues depend only on l,

$$\lambda_l = l(l+1),$$

and each has multiplicity $2l + 1$.

Proof By (5.29), the substitution $u = vw$ leads to the separated equation

$$\frac{\sin\varphi}{v}\frac{\partial}{\partial\varphi}\left(\sin\varphi\frac{\partial v}{\partial\varphi}\right) + \lambda = -\frac{1}{w}\frac{\partial^2 w}{\partial\theta^2}.$$

The continuity of u requires that w be 2π-periodic. Hence, for the θ equation

$$-\frac{\partial^2 w}{\partial\theta^2} = \kappa w,$$

the full set of solutions is represented by $w(\theta) = e^{im\theta}$ with $\kappa = m^2$ for $m \in \mathbb{Z}$.

With $u(\theta, \varphi) = v_m(\varphi)e^{im\theta}$, the eigenvalue equation (5.33) reduces to

$$\frac{1}{\sin\varphi}\frac{d}{d\varphi}\left(\sin\varphi\frac{dv_m}{d\varphi}\right) + \left(\lambda - \frac{m^2}{\sin^2\varphi}\right)v_m = 0.$$

Under the substitutions $z = \cos\varphi$ and $v_m(\varphi) = f(\cos\varphi)$, this becomes

$$(1 - z^2)f'' - 2zf' + \left(\lambda - \frac{m^2}{1 - z^2}\right)f = 0,$$

which is recognizable as the Legendre equation (5.30) with parameters $m = \mu$ and $\lambda = \nu(\nu + 1)$.

Although $\mathbb{S}^2$ does not have a boundary, use of the coordinate $\varphi \in [0, \pi]$ creates artificial boundaries at the endpoints, i.e., at the poles of the sphere. This is analogous to the boundary at $r = 0$ in Lemma 5.4. We need to find solutions which will be smooth at the poles.

It turns out that for $m \in \mathbb{Z}$, the function $Q_\nu^m(z)$ diverges as $z \to 1$ for any $\nu \in \mathbb{C}$. Similarly, the functions $P_\nu^m(z)$ diverge as $z \to -1$ except for the special cases $P_l^m(z)$ given by (5.31). In other words, up to a multiplicative constant $v_m(\varphi)$ must be equal to $P_l^m(\cos\varphi)$ for some $l \in \mathbb{N}_0$ with $l \geq |m|$. The corresponding solution u is proportional to the spherical harmonic Y_l^m.

By the identification $\nu = l$, the eigenvalue is given by $\lambda = l(l+1)$. The corresponding multiplicity is the number of possible choices of $m \in \{-l, \dots, l\}$, namely $2l + 1$. □

The spherical harmonics appearing in Lemma 5.6 give a complete set of eigenfunctions for $\Delta_{\mathbb{S}^2}$, in the sense that the only possible eigenvalues are $l(l+1)$ for $l \in \mathbb{N}_0$ and an eigenfunction with eigenvalue $l(l+1)$ is a linear combination of the Y_l^m for $m \in \{-l, \dots, l\}$. To prove this requires more advanced methods than we have available here.

Example 5.7 In 1925, Erwin Schrödinger developed a quantum model for the hydrogen atom in which the electron energy levels are given by the eigenvalues of the equation

$$\left(-\Delta - \frac{1}{r}\right)\phi = \lambda\phi \tag{5.34}$$

on $\mathbb{R}^3$. (We have omitted the physical constants.) The eigenfunctions ϕ are assumed to be bounded near $r = 0$ and decaying to zero as $r \to \infty$.

Since the term $1/r$ is radial, separation of the radial and angular variables is possible in (5.34). By Lemma 5.6, the angular components are given by spherical harmonics. A corresponding full solution has the form

$$\phi(r, \varphi, \theta) = h(r) Y_l^m(\varphi, \theta). \tag{5.35}$$

Substituting this into (5.34) and using the spherical form of the Laplacian (5.28) gives the radial equation

$$\left[-\frac{1}{r^2}\frac{d}{dr}\left(r^2\frac{d}{dr}\right) + \frac{l(l+1)}{r^2} - \frac{1}{r}\right] h(r) = \lambda h(r). \tag{5.36}$$

One strategy used to analyze an ODE such as (5.36) is to first consider the asymptotic behavior of solutions as $r \to 0$ or ∞.

Suppose we assume $h(r) \sim r^\alpha$ as $r \to 0$. Plugging this into (5.36) and comparing the two sides gives a leading term

$$-\alpha(\alpha+1)r^{\alpha-2} + l(l+1)r^{\alpha-2}$$

on the left side, with all other terms of order $r^{\alpha-1}$ or less. This shows that $h(r) \sim r^\alpha$ as $r \to 0$ is possible only if

$$\alpha(\alpha + 1) = l(l + 1).$$

The two solutions are $\alpha = l$ or $\alpha = -l - 1$. Taking $\alpha < 0$ would cause $h(r)$ to diverge as $r \to 0$. Therefore to obtain a solution bounded at the origin, we will assume that

$$h(r) \sim r^l$$

as $r \to 0$.

As $r \to \infty$, if we consider the terms in (5.36) with coefficients of order r^0 and drop the rest, the equation becomes

$$-h''(r) \sim \lambda h(r). \tag{5.37}$$

If $\lambda \geq 0$ then this shows that $h(r)$ could not possibly decay at infinity. Hence we assume that $\lambda < 0$ and set

$$\sigma^2 := -\lambda,$$

with $\sigma \in \mathbb{R}$. The asymptotic equation (5.37) implies the behavior

$$h(r) \sim ce^{-\sigma r}$$

as $r \to \infty$.

Determining these asymptotics allows us to make an educated guess for the form of the solution. For an as yet undetermined function $q(r)$, we set

$$h(r) = q(r)r^l e^{-\sigma r}, \tag{5.38}$$

with the conditions that $q(0) = 1$ and $q(r)$ has subexponential growth as $r \to \infty$. The goal of setting up the solution this way is that the equation for $q(r)$ will simplify. Substituting (5.38) into (5.36) leads to the equation

$$rq'' + 2(1 + l - r\sigma)q' + (1 - 2\sigma(l + 1))q = 0. \tag{5.39}$$

To find solutions, we suppose $q(r)$ is given by a power series

$$q(r) = \sum_{k=0}^{\infty} a_k r^k,$$

with $a_0 = 1$. Plugging this into (5.39) gives

$$0 = \sum_{k=0}^{\infty} \left[k(k-1)a_k r^{k-1} + 2(1 + l - r\sigma)ka_k r^{k-1} + (1 - 2\sigma(l + 1))a_k r^k \right].$$

Equating the coefficient of r^k to zero then gives a recursive relation

$$a_{k+1} = \frac{2\sigma(k+l+1)-1}{(k+1)(k+2l+2)} a_k. \tag{5.40}$$

If we assume that the numerator of (5.40) never vanishes, then the recursion relation implies that

$$a_k \sim \frac{(2\sigma)^k}{k!}$$

as $k \to \infty$. This would give $q(r) \sim ce^{2\sigma r}$ as $r \to \infty$, making $h(r)$ also grow exponentially as $r \to \infty$.

The only way to avoid this exponential growth is for the sequence of a_k to terminate at some point, so that q is a polynomial. The numerator on the right side of (5.40) will eventually vanish if and only if

$$\sigma = \frac{1}{2n},$$

for some integer $n \geq l+1$. Under this assumption the sequence a_k terminates at $k = n-l-1$. Since $\lambda = -\sigma^2$, this restriction on σ gives the set of eigenvalues

$$\lambda_n := -\frac{1}{4n^2}, \quad n \in \mathbb{N}.$$

This is in fact the complete set of eigenvalues for this problem, given the conditions we have imposed at $r = 0$ and $r \to \infty$. With this eigenvalue calculation, Schrödinger was able to give the first theoretical explanation of the emission spectrum of hydrogen gas (i.e., the set of wavelengths observed when the gas is excited electrically). The origin of these emission lines had been a mystery since their discovery by Anders Jonas Ångström in the mid-19th century.

Each value of n corresponds to a family of eigenfunctions given by

$$\phi_{n,l,m}(r,\varphi,\theta) = r^l q_{n,l}(r) e^{-\frac{r}{2n}} Y_l^m(\varphi,\theta),$$

for $l \in \{0,\ldots,n-1\}$, $m \in \{-l,\ldots,l\}$. Here $q_{n,l}(r)$ denotes the polynomial of degree $n-l-1$ with coefficients specified by (5.40). To compute the multiplicity of λ_n, we count $n-1$ choices for l and then $2l+1$ choices of m for each l. The total multiplicity is

$$\sum_{l=0}^{n-1} (2l+1) = n^2.$$

◇

5.5 Exercises

5.1 On the half-strip $\Omega = (0, 1) \times (0, \infty) \subset \mathbb{R}^2$, find the solutions of

$$\Delta u = 0$$

that factor as a product $u(x_1, x_2) = g(x_1)h(x_2)$, under the boundary conditions

$$u(0, x_2) = u(1, x_2) = u(x_1, 0) = 0.$$

5.2 The linear model for vibrations of a rectangular drumhead is the wave equation (4.30) with Dirichlet boundary conditions on a rectangle $\mathcal{R} := [0, \ell_1] \times [0, \ell_2] \subset \mathbb{R}^2$. Separation of variables leads to the corresponding Helmholtz problem

$$-\Delta\phi = \lambda\phi, \qquad \phi|_{\partial\mathcal{R}} = 0.$$

Find the eigenfunctions of product type, $\phi(x_1, x_2) = \phi_1(x_1)\phi_2(x_2)$, and the associated frequencies of vibration. For $\ell_1 = \ell_2$, compare the ratios of these frequencies to Table 5.2. Would a square drum do a better job of producing a definite pitch?

5.3 The one-dimensional *heat equation* for the temperature $u(t, x)$ of a metal bar of length ℓ is

$$\frac{\partial u}{\partial t} - \frac{\partial^2 u}{\partial x^2} = 0,$$

for $t \geq 0$ and $x \in (0, \ell)$. (We will derive this in Sect. 6.1.) If the ends of the bar are insulated, then u should satisfy Neumann boundary conditions

$$\frac{\partial u}{\partial x}(t, 0) = \frac{\partial u}{\partial x}(t, \ell) = 0.$$

Find the product solutions $u(t, x) = v(t)\phi(x)$.

5.4 The *damped wave equation* on $\Omega \subset \mathbb{R}^n$ is

$$\frac{\partial^2 u}{\partial t^2} + \gamma\frac{\partial u}{\partial t} - \Delta u = 0, \tag{5.41}$$

where $u \in C^2([0, \infty) \times \Omega)$ and $\gamma \geq 0$ is a constant called the coefficient of friction. Suppose that $\phi \in C^2(\Omega)$ satisfies the Helmholtz equation (5.3) on Ω with eigenvalue $\lambda > 0$, for some appropriate choice of boundary conditions. Show that that (5.41) has solutions of the form

$$u(t, \boldsymbol{x}) = \phi(\boldsymbol{x})e^{i\omega t},$$

and find the set of possible values of ω. In particular, show that Im $\omega > 0$ if $\gamma > 0$, which implies that the solutions decay exponentially in time. Does this decay rate depend on the oscillation frequency?

5.5 Consider this example of a nonlinear diffusion equation:

$$\frac{\partial u}{\partial t} - \Delta(u^2) = 0,$$

for $t \geq 0$, $\boldsymbol{x} \in \mathbb{R}^n$.

(a) Assuming a product solution of the form $u(t,x) = v(t)\phi(\boldsymbol{x})$, separate variables and find the equations for $v(t)$ and $\phi(\boldsymbol{x})$.
(b) Show that $\phi(\boldsymbol{x}) = |\boldsymbol{x}|^2$ solves the spatial equation, and find the corresponding function $v(t)$ given the initial condition $v(0) = a > 0$. (Observe that the solution "blows up" at a finite time that depends on a.)

5.6 In polar coordinates for $\mathbb{R}^2$, define the domain

$$\Omega = \left\{(r,\theta);\ 0 < r < 1,\ 0 < \theta < \pi/3\right\},$$

which is a sector within the unit disk. Find the eigenvalues of Δ on Ω with Dirichlet boundary conditions.

5.7 The quantum energy levels of a harmonic oscillator in $\mathbb{R}^n$ are the eigenvalues of the equation

$$\left(-\Delta + |\boldsymbol{x}|^2\right)\phi = \lambda\phi, \tag{5.42}$$

under the condition that $\phi \in C^2(\mathbb{R}^n)$ and $\phi \to 0$ at infinity.

(a) First consider the case $n = 1$:

$$\left(-\frac{\partial^2}{\partial x^2} + x^2\right)\phi = \kappa\phi, \tag{5.43}$$

Substitute

$$\phi(x) = q(x)e^{-x^2/2}$$

into (5.43) and find the corresponding ODE for q.
(b) Assume that the function q from (a) is given by a power series in x,

$$q(x) = \sum_{k=0}^{\infty} a_k x^k,$$

and find a recursive equation for a_{k+2} in terms of a_k.
(c) Find the values of κ for which the power series for q from (b) truncates to a polynomial. (The resulting functions q are called *Hermite polynomials*.)

(d) Returning to the original problem, by reducing (5.42) to n copies of the case (5.43), deduce the set of eigenvalues λ.

5.8 Let $\mathbb{B}^3 \subset \mathbb{R}^3$ be the unit ball $\{r < 1\}$. Consider the Helmholtz problem

$$-\Delta\phi = \lambda\phi,$$

with Dirichlet boundary conditions at $r = 1$.

(a) Assume that

$$\phi(r, \varphi, \theta) = h_l(r) Y_l^m(\varphi, \theta),$$

where Y_l^m is the spherical harmonic introduced in Sect. 5.4. Find the radial equation for $h_l(r)$.

(b) For $l = 0$ show that the radial equation is solved by

$$h_0(r) = \frac{\sin(\sqrt{\lambda} r)}{r}.$$

What set of eigenvalues λ does this give?

(c) Show that the substitution,

$$h_l(r) = r^{-\frac{1}{2}} f_l(\sqrt{\lambda} r),$$

reduces the equation from (a) to a Bessel equation (5.15) for $f_l(z)$, with a fractional value of k. Use this to write the solution $h_l(r)$ in terms of J_k.

(d) Express the eigenvalues λ in terms of Bessel zeros with fractional values of k.

Chapter 6
The Heat Equation

In physics, the term *heat* is used to describe the transfer of internal energy within a system of particles. When this transfer results from collective motion of particles in a gas or fluid, the process is called *convection*. The continuity equation developed in Sect. 3.1 describes convection by fluid flow, which is the special case called *advection*. Another form of convection is *conduction*, where the heat transfer caused by random collisions of individual particles.

The basic mathematical model for heat conduction is a PDE called the *heat equation*, developed by Joseph Fourier in the early 19th century. In this chapter we will discuss the derivation and develop some basic properties of this equation, our first example of a PDE of parabolic type.

6.1 Model Problem: Heat Flow in a Metal Rod

A metal rod that is sufficiently thin can be treated as one-dimensional system. Let $u(t, x)$ denote the temperature of the rod at time t and position x, with $x \in \mathbb{R}$ for now.

There are two physical principles that govern the flow of heat in the rod. The first is the relationship between thermal (internal) energy and temperature. Thermal energy is proportional to a product of density and temperature, by a constant c called the *specific heat* of the material. Thus, the total thermal energy in a segment $[a, b]$ is given by

$$\mathcal{U} = c \int_a^b \rho u \, dx. \tag{6.1}$$

We will assume that the density ρ is constant, although it could be variable in some applications.

The original version of the book was revised: Belated corrections from author have been incorporated. The erratum to the book is available at https://doi.org/10.1007/978-3-319-48936-0_14

D. Borthwick, *Introduction to Partial Differential Equations*, Universitext, DOI 10.1007/978-3-319-48936-0_6

The second principle is Fourier's law of heat conduction, which describes how heat flows from hotter regions to colder regions. In its one-dimensional form, Fourier's law says that the flux of thermal energy across a given point is given by

$$q = -k\frac{\partial u}{\partial x}, \tag{6.2}$$

where the constant $k > 0$ is the *thermal conductivity* of the material.

Assuming that the rod is thermally isolated, conservation of energy dictates the rate of change of the thermal energy within the segment is equal to the flux across its boundaries, i.e.,

$$\frac{d\mathcal{U}}{dt}(t) = q(t, a) - q(t, b). \tag{6.3}$$

As in our derivation of the local equation for conservation of mass, the combination of (6.1) and (6.3) yields an integral equation

$$\int_a^b \left(c\rho\frac{\partial u}{\partial t} + \frac{\partial q}{\partial x}\right) dx = 0.$$

Since a and b were arbitrary, this implies a local conservation law,

$$c\rho\frac{\partial u}{\partial t} + \frac{\partial q}{\partial x} = 0.$$

Using the formula for q from Fourier's law (6.2), we obtain the one-dimensional *heat equation*:

$$\frac{\partial u}{\partial t} - \frac{k}{c\rho}\frac{\partial^2 u}{\partial x^2} = 0. \tag{6.4}$$

For a rod of finite length, the solution u will satisfy boundary conditions that depend on how the rod interacts with its environment. If the rod is parametrized by $x \in [0, \ell]$ and we assume that each end is held at a fixed temperature, then this fixes the values at the endpoints,

$$u(t, 0) = T_0, \quad u(t, \ell) = T_1 \tag{6.5}$$

for all t. These are inhomogeneous Dirichlet boundary conditions. In one dimension the inhomogeneous problem can be reduced very simply to the homogeneous case by noting that

$$u_0(x) := T_0\left(1 - \frac{x}{\ell}\right) + T_1\frac{x}{\ell}$$

gives an equilibrium solution to the heat equation satisfying the boundary conditions (6.5). By the superposition principle, $u - u_0$ satisfies the heat equation with homogeneous Dirichlet conditions.

Another possible boundary assumption is that the ends are insulated, so that no thermal energy flows in or out. This means that q vanishes at the boundary, yielding the Neumann boundary conditions

$$\frac{\partial u}{\partial x}(t,0) = \frac{\partial u}{\partial x}(t,\ell) = 0.$$

Example 6.1 On the bounded interval $[0, \pi]$, we can find product solutions to the heat equation using Lemma 5.1. For the Dirichlet boundary conditions $u(0) = u(\pi) = 0$, Theorem 5.2 gives the set of sine eigenfunctions (5.5). The corresponding heat equation solutions are

$$u(t,x) = e^{-n^2 t} \sin(nx)$$

for $n \in \mathbb{N}$. Note that all of these solutions decay exponentially to 0 as $t \to \infty$.

For insulated ends we switch to Neumann conditions and obtain the cosine modes. The resulting set of solutions is

$$u(t,x) = e^{-n^2 t} \cos(nx)$$

for $n \in \mathbb{N}_0$. In this case the $n = 0$ mode yields a constant solution.

In Chap. 8 we will discuss the construction of series solutions from these trigonometric families. ♢

The higher dimensional form of the heat equation can be derived by an argument similar to that given above. In $\mathbb{R}^n$, the thermal flux $\boldsymbol{q}$ is vector valued, and Fourier's law becomes the gradient formula

$$\boldsymbol{q} = -k\nabla u.$$

Local conservation of energy is expressed by the continuity equation (3.20),

$$c\rho\frac{\partial u}{\partial t} + \nabla \cdot \boldsymbol{q} = 0.$$

In combination, these yield the n-dimensional heat equation,

$$\frac{\partial u}{\partial t} - \frac{k}{c\rho}\Delta u = 0. \tag{6.6}$$

The importance of the heat equation as a model extends well beyond its original thermodynamic context. One of the most prominent examples of this is Albert Einstein's probabilistic derivation of the heat equation as a model for Brownian motion in 1905, in one of the set of papers for which he was later awarded the Nobel prize. Brownian motion is named for the botanist Robert Brown, who observed in 1827 that minute particles ejected by pollen grains drifted erratically when suspended in water, with a jittery motion for which no explanation was available at the time. Einstein theorized that the motion was caused by collisions with a large number of molecules whose velocities were distributed randomly. The existence of atoms and molecules was still unconfirmed in 1905, and Einstein's model provided crucial supporting evidence.

To summarize Einstein's argument, suppose that a total of n particles are distributed on the real line. In an interval of time τ, the position of each particle is assumed to change by a random amount according to a distribution function ϕ. To be more precise, the number of particles experiencing a displacement between σ and $\sigma + d\sigma$ is

$$dn = n\phi(\sigma)\, d\sigma.$$

The total number of particles is conserved, which imposes the condition

$$\int_{-\infty}^{\infty} \phi(\sigma)\, d\sigma = 1. \tag{6.7}$$

The distribution of displacements is assumed to be symmetric, $\phi(\sigma) = \phi(-\sigma)$, meaning that particles are equally likely to move left or right.

Suppose the distribution of particles at time t is given by a density function $\rho(t, x)$. Under the displacement hypothesis, the values of this density at times t and $t + \tau$ are related by

$$\rho(t + \tau, x) = \int_{-\infty}^{\infty} \rho(t, x - \sigma)\phi(\sigma)\, d\sigma. \tag{6.8}$$

To find an equation for ρ, Einstein takes the Taylor expansions of the density on both sides of (6.8), obtaining

$$\rho(t + \tau, x) = \rho(t, x) + \frac{\partial \rho}{\partial t}(t, x)\tau + \dots \tag{6.9}$$

on the left, and

$$\rho(t, x - \sigma) = \rho(t, x) - \frac{\partial \rho}{\partial x}(t, x)\sigma + \frac{1}{2}\frac{\partial^2 \rho}{\partial x^2}(t, x)\sigma^2 + \dots$$

inside the integral. Integrating the latter expansion against ϕ gives, by (6.7) and the assumption that ϕ is even (which knocks out the linear term),

$$\int_{-\infty}^{\infty} \rho(t, x - \sigma)\phi(\sigma)\, d\sigma = \rho(t, x) + \frac{1}{2}\frac{\partial^2 \rho}{\partial x^2}(t, x)\int_{-\infty}^{\infty} \sigma^2 \phi(\sigma)\, d\sigma + \dots.$$

Substituting this formula together with (6.9) into (6.8) and keeping the leading terms gives

$$\frac{\partial \rho}{\partial t}(t, x)\tau = \frac{1}{2}\frac{\partial^2 \rho}{\partial x^2}(t, x)\int_{-\infty}^{\infty} \sigma^2 \phi(\sigma)\, d\sigma.$$

Einstein then assumes that the value

$$D = \frac{1}{2\tau} \int_{-\infty}^{\infty} \sigma^2 \phi(\sigma)\, d\sigma \tag{6.10}$$

is a fixed constant, so that the equation for ρ becomes

$$\frac{\partial \rho}{\partial t} - D \frac{\partial^2 \rho}{\partial x^2} = 0,$$

i.e., the heat equation. Remarkably, the function ϕ representing the random distribution of displacements plays no role in the final equation, except in the value of the constant D. This fact is related to a fundamental result in probability called the central limit theorem.

Diffusion models involving random motions of particles are prevalent in physics, biology, and chemistry. The same statistical principles appear in other applications as well, for example in models of the spread of infection in medicine, or in the study of fluctuating financial markets. The heat equation plays a fundamental role in all of these applications.

6.2 Scale-Invariant Solution

Let us consider the heat equation on $\mathbb{R}$, with physical constants normalized to 1,

$$\frac{\partial u}{\partial t} - \frac{\partial^2 u}{\partial x^2} = 0. \tag{6.11}$$

Note that the equation is invariant under the rescaling $(t, x) \mapsto (\lambda^2 y, \lambda x)$, with λ a nonzero constant. This suggests a change of variables to the scale-invariant ratio $y := x/\sqrt{t}$ might simplify the equation.

Let us try to find a solution of the form $u(t, x) = q(y)$ for $t > 0$. By the chain rule,

$$\frac{\partial u}{\partial t} = -\frac{y}{2t} q', \qquad \frac{\partial^2 u}{\partial x^2} = \frac{1}{t} q''.$$

Thus, as an equation for q, (6.11) reduces to an ODE,

$$q'' = -\frac{y}{2} q'.$$

This can be solved for q' by separation of variables for ODE, as described in Sect. 2.5,

$$q'(y) = q'(0) e^{-y^2/4}.$$

A further integration yields

$$q(y) = q'(0) \int_0^y e^{-y'^2/4}\, dy' + q(0).$$

In the original coordinates this translates to

$$u(t,x) = C_1 \int_0^{\frac{x}{\sqrt{t}}} e^{-y^2/4}\, dy + C_2.$$

It is easy to confirm that this solves (6.11) for $t > 0$. To see what happens as $t \to 0$, note that

$$\int_0^\infty e^{-y^2/4}\, dy = \sqrt{\pi},$$

by the computations from Exercise 2.5. Thus

$$\lim_{t\to 0} u(t,x) = \begin{cases} C_1\sqrt{\pi} + C_2, & x > 0, \\ 0, & x = 0, \\ -C_1\sqrt{\pi} + C_2, & x < 0. \end{cases} \tag{6.12}$$

In view of this limiting behavior, let us consider the particular solution U defined by setting $C_1 = \frac{1}{\sqrt{4\pi}}$, $C_2 = \frac{1}{2}$,

$$U(t,x) := \frac{1}{\sqrt{4\pi}} \int_0^{\frac{x}{\sqrt{t}}} e^{-y^2/4}\, dy + \frac{1}{2}. \tag{6.13}$$

This solution is plotted for some small values of t in Fig. 6.1. By (6.12), $\lim_{t\to 0} U(t,x) = \Theta(x)$, the *Heaviside step function* defined by

$$\Theta(x) := \begin{cases} 1, & x > 0, \\ \frac{1}{2}, & x = 0, \\ 0, & x < 0. \end{cases}$$

The fact that $U(t,x)$ has such a simple limit as $t \to 0$ can be used to derive a more general integral formula. Suppose we want to solve (6.11) for the initial condition

$$u(t,x) = \varphi(x)$$

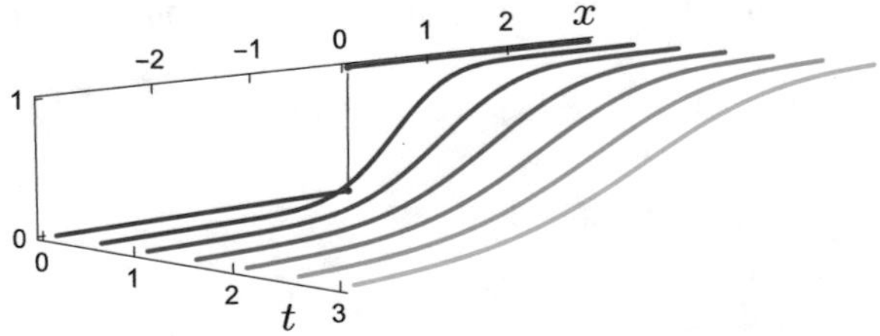

Fig. 6.1 The heat solution $U(t,x)$ for times from $t = 0$ to 3

with $\varphi \in C^{\infty}_{\text{cpt}}(\mathbb{R})$. The key observation is that φ can be reproduced by integrating its derivative against the Heaviside function,

$$\begin{aligned}\int_{-\infty}^{\infty} \varphi'(z)\Theta(x-z)\,dz &= \int_{-\infty}^{x} \varphi(z)\,dx' \\ &= \varphi(x).\end{aligned}$$

This suggests that we could solve the heat equation with initial condition φ by setting

$$u(t,x) := \int_{-\infty}^{\infty} \varphi'(z)U(t,x-z)\,dz.$$

For $t > 0$ the function $U(t, x - z)$ is continuously differentiable, so we can integrate by parts to rewrite this as

$$\begin{aligned}u(t,x) &= -\int_{-\infty}^{\infty} \varphi(z)\frac{\partial U}{\partial z}(t,x-z)\,dz \\ &= \frac{1}{\sqrt{4\pi t}}\int_{-\infty}^{\infty} \varphi(z)e^{-(x-z)^2/4t}\,dz.\end{aligned} \tag{6.14}$$

In terms of the function

$$H_t(x) := \frac{1}{\sqrt{4\pi t}}e^{-x^2/4t},$$

the solution is

$$u(t,x) = \int_{-\infty}^{\infty} H_t(x-z)\varphi(z)\,dz.$$

This integral is called the *convolution* of H_t with φ and denoted $u = H_t * \varphi$.

In the next section, we will generalize this convolution formula to higher dimension, and check that it does yield a solution of the heat equation satisfying the desired initial condition.

6.3 Integral Solution Formula

Consider the heat equation on $\mathbb{R}^n$,

$$\frac{\partial u}{\partial t} - \Delta u = 0 \tag{6.15}$$

for $t > 0$, with initial condition

$$u(0,\boldsymbol{x}) = g(\boldsymbol{x})$$

for $\boldsymbol{x} \in \mathbb{R}^n$.

Inspired by the calculations in Sect. 6.2, we define

$$H_t(\boldsymbol{x}) := (4\pi t)^{-\frac{n}{2}} e^{|\boldsymbol{x}|^2/4t} \tag{6.16}$$

for $t > 0$. The normalizing factor $(4\pi t)^{-\frac{n}{2}}$ is chosen so that

$$\int_{\mathbb{R}^n} H_t(\boldsymbol{x})\, d^n\boldsymbol{x} = 1. \tag{6.17}$$

(See Exercise 2.5 for the computation of integrals of this type.)

Direct differentiation shows that for $t > 0$,

$$\frac{\partial}{\partial t} H_t - \Delta H_t = 0, \tag{6.18}$$

so that $H_t(\boldsymbol{x})$ is a solution of the heat equation. However, the limit of $H_t(\boldsymbol{x})$ as $t \to 0$ is 0 for $\boldsymbol{x} \neq 0$ and ∞ for $\boldsymbol{x} = 0$, which does not seem to make sense as a distribution of temperatures. (We will return to discuss the interpretation of this initial condition in Chap. 12.)

With (6.14) as motivation, our goal in this section is to show that the convolution $u = H_t * g$ satisfies the heat equation on $\mathbb{R}^n$ for a continuous and bounded initial condition g. A function that acts on other functions by convolution is an *integral kernel*, and H_t is specifically called the *heat kernel* on $\mathbb{R}^n$. It is also called the *fundamental solution* of the heat equation, for reasons we will explain in Chap. 12.

Theorem 6.2 *For a bounded function $g \in C^0(\mathbb{R}^n)$, the heat equation*

$$\left(\frac{\partial}{\partial t} - \Delta\right) u = 0, \qquad u|_{t=0} = g, \tag{6.19}$$

admits a classical solution given by

$$u(t, \boldsymbol{x}) = H_t * g(\boldsymbol{x}). \tag{6.20}$$

Proof Explicitly, the formula (6.20) says that

$$u(t, \boldsymbol{x}) = (4\pi t)^{-\frac{n}{2}} \int_{\mathbb{R}^n} e^{-|\boldsymbol{x}-\boldsymbol{y}|^2/4t} g(\boldsymbol{y})\, d^n\boldsymbol{y}. \tag{6.21}$$

Because the domain is infinite here, we should treat differentiation under the integral with some care. To justify this, the key point is that the partial derivatives of H_t can be estimated by expressions of the form $c_1(t, \boldsymbol{x})e^{-c_2(t,\boldsymbol{x})|\boldsymbol{y}|^2}$, where the dependence of the constants c_1 and c_2 on t and $\boldsymbol{x}$ is continuous. We will not go into the technical details, but this makes it relatively straightforward to check that differentiating under the integral works in this case. In particular, the fact that $H_t(\boldsymbol{x})$ solves the heat equation implies (6.19).

To check the initial condition, fix some $\boldsymbol{x} \in \mathbb{R}^n$. A change of variables to $\boldsymbol{w} = (\boldsymbol{y} - \boldsymbol{x})/\sqrt{t}$ in (6.21) gives

$$\begin{aligned} u(t, \boldsymbol{x}) &= (4\pi)^{-\frac{n}{2}} \int_{\mathbb{R}^n} e^{-|\boldsymbol{w}|^2} g\left(\boldsymbol{x} + t^{\frac{1}{2}} \boldsymbol{w}\right) d^n \boldsymbol{w} \\ &= \int_{\mathbb{R}^n} H_1(\boldsymbol{w}) g\left(\boldsymbol{x} + t^{\frac{1}{2}} \boldsymbol{w}\right) d^n \boldsymbol{w} \end{aligned}$$

By (6.17) we can also write

$$g(\boldsymbol{x}) = \int_{\mathbb{R}^n} H_1(\boldsymbol{w}) g(\boldsymbol{x})\, d^n \boldsymbol{w}.$$

Thus the difference we are trying to estimate is

$$u(t, \boldsymbol{x}) - g(\boldsymbol{x}) = \int_{\mathbb{R}^n} H_1(\boldsymbol{w}) \left[g\left(\boldsymbol{x} + t^{\frac{1}{2}} \boldsymbol{w}\right) - g(\boldsymbol{x}) \right] d^n \boldsymbol{w}. \tag{6.22}$$

Given $\varepsilon > 0$, we can use the exponential decay of $H_1(\boldsymbol{w})$ as $|\boldsymbol{w}| \to \infty$ to choose R so that

$$\int_{|\boldsymbol{w}| \geq R} H_1(\boldsymbol{w})\, d^n \boldsymbol{w} < \varepsilon.$$

Since g is bounded by assumption, there exists a constant M so that $|g| \leq M$. The "large-$\boldsymbol{w}$" piece of (6.22) can thus be estimated by

$$\int_{|\boldsymbol{w}| \geq R} H_1(\boldsymbol{w}) \left| g\left(\boldsymbol{x} + t^{\frac{1}{2}} \boldsymbol{w}\right) - g(\boldsymbol{x}) \right| d^n \boldsymbol{w} \leq 2M\varepsilon. \tag{6.23}$$

On the other hand, by the continuity of g we can choose $\delta > 0$ so that for $\boldsymbol{y}$ such that for $|\boldsymbol{x} - \boldsymbol{y}| < \delta$, we have

$$|g\,(\boldsymbol{x} - \boldsymbol{y}) - g(\boldsymbol{x})| \leq \varepsilon.$$

Thus for $|\boldsymbol{w}| < R$ and $t < \delta^2/R^2$,

$$\left| g\left(\boldsymbol{x} + t^{\frac{1}{2}} \boldsymbol{w}\right) - g(\boldsymbol{x}) \right| \leq \varepsilon.$$

It follows that for $t < \delta^2/R^2$,

$$
\begin{aligned}
\int_{|w|<R} H_1(w) \left| g\left(x + t^{\frac{1}{2}} w\right) - g(x) \right| d^n w &\le \varepsilon \int_{|w|<R} H_1(w)\, d^n w \\
&\le \varepsilon \int_{\mathbb{R}^n} H_1(w)\, d^n w \\
&= \varepsilon.
\end{aligned}
\tag{6.24}
$$

Combining the estimates (6.23) and (6.24) gives

$$|u(t, x) - g(x)| \le (2M + 1)\varepsilon$$

for $0 < t < \delta^2/R^2$. Since ε was arbitrary, this shows that

$$\lim_{t\to 0} u(t, x) = g(x). \qquad \square$$

Without extra restrictions on u, the solution of (6.19) is not necessarily unique. However, if we start from a bounded initial condition g, then it is physically reasonable to assume that u is bounded over finite time intervals.

Theorem 6.3 *Under the assumption that* $u(t, \cdot)$ *is bounded on* $[0, T] \times \mathbb{R}^n$ *for each* $T > 0$*, the solution of the heat equation* (6.19) *is unique.*

We will develop tools to prove this result (maximum principles) in Chap. 9. The statement can be improved by weakening the boundedness hypothesis to an assumption of exponential growth. The counterexamples to uniqueness exhibit superexponential growth and are not considered valid as physical solutions.

In combination, Theorems 6.2 and 6.3, show that a bounded solution of the heat equation on $\mathbb{R}^n$ with continuous initial data satisfies (6.21). The function $H_t(x)$ is C^∞ in both variables for $t > 0$. As we noted in the proof of Theorem 6.2, differentiation under the integral is justified in (6.21), so this regularity can be extended to general solutions.

Theorem 6.4 *Suppose that* u *is a bounded solution of the heat equation* (6.19) *for a bounded initial condition* $g \in C^0(\mathbb{R}^n)$*. Then* $u \in C^\infty((0, \infty) \times \mathbb{R}^n)$.

Similar regularity results hold for the heat equation in other contexts, for example on a bounded domain. We will discuss some of these cases later in Sect. 8.6. This behavior, i.e., smoothness of solutions that does not depend on the regularity of the initial data, is characteristic of parabolic equations.

Another interesting feature of the heat kernel is the fact that $H_t(x)$ is strictly positive for all $t > 0$ and $x \in \mathbb{R}^n$. This means that if g is nonnegative and not identically zero, then u is nonzero at all points $x \in \mathbb{R}^n$ for $t > 0$. Compare this to the Huygens principle that we observed in Chap. 4, which says that for solutions of the wave equation the range of influence of a point is limited by the (finite) propagation speed. The heat equation exhibits *infinite propagation speed*.

We can see the origin of the infinite propagation speed in Einstein's diffusion model from Sect. 6.1. In (6.10) the value D, which is assumed to be constant, is the average squared displacement per unit time. The fact that D is fixed implies that in the continuum limit $\tau \to 0$, the average absolute displacement per unit time diverges. Hence the infinite propagation speed is built in to the construction of the model. It reflects the fact that models of diffusion are inherently statistical, and not expected to be accurate on a microscopic scale.

6.4 Inhomogeneous Problem

Duhamel's method, which was used to construct solutions of the inhomogeneous wave equation in Sect. 4.4, was originally developed in the context of the heat equation. There are slight differences in the setup, but the basic idea of translating a forcing term into an initial condition applies in both settings.

Consider the equation on $\mathbb{R}^n$,

$$\frac{\partial u}{\partial t} - \Delta u = f \tag{6.25}$$

for $t > 0$, with initial condition $u(0, \boldsymbol{x}) = 0$. For $s \geq 0$, let $\eta_s(t, \boldsymbol{x})$ be the solution of the homogeneous heat equation (6.19) for $t \geq s$, subject to the initial condition

$$\eta_s(t, \boldsymbol{x})\big|_{t=s} = f(s, \boldsymbol{x}). \tag{6.26}$$

We claim that the solution is then given by the integral

$$u(t, \boldsymbol{x}) := \int_0^t \eta_s(t, \boldsymbol{x})\, ds.$$

Using the formula for η_s provided by Theorem 6.2, the proposed solution can be written

$$u(t, \boldsymbol{x}) = \int_0^t \int_{\mathbb{R}^n} H_{t-s}(\boldsymbol{x} - \boldsymbol{y}) f(s, \boldsymbol{y})\, d^n\boldsymbol{y}\, ds. \tag{6.27}$$

To justify this formula, we must investigate carefully what happens near the point $t = s$.

Theorem 6.5 *Assuming that $f \in C^2([0, \infty) \times \mathbb{R}^n)$ and is compactly supported, the formula* (6.27) *yields a classical solution to the inhomogeneous heat equation* (6.25).

Proof We can see that u is at least C^2 by changing variables in the integral formula to obtain

$$u(t, \boldsymbol{x}) = \int_0^t \int_{\mathbb{R}^n} H_s(\boldsymbol{y}) f(t - s, \boldsymbol{x} - \boldsymbol{y})\, d^n\boldsymbol{y}\, ds.$$

Since $H_s(\mathbf{y})$ is smooth near $s = t$ and f is compactly supported, differentiation under the integral is justified. This gives

$$\begin{aligned}\frac{\partial u}{\partial t}(t, \mathbf{x}) = &\int_0^t \int_{\mathbb{R}^n} H_s(\mathbf{y}) \frac{\partial f}{\partial t}(t - s, \mathbf{x} - \mathbf{y})\, d^n\mathbf{y}\, ds \\ &+ \int_{\mathbb{R}^n} H_t(\mathbf{y}) \frac{\partial f}{\partial t}(0, \mathbf{x} - \mathbf{y})\, d^n\mathbf{y},\end{aligned} \tag{6.28}$$

and

$$\Delta u(t, \mathbf{x}) = \int_0^t \int_{\mathbb{R}^n} H_s(\mathbf{y}) \Delta_{\mathbf{x}} f(t - s, \mathbf{x} - \mathbf{y})\, d^n\mathbf{y}\, ds. \tag{6.29}$$

Our goal is to integrate by parts in these formulas, to exploit the fact that H_s solves the heat equation. Here we must be careful, because of the singular behavior of H_s at $s = 0$.

To deal with this singularity, we split the integral at $s = \varepsilon$. For the first integral in (6.28), switching the t derivative to an s derivative and integrating by parts gives

$$\begin{aligned}&\int_\varepsilon^t \int_{\mathbb{R}^n} H_s(\mathbf{y}) \frac{\partial f}{\partial t}(t - s, \mathbf{x} - \mathbf{y})\, d^n\mathbf{y}\, ds \\ &\quad = -\int_\varepsilon^t \int_{\mathbb{R}^n} H_s(\mathbf{y}) \frac{\partial f}{\partial s}(t - s, \mathbf{x} - \mathbf{y})\, d^n\mathbf{y}\, ds \\ &\quad = \int_\varepsilon^t \int_{\mathbb{R}^n} \frac{\partial H_s}{\partial s}(\mathbf{y}) f(t - s, \mathbf{x} - \mathbf{y})\, d^n\mathbf{y}\, ds \\ &\qquad - \int_{\mathbb{R}^n} H_t(\mathbf{y}) f(0, \mathbf{x} - \mathbf{y})\, d^n\mathbf{y} + \int_{\mathbb{R}^n} H_\varepsilon(\mathbf{y}) f(t - \varepsilon, \mathbf{x} - \mathbf{y})\, d^n\mathbf{y}\end{aligned}$$

The corresponding result for (6.29) has no boundary terms because of the compact support of f,

$$\begin{aligned}&\int_\varepsilon^t \int_{\mathbb{R}^n} H_s(\mathbf{y}) \Delta_{\mathbf{x}} f(t - s, \mathbf{x} - \mathbf{y})\, d^n\mathbf{y}\, ds \\ &\quad = \int_\varepsilon^t \int_{\mathbb{R}^n} \Delta_{\mathbf{y}} H_s(\mathbf{y}) f(t - s, \mathbf{x} - \mathbf{y})\, d^n\mathbf{y}\, ds\end{aligned}$$

Applying these integrations by parts to (6.28) and (6.29), and using the fact that $(\frac{\partial}{\partial s} - \Delta) H_s = 0$, we obtain

$$\begin{aligned}\left(\frac{\partial}{\partial t} - \Delta\right) u(t, \mathbf{x}) = &\int_{\mathbb{R}^n} H_\varepsilon(\mathbf{y}) f(t - \varepsilon, \mathbf{x} - \mathbf{y})\, d^n\mathbf{y} \\ &+ \int_0^\varepsilon \int_{\mathbb{R}^n} H_s(\mathbf{y}) \left(\frac{\partial}{\partial t} - \Delta_{\mathbf{x}}\right) f(t - s, \mathbf{x} - \mathbf{y})\, d^n\mathbf{y}\, ds.\end{aligned} \tag{6.30}$$

Since $H_s > 0$, the second term can be estimated by

$$\begin{aligned}\left|\int_0^{\varepsilon}\int_{\mathbb{R}^n} H_s(\mathbf{y})\left(\frac{\partial}{\partial t} - \Delta_{\mathbf{x}}\right) f(t-s, \mathbf{x}-\mathbf{y})\, d^n\mathbf{y}\, ds\right| \\ \le C \int_0^{\varepsilon}\int_{\mathbb{R}^n} H_s(\mathbf{y})\, d^n\mathbf{y}\, ds,\end{aligned}$$

where C is the maximum value of $\left|(\frac{\partial}{\partial t} - \Delta)f\right|$ (which exists by the assumption of compact support). By (6.17), the integral of $H_s(\mathbf{y})$ over $\mathbf{y} \in \mathbb{R}^n$ evaluates to 1, yielding the estimate

$$\left|\int_0^{\varepsilon}\int_{\mathbb{R}^n} H_s(\mathbf{y})\left(\frac{\partial}{\partial t} - \Delta_{\mathbf{x}}\right) f(t-s, \mathbf{x}-\mathbf{y})\, d^n\mathbf{y}\, ds\right| \le C\varepsilon.$$

We can therefore take $\varepsilon \to 0$ in (6.30) to obtain

$$\left(\frac{\partial}{\partial t} - \Delta\right) u(t, \mathbf{x}) = \lim_{\varepsilon\to 0}\int_{\mathbb{R}^n} H_{\varepsilon}(\mathbf{y}) f(t-\varepsilon, \mathbf{x}-\mathbf{y})\, d^n\mathbf{y}.$$

The remaining limit is very close to the limit computed in the proof of Theorem 6.2, except that t is replaced by $t-\varepsilon$. A simple modification of that argument shows that

$$\lim_{\varepsilon\to 0}\int_{\mathbb{R}^n} H_{\varepsilon}(\mathbf{y}) f(t-\varepsilon, \mathbf{x}-\mathbf{y})\, d^n\mathbf{y} = f(t, \mathbf{x}).$$

This completes the proof that u satisfies (6.25). □

6.5 Exercises

6.1 Biological processes and chemical reactions are frequently described by *reaction-diffusion equations*, consisting of the heat equation modified by a reaction term. Consider the simplest such equation,

$$\frac{\partial u}{\partial t} + \gamma u - \Delta u = 0.$$

on $\mathbb{R}^n$ with initial condition $u(0, \mathbf{x}) = f(\mathbf{x})$. Assuming f is continuous and bounded, find a formula for the solution. Hint: use a substitution of the form $u \to e^{-at}u$ to reduce this to the ordinary heat equation.

6.2 For $t \ge 0$, $x \ge 0$, suppose that $u(t, x)$ satisfies the one-dimensional heat equation (6.11) with the initial condition $u(0, x) = 0$ for $x \ge 0$ and the boundary condition

$$u(t, 0) = A\cos(\omega t)$$

for $t \geq 0$. Under the additional requirement that $u(t, \cdot)$ is bounded, find a solution $u(t, x)$. Hint: use separation of variables and assume that the temporal components have the form $e^{\pm i\omega t}$.

6.3 Let $\Omega \subset \mathbb{R}^n$ be a bounded domain with piecewise C^1 boundary. Suppose that $u(t, \boldsymbol{x})$ satisfies the heat equation

$$\frac{\partial u}{\partial t} - \Delta u = 0,$$

on $(0, \infty) \times \Omega$. Following the discussion from Sect. 6.1, we define the total thermal energy at time t by

$$\mathcal{U}[t] = \int_\Omega u(t, \boldsymbol{x})\, d^n\boldsymbol{x}.$$

(a) Assume that u satisfies Neumann boundary conditions,

$$\left.\frac{\partial u}{\partial \nu}\right|_{\partial\Omega} = 0,$$

(the insulated case). Show that $\mathcal{U}$ is constant.

(b) Assume that u is positive in the interior of Ω and equals 0 on the boundary. Show that $\mathcal{U}(t)$ is decreasing in this case.

6.4 Let $\Omega \subset \mathbb{R}^n$ be a bounded domain with piecewise C^1 boundary. Suppose that $u(t, \boldsymbol{x})$ is real-valued and satisfies the heat equation

$$\frac{\partial u}{\partial t} - \Delta u = 0,$$

on $(0, \infty) \times \Omega$. Define

$$\eta(t) := \int_\Omega u(t, \boldsymbol{x})^2\, d^n\boldsymbol{x}. \tag{6.31}$$

(a) Assume that u satisfies the Dirichlet boundary conditions:

$$u(t, \boldsymbol{x})|_{\boldsymbol{x} \in \partial\Omega} = 0$$

for $t \geq 0$. Show that η decreases as a function of t.

(b) Use (a) to show that a solution u satisfying boundary and initial conditions

$$u|_{t=0} = g, \quad u|_{\boldsymbol{x} \in \partial\Omega} = h,$$

for some continuous functions g on Ω and h on $\partial\Omega$, is uniquely determined by g and h.

Chapter 7
Function Spaces

In the preceding chapters we have seen that separation of variables can generate families of product solutions for certain PDE. For example, we found families of trigonometric solutions of the wave equation in Sect. 5.2 and the heat equation in Sect. 6.1. By the superposition principle, finite linear combinations of these functions give more general solutions.

It is natural to hope that we could push this construction farther and obtain solutions by infinite series. Solutions of PDE by trigonometric series were studied extensively in the 18th century by d'Alembert, Euler, Bernoulli, and others. However, notions of convergence were not well developed at that time, and many fundamental questions were left open.

In this chapter we will introduce some basic concepts of functional analysis, which will give us the tools to address some of these fundamental issues.

7.1 Inner Products and Norms

We assume that the reader has had a basic course in linear algebra and is familiar with the notion of a *vector space*, i.e., a set equipped with the operations of addition and scalar multiplication. The basic finite-dimensional example is the vector space $\mathbb{R}^n$. This space comes equipped with a natural inner product given by the dot product $\boldsymbol{v} \cdot \boldsymbol{w}$ for $\boldsymbol{v}, \boldsymbol{w} \in \mathbb{R}^n$. The Euclidean length of a vector $\boldsymbol{v} \in \mathbb{R}^n$ is $\|\boldsymbol{v}\| := \sqrt{\boldsymbol{v} \cdot \boldsymbol{v}}$.

In this section we will review the corresponding definitions for general real or complex vector spaces, which include function spaces. One important set of examples are the spaces $C^m(\Omega)$ introduced in Sect. 2.4, consisting of m-times continuously differentiable complex-valued functions on a domain $\Omega \subset \mathbb{R}^n$. Because differentiability and continuity of functions are preserved under linear combination and scalar multiplication, $C^m(\Omega)$ is naturally a complex vector space.

An *inner product* on a complex vector space V is a function of two variables,

$$u, v \in V \mapsto \langle u, v \rangle \in \mathbb{C},$$

D. Borthwick, *Introduction to Partial Differential Equations*,
Universitext, DOI 10.1007/978-3-319-48936-0_7

satisfying the following properties:

(I1) Positive definiteness: $\langle v, v\rangle \geq 0$ for $v \in V$, with equality only if $v = 0$.
(I2) Symmetry: for $v, w \in V$,

$$\langle v, w\rangle = \overline{\langle w, v\rangle}.$$

(I3) Linearity in the first variable: for $c_1, c_2 \in \mathbb{C}$ and $v_1, v_2, w \in V$,

$$\langle c_1 v_1 + c_2 v_2, w\rangle = c_1 \langle v_1, w\rangle + c_2 \langle v_2, w\rangle .$$

Together, (I2) and (I3) imply conjugate linearity in the second variable,

$$\langle w, c_1 v_1 + c_2 v_2\rangle = \overline{c_1} \langle w, v_1\rangle + \overline{c_2} \langle w, v_2\rangle .$$

The combination of linearity and conjugate linearity in the respective variables is called *sesquilinearity*. In the real case, the complex conjugation can be omitted, reducing sesquilinearity to *bilinearity*.

An *inner product space* is a real or complex vector space V equipped with an inner product $\langle \cdot, \cdot\rangle$. The *Euclidean* inner product on $\mathbb{C}^n$ is defined by including a conjugation in the dot product,

$$\langle \boldsymbol{v}, \boldsymbol{w}\rangle := \boldsymbol{v} \cdot \overline{\boldsymbol{w}}. \tag{7.1}$$

One way to define an inner product on function spaces is by integration. For example, on $C^0[0, 1]$ we could take

$$\langle f, g\rangle := \int_0^1 f\bar{g}\, dx.$$

Certain geometric notions are carried over from Euclidean geometry to inner product spaces. For example, vectors u, v in an inner product space V are called *orthogonal* if

$$\langle u, v\rangle = 0.$$

The analog of length for vectors in V is called a *norm*. A norm is a function $\|\cdot\| : V \to \mathbb{R}$ satisfying the following properties: for all $u, v \in V$ and scalar λ,

(N1) Positive definiteness: $\|u\| \geq 0$ with equality only if $u = 0$.
(N2) Homogeneity: $\|\lambda u\| = |\lambda|\, \|u\|$.
(N3) Triangle inequality: $\|u + v\| \leq \|u\| + \|v\|$.

For an inner product space, the definition of the Euclidean length in terms of the dot product suggests that the function

$$\|u\| := \sqrt{\langle u, u\rangle} \tag{7.2}$$

should yield a norm.

Positive definiteness of (7.2) clearly follows from positive definiteness of the inner product, and homogeneity follows from sesquilinearity. To see that (7.2) also satisfies the triangle inequality, we present a relation first derived in the Euclidean case by the great 19th century analyst Augustin-Louis Cauchy; Hermann Schwarz later generalized the result to inner product spaces.

Theorem 7.1 (Cauchy-Schwarz inequality) *For an inner product space V with $\|\cdot\|$ defined by* (7.2),

$$|\langle v, w\rangle| \leq \|v\| \, \|w\|$$

for all $v, w \in V$.

Proof For $v, w \in V$ and $t \in \mathbb{R}$, consider the function

$$q(t) := \left\| v + t \, \langle v, w\rangle \, w \right\|^2,$$

The claimed inequality is trivial if $w = 0$, so assume $w \neq 0$. By (I2), (I3), and (7.2),

$$\begin{aligned} q(t) &= \Big\langle v + t \, \langle v, w\rangle \, w, \, v + t \, \langle v, w\rangle \, w \Big\rangle \\ &= \|v\|^2 + 2t \, |\langle v, w\rangle|^2 + t^2 \, |\langle v, w\rangle|^2 \, \|w\|^2 . \end{aligned}$$

The minimum of this quadratic polynomial occurs at $t_0 = -\|w\|^{-2}$. Since $q \geq 0$,

$$0 \leq q(t_0) = \|v\|^2 - \frac{|\langle v, w\rangle|^2}{\|w\|^2},$$

which gives the claimed inequality. □

The triangle inequality for (7.2) follows from the Cauchy-Schwarz inequality by

$$\begin{aligned} \|u + v\|^2 &= \langle u + v, u + v\rangle \\ &= \|u\|^2 + 2 \, \mathrm{Re} \, \langle u, v\rangle + \|v\|^2 \\ &\leq \|u\|^2 + 2 \, |\langle u, v\rangle| + \|v\|^2 \\ &\leq \|u\|^2 + 2 \, \|u\| \, \|v\| + \|v\|^2 \\ &= (\|u\| + \|v\|)^2 . \end{aligned}$$

Thus (7.2) defines a norm associated to the inner product. This definition of the norm is used by default on an inner product space.

It is possible to have a norm that is not associated to an inner product. For example, this is the case for the *sup norm*, defined for $f \in C^0(\overline{\Omega})$, with $\Omega \subset \mathbb{R}^n$ bounded, by

$$\sup_{x \in \Omega} |f(x)| := \sup\left\{|f(x)|\,;\ x \in \Omega\right\}. \tag{7.3}$$

We will explain how to tell that a norm does not come from an inner product in the exercises.

7.2 Lebesgue Integration

In the early 20th century, Henri Lebesgue developed an extension of the classic definition of the integral introduced by Bernhard Riemann in 1854. (Riemann's is the version commonly taught in calculus courses.) Lebesgue's definition agrees with the Riemann integral when the latter exists, but extends to a broader class of integrable functions.

A full course would be needed to develop this integration theory properly. In this section, we present only a brief sketch of the Lebesgue theory, with the focus on the features most relevant for applications to PDE.

The Lebesgue integral is based on a generalized notion of volume for subsets of $\mathbb{R}^n$, which can be defined in terms of approximation by rectangles. For a rectangular subset in $\mathcal{R} \subset \mathbb{R}^n$, let $\mathrm{vol}(\mathcal{R})$ denote the usual notion of volume, the product of the lengths of the sides. (It is conventional to use "volume" as a general term when the dimension is arbitrary.) The volume of a subset $A \subset \mathbb{R}^n$ can be overestimated by covering the set with rectangles, as illustrated in Fig. 7.1. The (n-dimensional) *measure* of A is defined by taking the infimum of these overestimates,

$$m_n(A) := \inf\left\{\sum_{j=1}^{\infty} \mathrm{vol}(\mathcal{R}_j);\ A \subset \bigcup_{j=1}^{\infty} \mathcal{R}_j\right\}. \tag{7.4}$$

For a bounded region with C^1 boundary, the definition (7.4) reproduces the notion of volume used in multivariable calculus. Note that the concept of measure is dependent on the dimension. The measure of a line segment in $\mathbb{R}^1$ is the length, but a line segment has measure zero in $\mathbb{R}^n$ for $n \geq 2$.

There is a major technicality in the application of (7.4). In order to make the definition of measure consistent with respect to basic set operations, we cannot apply it to all possible subsets of $\mathbb{R}^n$. Instead, the definition is restricted to a special class of *measurable* sets. Lebesgue gave a criterion for measurability that rules out certain exotic sets for which volume is ill-defined. Fortunately, these sets are so exotic that we are unlikely to encounter them in normal usage. All open and closed sets in $\mathbb{R}^n$ are included in the measurable category, as are any sets constructed from them by basic set operations of union and intersection.

Fig. 7.1 Covering a set with *rectangles*

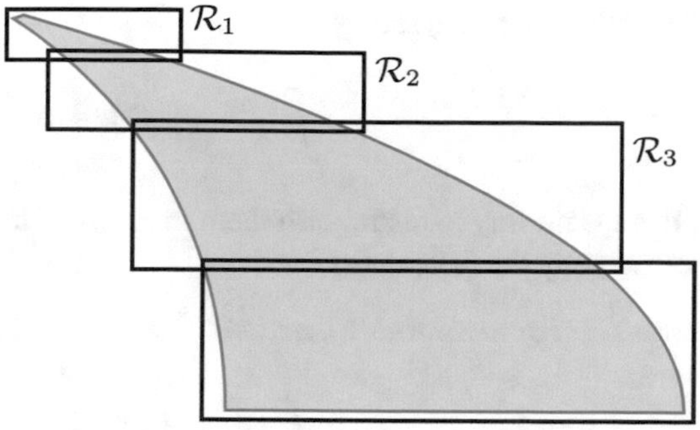

The *characteristic function* of a set $A \subset \mathbb{R}^n$ is defined by

$$\chi_A(\boldsymbol{x}) := \begin{cases} 1, & \boldsymbol{x} \in A, \\ 0, & \text{otherwise.} \end{cases} \tag{7.5}$$

The measure can be used to define the integral of a characteristic function,

$$\int_{\mathbb{R}^n} \chi_A \, d^n\boldsymbol{x} := m_n(A),$$

provided A is a measurable set. The integral of a general function is then built from approximations by linear combinations of characteristic functions. In order to construct these approximations, we need to use a restricted class of functions. A function $f : \Omega \to \mathbb{C}$ is called *measurable* if the preimage $f^{-1}(\mathcal{R})$ is a measurable subset of Ω for every rectangle $\mathcal{R} \subset \mathbb{C}$. Every Riemann-integrable function is measurable in the Lebesgue sense, so the measurable class includes all functions encountered in a traditional calculus class. Henceforth, whenever we write $f : \Omega \to \mathbb{C}$ or $\mathbb{R}$, we will assume implicitly that f is measurable.

With this basic picture in mind, we will ask the reader to accept certain important consequences of the Lebesgue definition without further justification. In examples and exercises we will confine our attention to functions for which ordinary Riemannian integrals exist.

It is standard practice when working with function spaces related to integration to make an equivalence:

$$f \equiv g \quad \Longleftrightarrow \quad f = g \text{ except on a set of measure zero.} \tag{7.6}$$

For example, in $\mathbb{R}$ the characteristic functions of the intervals (a, b) and $[a, b]$ are equivalent. In measure theory, a property is said to hold *almost everywhere* if it fails only on a set of measure zero. The equivalence (7.6) amounts to identifying functions that agree almost everywhere.

If functions f and g satisfy

$$\int |f - g| \, d^n\boldsymbol{x} = 0,$$

then there is no way to distinguish them in terms of integration. The definition (7.6) is motivated by the following:

Lemma 7.2 *For measurable functions* $f, g : \Omega \to \mathbb{C}$ *with* $\Omega \subset \mathbb{R}^n$,

$$\int_\Omega |f - g| \, d^n\boldsymbol{x} = 0$$

if and only if $f \equiv g$.

7.3 L^p Spaces

A function $f : \Omega \to \mathbb{C}$ is defined to be *integrable* if its integral converges absolutely, i.e.,

$$\int_\Omega |f| \, d^n\boldsymbol{x} < \infty.$$

For $p \geq 1$, we define the space of "p-integrable" functions by

$$L^p(\Omega) := \left\{ f : \Omega \to \mathbb{C}; \int_\Omega |f|^p \, d^n\boldsymbol{x} < \infty \right\}, \tag{7.7}$$

with the understanding that functions in L^p are identified according to the equivalence (7.6). The space $L^p(\Omega)$ is clearly closed under scalar multiplication. Closure under addition is a consequence of the convexity of the function $x \mapsto |x|^p$ for $p \geq 1$, which implies the inequality

$$\left| \frac{f + g}{2} \right|^p \leq \frac{|f|^p + |g|^p}{2}.$$

Hence $L^p(\Omega)$ is a complex vector space for $p \geq 1$.

The L^p norm is defined by

$$\|f\|_p := \left(\int_\Omega |f|^p \, d^n\boldsymbol{x} \right)^{\frac{1}{p}}.$$

To check that this is really a norm, we first note that Lemma 7.2 implies positive definiteness (N1) because of the equivalence relation (7.6). Homogeneity (N2) is satisfied because of cancellation between the powers p and $1/p$.

Fig. 7.2 Step function

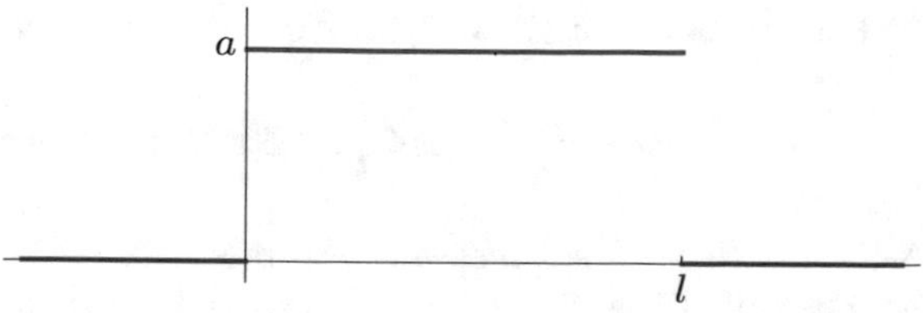

The triangle inequality (N3) is clear for $p = 1$ because $|f + g| \leq |f| + |g|$. And for $p = 2$ it follows from Cauchy-Schwarz inequality, because $\|\cdot\|_2$ is associated to the inner product

$$\langle f, g \rangle := \int_\Omega f \bar{g}\, d^n\boldsymbol{x}. \tag{7.8}$$

In general the L^p triangle inequality,

$$\|f + g\|_p \leq \|f\|_p + \|g\|_p\,,$$

is called the *Minkowski inequality* and holds for $p \geq 1$. We omit the proof because we are mainly concerned with the cases L^1 and L^2.

Example 7.3 To illustrate the distinction between the L^p norms, consider the function

$$h := a\chi_{[0,l]},$$

for $a, l > 0$, as shown in Fig. 7.2.

For general $p \geq 1$,

$$\begin{aligned}\|h\|_p &= \left[\int_0^l a^p\, dx\right]^{1/p} \\ &= a l^{1/p}.\end{aligned} \tag{7.9}$$

If we think of h as a density function, then the L^1 norm gives the total mass $\|h\|_1 = al$. The sensitivity of $\|\cdot\|_p$ to the spread of the function decreases as p increases, as illustrated by the fact that

$$\lim_{p\to\infty} \|h\|_p = a,$$

For large p, the L^p norms increasingly become measures of local concentration rather than mass. ◊

Example 7.3 suggests the possibility of defining a space L^∞ that is a limiting case of the L^p spaces, with a norm that generalizes the sup norm (7.3). The sup norm itself does not respect the equivalence (7.6), so we must modify the definition to define a norm consistent with the other L^p spaces.

For a function $h : \Omega \to \mathbb{R}$, the *essential supremum* is

$$\text{ess-sup}(h) := \inf \Big\{ a \in \mathbb{R};\ \{h > a\} \text{ has measure zero} \Big\}. \tag{7.10}$$

Note that $\{h > a\}$ has measure zero precisely when h is equivalent to a function bounded by a. The value ess-sup(h) is thus the least upper bound among all functions equivalent to h. For continuous functions the essential supremum reduces to the supremum.

For $f : \Omega \to \mathbb{C}$, we define

$$\|f\|_\infty := \text{ess-sup}\, |f|\,. \tag{7.11}$$

The normed vector space $L^\infty(\Omega)$ consists of functions which are "essentially bounded",

$$L^\infty(\Omega) := \Big\{ f : \Omega \to \mathbb{C};\ \|f\|_\infty < \infty \Big\}, \tag{7.12}$$

subject to the equivalence (7.6).

Collectively, the L^p spaces play a vital role in the analysis of PDE. The different norms can be thought of as a collection of measuring tools. Although the full toolkit is needed for many applications, for this book we will rely on the cases $p = 1, 2$, or ∞.

Example 7.4 The Schrödinger equation in $\mathbb{R}^n$ describes the evolution of a quantum-mechanical wave function $\psi(t, \boldsymbol{x})$:

$$-i\frac{\partial \psi}{\partial t} = \Delta \psi.$$

In Exercise 4.7 we saw that solutions have constant spatial L^2 norms,

$$\|\psi(t, \cdot)\|_2 = \|\psi(0, \cdot)\|_2\,,$$

which corresponds to the conservation of total probability. On the other hand, solutions also satisfy a *dispersive estimate*

$$\|\psi(t, \cdot)\|_\infty \le C t^{-n/2}\, \|\psi(0, \cdot)\|_1\,,$$

for all $t > 0$, with C a dimensional constant. The norm on the left measures the peak amplitude of the wave. By the estimate on the right, this amplitude is bounded in terms of the mass and decays as a function of time. In general, dispersive estimates describe the spreading of solutions as a function of time. ◊

It is conventional to represent elements of L^p as ordinary functions, even though each element is actually an equivalence class of functions identified under (7.6). This usually causes no trouble because equivalent functions give the same results in integrals.

One point that requires clarification, however, is the issue of continuity or differentiability of functions in L^p. Under (7.6), a C^m function is equivalent to a class of functions which are not even continuous. To account for this technicality, we adopt the convention that if a function in L^p is equivalent to a continuous function, then the continuous representative is used by default. This is unambiguous because the continuous representative is unique when it exists. Under this convention, the statement that $f \in L^p$ is a C^m function really means that f admits a continuous representative which is C^m.

7.4 Convergence and Completeness

In a normed vector space V, convergence of a sequence $v_n \to v$ means

$$\lim_{n\to\infty} \|v_n - v\| = 0. \tag{7.13}$$

We might also write this as

$$v = \lim v_n,$$

provided the choice of norm is clear.

It frequently proves useful to approximate L^p functions by smooth functions. For $p \geq 1$ there is a natural inclusion

$$C^\infty_{\text{cpt}}(\Omega) \subset L^p(\Omega),$$

because continuous functions on a compact set are bounded. The Lebesgue theory gives the following:

Theorem 7.5 *Assume* $1 \leq p < \infty$. *For a function* $f \in L^p(\Omega)$ *there exists an approximating sequence* $\{\psi_k\} \subset C^\infty_{\text{cpt}}(\Omega)$, *such that*

$$\lim_{k\to\infty} \|\psi_k - f\|_p = 0.$$

A subset W of a normed vector space V is called *dense* if every $v \in V$ can be obtained as a limit of a sequence in W. Theorem 7.5 thus states that $C^\infty_{\text{cpt}}(\Omega)$ is dense in $L^p(\Omega)$ for $p \in [1, \infty)$.

In PDE applications, a common method of proving the existence of a solution is to construct a sequence of approximate solutions, and then establish convergence of this sequence with respect to an appropriate norm. We cannot simply use the definition (7.13) to check convergence in this situation, because the limiting function may not exist. It is therefore crucial to be able to deduce convergence using only the sequence itself.

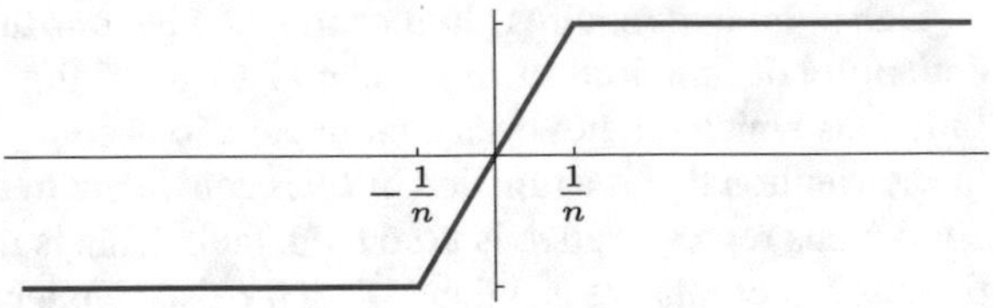

Fig. 7.3 A Cauchy sequence with respect to $\|\cdot\|_1$

The most useful tool for this purpose is a slightly weaker form of convergence. A sequence $\{v_k\} \subset V$ is said to be *Cauchy* if the difference between elements converges to zero: given $\varepsilon > 0$ there exists an N such that $k, m \geq N$ implies

$$\|v_k - v_m\| < \varepsilon.$$

This Cauchy condition is sometimes written as a double limit,

$$\lim_{k,m\to\infty} \|v_k - v_m\| = 0.$$

Every convergent sequence is Cauchy. This is because the triangle inequality implies

$$\begin{aligned}\|v_k - v_m\| &= \|v_k - v + v - v_m\| \\ &\leq \|v_k - v\| + \|v - v_m\|.\end{aligned}$$

If the sequence converges then the terms on the right are arbitrarily small for k and m sufficiently large.

In $\mathbb{R}^n$, it follows from the completeness axiom for real numbers that all Cauchy sequences are convergent. (See Theorem A.3.) This property does not necessarily hold in a general normed vector space, as the following demonstrates.

Example 7.6 Consider the space $C^0[-1, 1]$ equipped with the L^1 norm $\|\cdot\|_1$. For $n \in \mathbb{N}$ define the functions

$$f_n(x) = \begin{cases} -1, & x < -\frac{1}{n}, \\ nx & -\frac{1}{n} \leq x \leq \frac{1}{n}, \\ 1, & x > \frac{1}{n}, \end{cases}$$

as illustrated in Fig. 7.3.

We can see that the sequence $\{f_n\}$ is Cauchy by computing

$$\begin{aligned}\|f_k - f_m\|_1 &= \int_{-1}^{1} |f_k - f_m| \, dx \\ &= \left|\frac{1}{k} - \frac{1}{m}\right|.\end{aligned}$$

However, for $f \in C^0[-1, 1]$,

$$\lim_{k\to\infty} \|f_k - f\|_1 = \int_{-1}^{0} |f + 1|\, dx + \int_0^1 |f - 1|\, dx.$$

This limit equals 0 only if $f(x) = -1$ for $x < 0$ and $f(x) = 1$ for $x > 0$. That is not possible for f continuous. Therefore the sequence $\{f_n\}$ does not converge in $C^0[-1, 1]$. ♢

A normed vector space V is *complete* if all Cauchy sequences in V converge within V. Theorem A.3 implies that Euclidean $\mathbb{R}^n$ is complete in this sense. For L^p spaces the Lebesgue integration theory gives the following result.

Theorem 7.7 *For a domain $\Omega \subset \mathbb{R}^n$, the normed vector space $L^p(\Omega)$ is complete for each $p \in [1, \infty]$.*

In functional analysis, a complete normed vector space is called a *Banach space* and a complete inner product space is called a *Hilbert space*. Thus Theorem 7.7 could be paraphrased as the statement that $L^p(\Omega)$ is a Banach space. The inner product space $L^2(\Omega)$ is a Hilbert space.

A subspace $W \subset V$ is *closed* if it contains the limit of every sequence in W that converges in V.

Lemma 7.8 *If V is a complete normed vector space and $W \subset V$ is a closed subspace, then W is complete with respect to the norm of V.*

Proof Suppose $\{w_k\} \subset W$ is a Cauchy sequence. The sequence is also Cauchy in V, and so converges to some $v \in V$ by the completeness of V. Since W is closed, $v \in W$. □

The L^p function spaces have discrete counterparts, denoted by ℓ^p, whose elements are sequences. To a sequence $(a_1, a_2, \dots)$ of complex numbers we associate the function $a : \mathbb{N} \to \mathbb{C}$ defined by $j \mapsto a_j$. The ℓ^p norm of this function is

$$\|a\|_{\ell^p} := \left[\sum_{j=1}^{\infty} |a_j|^p\right]^{\frac{1}{p}},$$

The corresponding vector spaces are

$$\ell^p(\mathbb{N}) := \left\{a : \mathbb{N} \to \mathbb{C};\ \|a\|_{\ell^p} < \infty\right\}, \tag{7.14}$$

for $p \geq 1$. It is possible to prove directly that $\ell^p(\mathbb{N})$ is complete, but this can also be deduced easily from Lemma 7.8. We interpret $\ell^p(\mathbb{N})$ as a closed subspace of $L^p(\mathbb{R})$ consisting of functions which are constant on each interval $[j, j + 1)$ for $j \in \mathbb{N}$ and zero on $(-\infty, 0)$. On this subspace the L^p norm reduces to the ℓ^p norm, so that Lemma 7.8 implies that $\ell^p(\mathbb{N})$ is complete. In particular, $\ell^2(\mathbb{N})$ is a Hilbert space with the inner product

$$\langle a, b \rangle_{\ell^2} := \sum_{j=1}^{\infty} a_j \overline{b_j}.$$

7.5 Orthonormal Bases

Let H be an infinite-dimensional complex Hilbert space. A sequence of vectors $\{e_1, e_2, \dots\} \subset H$ is *orthonormal* if

$$\langle e_j, e_k \rangle = \begin{cases} 1, & j = k, \\ 0, & j \neq k, \end{cases} \tag{7.15}$$

for all $j, k \in \mathbb{N}$. An *orthonormal basis* for H is an orthonormal sequence such that each $v \in H$ admits a unique representation as a convergent series,

$$v = \sum_{j=1}^{\infty} c_j e_j, \tag{7.16}$$

with $c_j \in \mathbb{C}$.

As we will see in Sect. 7.6, the sets of eigenfunctions of certain differential operators naturally form orthonormal sequences with respect to the L^2 inner product. For example the sine eigenfunctions appearing in Theorem 5.2 have this property. If a sequence of eigenfunctions forms a basis, then we can expand general functions in terms of eigenfunctions.

Suppose we are given an orthonormal sequence $\{e_j\} \subset H$, and we would like to show that this forms a basis. To represent an element $v \in H$ in the form (7.16), we must decide how to choose the coefficients c_j. This works in much the same way as it does in finite dimensions. By the orthonormality property (7.15), we can compute that

$$\left\langle \sum_{j=1}^{n} c_j e_j, e_k \right\rangle = c_k \tag{7.17}$$

for all $n \geq k$. Assuming that $\sum c_j e_j$ converges to v in H, we can take the limit $n \to \infty$ in (7.17) to compute

$$\langle v, e_k \rangle = c_k. \tag{7.18}$$

Based on this calculation, we assign coefficients to v by setting

$$c_j[v] := \langle v, e_j \rangle. \tag{7.19}$$

The corresponding partial sums for $n \in \mathbb{N}$ are denoted by

$$S_n[v] := \sum_{j=1}^{n} c_j[v] e_j. \tag{7.20}$$

The condition that $\{e_j\}$ is a basis is equivalent to the convergence of $S_n[v] \to v$ in H for every $v \in H$.

Theorem 7.9 (Bessel's inequality) *Assume that $\{e_j\}$ is an orthonormal sequence in an infinite-dimensional Hilbert space H. For $v \in H$, the series $\sum |c_j[v]|^2$ converges and the limit satisfies*

$$\sum_{j=1}^{\infty} |c_j[v]|^2 \leq \|v\|^2 .$$

Equality holds if and only if $S_n[v] \to v$ in H.

Proof Using the sesquilinearity (I3) of the inner product, we can expand

$$\begin{aligned} \|v - S_n[v]\|^2 &:= \langle v - S_n[v], v - S_n[v] \rangle \\ &= \langle v, v \rangle - \langle S_n[v], v \rangle - \langle v, S_n[v] \rangle + \langle S_n[v], S_n[v] \rangle , \end{aligned}$$

for $n \in \mathbb{N}$. By the definition (7.20) of $S_n[v]$ and the orthonormality condition (7.15),

$$\langle S_n[v], v \rangle = \langle v, S_n[v] \rangle = \langle S_n[v], S_n[v] \rangle = \sum_{j=1}^{n} |c_j[v]|^2.$$

We thus conclude that

$$\|v - S_n[v]\|^2 = \|v\|^2 - \sum_{j=1}^{n} |c_j[v]|^2. \tag{7.21}$$

Since the left-hand side is positive, the identity (7.21) shows that

$$\sum_{j=1}^{n} |c_j[v]|^2 \leq \|v\|^2 ,$$

for all $n \in \mathbb{N}$. The partial sums of the series $\sum |c_j[v]|^2$ are thus bounded and the terms are all positive. Hence the series converges by the monotone sequence theorem, to a limit satisfying the claimed bound,

$$\sum_{j=1}^{\infty} \left| c_j[v] \right|^2 \leq \|v\|^2 .$$

To complete the proof, note that $S_n[v] \to v$ in H means that the limit as $n \to \infty$ of the left-hand side of (7.21) is zero. Hence $S_n[v] \to v$ if and only if

$$\lim_{n \to \infty} \sum_{j=1}^{n} \left| c_j[v] \right|^2 = \|v\|^2 .$$

□

The combination of completeness and Bessel's inequality leads to an alternative characterization of a basis that is easier to apply.

Theorem 7.10 *Suppose H is an infinite-dimensional Hilbert space. An orthonormal sequence in H forms a basis if and only if 0 is the only element of H that is orthogonal to all vectors in the sequence.*

Proof Assume first that $\{e_j\}$ forms a basis, so that every $v \in H$ can be written as a convergent sum $\sum c_j[v]e_j$. If v is orthogonal to all of the vectors e_j, then $c_j[v] = 0$ for all j by (7.19). Hence $v = 0$.

To establish the converse statement, let $\{e_j\}$ be an orthonormal sequence. For $v \in H$, Bessel's inequality implies

$$\sum_{j=1}^{\infty} \left| c_j[v] \right|^2 \leq \|v\|^2 < \infty. \tag{7.22}$$

For $n \leq m$,

$$\begin{aligned} \left\| S_m[v] - S_n[v] \right\|^2 &= \left\| \sum_{j=n+1}^{m} c_j[v]e_j \right\|^2 \\ &= \sum_{j=n+1}^{m} \left| c_j[v] \right|^2 . \end{aligned}$$

Hence (7.22) implies that

$$\lim_{m,n \to \infty} \left\| S_m[v] - S_n[v] \right\|^2 = 0,$$

meaning that the sequence $\{S_n[v]\}$ is Cauchy in H. By completeness of H this implies that $S_n[v] \to \tilde{v}$ for some $\tilde{v} \in H$.

Now assume that 0 is the only vector orthogonal to e_j for all j. For $n \geq j$ we have

$$\begin{aligned}\langle v - S_n[v], e_j\rangle &= \langle v, e_j\rangle - \langle S_n[v], e_j\rangle \\ &= c_j[v] - c_j[v] \\ &= 0.\end{aligned}$$

Taking the limit as $n \to \infty$ with j fixed gives

$$\langle v - \tilde{v}, e_j\rangle = 0.$$

Thus $v - \tilde{v}$ is orthogonal to every e_j, implying that $v = \tilde{v}$. This proves that $S_n[v] \to v$ in H for each $v \in H$, and thus $\{e_j\}$ is a basis. □

7.6 Self-adjointness

The process of forming a basis from eigenvectors of an operator should be familiar from linear algebra; for a finite-dimensional matrix this is called *diagonalization*. Let us briefly recall the basic facts for the finite-dimensional case. A complex $n \times n$ matrix A is *self-adjoint* (also called Hermitian) if the matrix is equal to its conjugate transpose. In terms of the Euclidean inner product (7.1) this means precisely that

$$\langle Au, v\rangle = \langle u, Av\rangle \tag{7.23}$$

for all $u, v \in \mathbb{C}^n$. (In the real case self-adjoint is the same as *symmetric*.)

The spectral theorem in linear algebra says that for a self-adjoint matrix A there exists an orthonormal basis for $\mathbb{C}^n$ consisting of eigenvectors for A, with real eigenvalues. Functional analysis allows a powerful extension of this result, that applies in particular to certain differential operators acting on L^2 spaces. The full spectral theorem for Hilbert spaces is too technical for us to state here, but we will prove a version of this for the Laplacian on bounded domains later in Sect. 11.5.

Self-adjointness remains important as a hypothesis for the more general spectral theorem, but even this condition becomes rather technical in the Hilbert space setting. The issues arise from the fact that differentiable operators cannot act on the whole space $L^2(\Omega)$ because L^2 functions need not be differentiable. We will avoid these complexities, by focusing on the Laplacian and restricting our attention to C^2 functions.

Lemma 7.11 *Suppose that $\Omega \in \mathbb{R}^n$ is a bounded domain with C^1 boundary. If $u, v \in C^2(\overline{\Omega})$ both satisfy either Dirichlet or Neumann boundary conditions on $\partial\Omega$, then*

$$\langle \Delta u, v\rangle = \langle u, \Delta v\rangle\,. \tag{7.24}$$

Proof By Green's first identity (Theorem 2.10),

$$\int_{\Omega}\left[u\overline{\Delta v}-\bar{v}\Delta u\right]d^n\mathbf{x}=\int_{\partial\Omega}\left[u\frac{\partial\bar{v}}{\partial\nu}-\bar{v}\frac{\partial u}{\partial\nu}\right]dS. \tag{7.25}$$

The Dirichlet conditions require that

$$u|_{\partial\Omega}=v|_{\partial\Omega}=0,$$

implying the vanishing of the right-hand side of (7.25). Similarly, the Neumann conditions

$$\left.\frac{\partial u}{\partial\nu}\right|_{\partial\Omega}=\left.\frac{\partial v}{\partial\nu}\right|_{\partial\Omega}=0 \tag{7.26}$$

also imply that the integrand on the right vanishes. □

Boundary conditions for which (7.24) holds are called *self-adjoint* boundary conditions (for the Laplacian). Formally, (7.24) resembles the matrix condition (7.23), but of course there is no analog of boundary conditions in the matrix case. The proper definition of self-adjointness in functional analysis involves a more precise specification of the domain on which Δ acts and (7.24) holds. Even without going into these details, we can still draw some meaningful conclusions from Lemma 7.11.

Lemma 7.12 *Suppose $\{\lambda_j\}$ is a sequence of eigenvalues of $-\Delta$ on a bounded domain $\Omega\subset\mathbb{R}^n$, with eigenvectors in $C^2(\overline{\Omega})$ subject to a self-adjoint boundary condition. Then $\lambda_j\in\mathbb{R}$ and, after possible rearrangement, the eigenvectors form an orthonormal sequence in $L^2(\Omega)$.*

Furthermore, $\lambda_j>0$ for Dirichlet conditions, and $\lambda_j\geq 0$ for Neumann.

Proof Suppose we have a sequence $\{\phi_j\}\subset C^2(\overline{\Omega})$ satisfying

$$-\Delta\phi_j=\lambda_j\phi_j.$$

The condition (7.24) implies that for $j,k\in\mathbb{Z}$,

$$\langle\Delta\phi_j,\phi_k\rangle=\langle\phi_j,\Delta\phi_k\rangle.$$

By the eigenvalue property this reduces to

$$\left(\lambda_j-\overline{\lambda_k}\right)\langle\phi_j,\phi_k\rangle=0. \tag{7.27}$$

For $j=k$ the inner product equals $\|\phi_j\|_2^2>0$, implying that $\lambda_j\in\mathbb{R}$ for all j. We can thus drop the conjugation in (7.27). If $\lambda_j\neq\lambda_k$, then this now implies that $\langle\phi_j,\phi_k\rangle=0$.

If some of the λ_j's are equal, then every linear combination of the corresponding of eigenfunctions will still be an eigenfunction for the same value of λ_j. Hence we can rearrange the eigenfunctions sharing a common eigenvalue into an orthogonal set using the Gram-Schmidt procedure from linear algebra.

By multiplying the eigenfunctions by constants we can normalize so that $\|\phi_j\|_2 = 1$. The divergence theorem (Theorem 2.6) then implies

$$\begin{aligned}\lambda_j &= \langle -\Delta\phi_j, \phi_j\rangle \\ &= \int_\Omega |\nabla\phi_j|^2 \, d^n\mathbf{x} - \int_{\partial\Omega} \overline{\phi_j}\frac{\partial\phi_j}{\partial\nu}\, dS.\end{aligned}$$

Either Dirichlet or Neumann conditions will cause the second term to vanish, implying that $\lambda_j \geq 0$. If $\lambda_j = 0$ then the equation also shows that $\nabla\phi_j \equiv 0$, implying that ϕ_j is constant. In the Dirichlet case the only constant solution is trivial, $\phi_j \equiv 0$, but for Neumann conditions a nonzero constant is possible. □

Example 7.13 In Example 5.5 we found a set of eigenfunctions for a circular drumhead modeled by the unit disk, with Dirichlet boundary conditions. The eigenfunctions were given in polar coordinates by

$$\phi_{k,m}(r,\theta) := e^{ik\theta} J_k(j_{k,m}r), \quad k \in \mathbb{Z}, m \in \mathbb{N},$$

where $j_{k,m}$ is the mth positive zero of the Bessel function J_k. The eigenvalues of $-\Delta$ in this case are the values $j_{k,m}$. Since the only possible matches among the Bessel zeros are $j_{k,m} = j_{-k,m}$, these are the only potential non-orthogonal pairs.

Let us examine the orthogonality condition more explicitly. In polar coordinates, the L^2 inner product of two eigenfunctions is given by

$$\begin{aligned}\langle \phi_{k,m}, \phi_{k',m'}\rangle_{L^2} &= \int_0^1\int_0^{2\pi} \phi_{k,m}(r,\theta)\overline{\phi_{k',m'}(r,\theta)}\, r\, d\theta\, dr \\ &= \int_0^1\int_0^{2\pi} e^{i(k-k')\theta} J_k(j_{k,m}r) J_{k'}(j_{k',m'}r)\, r\, d\theta\, dr.\end{aligned}$$

Note that the eigenfunctions are clearly orthogonal when $k \neq k'$, because the θ integral vanishes in this case. If we set $k = k'$, then the θ integral is trivial and the inner product becomes

$$\langle \phi_{k,m}, \phi_{k,m'}\rangle_{L^2} = 2\pi \int_0^1 r J_k(j_{k,m}r) J_k(j_{k,m'}r)\, dr.$$

By Lemma 7.12 this integral vanishes for $m \neq m'$. The cancellations occur because of the oscillations, just as for sine functions, as Fig. 7.4 illustrates.

Fig. 7.4 Radial components $J_1(j_{1,m}r)$ of orthogonal eigenfunctions on the disk

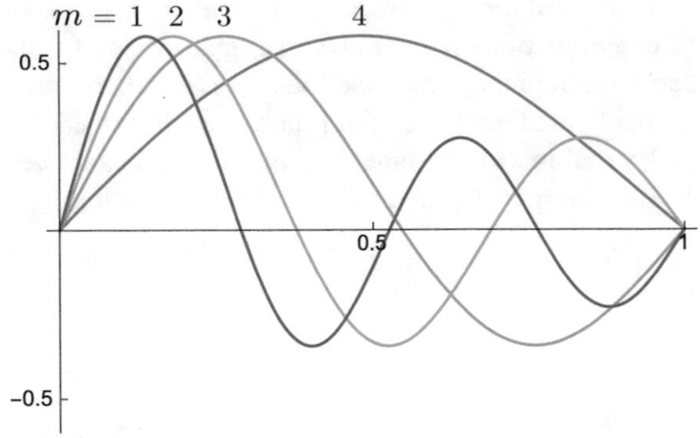

7.7 Exercises

7.1 A norm $\|\cdot\|$ on a vector space V satisfies the *parallelogram law* if

$$\|v + w\|^2 + \|v - w\|^2 = 2\,\|v\|^2 + 2\,\|w\|^2,$$

for all $v, w \in V$.

(a) Show that a norm defined by an inner product as in (7.2) satisfies the parallelogram law.

(b) In $L^p(\mathbb{R})$, define the functions

$$f(x) = \chi_{[0,2]}, \qquad g(x) = \chi_{[0,1]} - \chi_{[1,2]}.$$

Use these to show that the parallelogram law fails for $\|\cdot\|_p$ if $p \neq 2$.

(c) Find an example to show that the parallelogram law fails for the sup norm (7.3).

7.2 Consider the sequence of functions on $\mathbb{R}$ defined by

$$f_n(x) = \begin{cases} ne^{-n^2 x}, & x \geq 0, \\ 0, & x < 0. \end{cases}$$

Show that $f_n \to 0$ in $L^1(\mathbb{R})$ but not in $L^2(\mathbb{R})$.

7.3 Consider the sequence of functions on $\mathbb{R}$ defined by

$$g_n(x) = n^{-1}\chi_{[0,n]}.$$

Show that $g_n \to 0$ in $L^2(\mathbb{R})$ but not in $L^1(\mathbb{R})$.

7.4 Consider the sequence $f_n(x) = x^n$ for $x \in (0, 1)$. Show that $f_n \to 0$ in $L^p(0, 1)$ for each $p \in [1, \infty)$, but not for $p = \infty$.

7.5 Assume that $\Omega \subset \mathbb{R}^n$ is a bounded domain. Show that there is a constant $C > 0$ such that for $f \in L^2(\Omega)$,

$$\|f\|_1 \le C \|f\|_2 .$$

This implies in particular that $L^2(\Omega) \subset L^1(\Omega)$. Find an example to show that this result does not hold for Ω unbounded.

7.6 As an application of the Cauchy-Schwarz inequality, we can use the quantity η defined in Exercise 6.4 to show that solutions of the heat equation with fixed boundary values are uniquely determined by the values at time $t = T > 0$. Under the hypotheses from that exercise, suppose that u solves the heat equation with

$$u|_{t=T} = 0, \quad u|_{x\in\partial\Omega} = 0.$$

The goal is to show that these assumptions imply $u = 0$ for all t.

(a) Use the Cauchy-Schwarz inequality to deduce that

$$\eta'(t)^2 \le 4\eta(t) \int_\Omega \left|\frac{\partial u}{\partial t}\right|^2 d^n\boldsymbol{x}.$$

where η is defined as in (6.31).

(b) Show that

$$\eta''(t) = 4 \int_\Omega \left|\frac{\partial u}{\partial t}\right|^2 d^n\boldsymbol{x},$$

so that the inequality from (a) becomes

$$\eta'(t)^2 \le \eta(t)\eta''(t). \tag{7.28}$$

(c) Suppose that $\eta(0) > 0$. Then by continuity $\log \eta(t)$ is defined at least in some neighborhood of $t = 0$. Using (7.28), show that

$$(\log \eta(t))'' \ge 0.$$

This implies that $\log \eta(t)$ is bounded below by its tangent lines. In particular

$$\log \eta(t) \ge \log \eta(0) + \frac{\eta'(0)}{\eta(0)} t,$$

which implies

$$\eta(t) \geq \eta(0)e^{-ct},$$

for $c = -\eta'(0)/\eta(0) > 0$. Thus if $\eta(0) > 0$ then η is strictly positive for all $t \geq 0$.

(d) Conclude from (c) that if $\eta(T) = 0$, then $\eta(t) = 0$ for all t, and deduce that $u = 0$.

7.7 Recall the radial decomposition formula (2.10). We can use this to get a basic picture of the degree of singularity or decay at infinity that is allowed in each L^p.

(a) For $\gamma \in \mathbb{R}$ consider the function

$$g(\boldsymbol{x}) := \begin{cases} r^\gamma, & r \leq 1, \\ 0, & r \geq 1. \end{cases}$$

For what values of γ and $p \in [1, \infty]$ is $g \in L^p(\mathbb{R}^n)$?

(b) For $\gamma \in \mathbb{R}$ consider the function

$$h(\boldsymbol{x}) := \begin{cases} 0, & r \leq 1, \\ r^\gamma, & r \geq 1. \end{cases}$$

For what values of γ and $p \in [1, \infty]$ is $h \in L^p(\mathbb{R}^n)$?

7.8 Consider the eigenfunctions given by (5.5) with $\ell = \pi$.

(a) Show that

$$\phi_n(x) := \sqrt{\frac{2}{\pi}} \sin(nx), \quad n \in \mathbb{N},$$

defines an orthonormal sequence in $L^2(0, \pi)$. (Hint: recall the trigonometric identity $\sin(\alpha)\sin(\beta) = \frac{1}{2}\big[\cos(\alpha - \beta) - \cos(\alpha + \beta)\big]$.)

(b) For the function $u \equiv 1$, compute the corresponding expansion coefficients,

$$c_k[1] := \langle 1, \phi_k \rangle .$$

Under Theorem 7.9, what explicit summation condition corresponds to the convergence $S_n[1] \to 1$ in $L^2(0, \pi)$?

Chapter 8
Fourier Series

In his study of heat flow in 1807, Fourier made the radical claim that it should be possible to represent all solutions of the one-dimensional heat equation by trigonometric series. As we noted in the introduction to Chap. 7, trigonometric series had been studied earlier by other mathematicians. Fourier's innovation was to suggest that the general solution could be obtained this way.

This claim proved difficult to resolve, because the tools of functional analysis that we discussed in Chap. 7 were not yet available in Fourier's time. Indeed, the difficult problem of Fourier series convergence provided some of the strongest motivation for the development of these tools.

In this chapter we will analyze Fourier series in more detail, and show that the Fourier approach yields a general solution for the one-dimensional heat equation. The primary significance of this approach to PDE is the philosophy of *spectral analysis* that it inspired. The decomposition of functions with respect to the spectrum of a differential operator is a tool with enormous applications, both theoretical and practical.

8.1 Series Solution of the Heat Equation

Consider the heat equation

$$\frac{\partial u}{\partial t} - \Delta u = 0, \tag{8.1}$$

on a domain $\Omega \subset \mathbb{R}^n$, with Dirichlet or Neumann boundary conditions. According to Lemma 5.1 the product solutions of (8.1) have the form

$$u(t, \boldsymbol{x}) = v(t)\phi(\boldsymbol{x}),$$

D. Borthwick, *Introduction to Partial Differential Equations*,
Universitext, DOI 10.1007/978-3-319-48936-0_8

where ϕ solves the Helmholtz equation

$$-\Delta\phi = \lambda\phi$$

on Ω. The temporal equation is a simple ODE,

$$\frac{\partial v}{\partial t} = -\lambda v,$$

with the family of solutions

$$v(t) = v(0)e^{-\lambda t}. \tag{8.2}$$

Let us assume that the equation for ϕ admits a sequence of solutions ϕ_k, with eigenvalues λ_k. We have seen specific examples of this in Chap. 5, including the trigonometric case in Theorem 5.2. By (8.2), the corresponding product solutions of the heat equation are

$$u_k(t, \boldsymbol{x}) := e^{-\lambda_k t}\phi_k(\boldsymbol{x}).$$

Fourier's strategy calls for us to express the general solution as a series,

$$u(t, x) = \sum_{n=1}^{\infty} a_n e^{-\lambda_n t}\phi_n(\boldsymbol{x}), \tag{8.3}$$

for some choice of coefficients a_n. To fix the coefficients a_n in (8.3) we assume an initial condition $u(0, \boldsymbol{x}) = h(\boldsymbol{x})$. Setting $t = 0$ gives

$$h(\boldsymbol{x}) = \sum_{n=1}^{\infty} a_n\phi_n(\boldsymbol{x}). \tag{8.4}$$

If we can show that $\{\phi_n\}$ forms an orthonormal basis of $L^2(\Omega)$, then this gives us a way to assign coefficients to h such that (8.4) holds, at least in the sense of L^2 convergence.

Even if the orthonormal basis property is established, some big issues still remain. The fact that each term u_k satisfies the heat equation does not guarantee that u does, because of the infinite series summation. Similarly, the limit of (8.3) as $t \to 0$ is not necessarily (8.4), because the limit cannot necessarily be taken inside the summation. In this chapter we will explain how to resolve these problems in the context of trigonometric series.

Example 8.1 Consider the case of a one-dimensional metal rod with insulated ends. For convenience take the length to be π, so that $x \in [0, \pi]$ and the Neumann boundary conditions are

$$\frac{\partial u}{\partial x}(t, 0) = \frac{\partial u}{\partial x}(t, \pi) = 0.$$

Let us consider the initial condition

$$h(x) = 3\pi x^2 - 2x^3, \tag{8.5}$$

as pictured on the left in Fig. 8.1.

The boundary conditions are satisfied by cosines,

$$\phi_n(x) = \cos(nx), \quad n \in \mathbb{N}_0.$$

Hence the strategy outlined above calls for us to represent the initial condition as a series

$$h(x) = \sum_{n=0}^{\infty} a_n \cos(nx). \tag{8.6}$$

To choose the coefficients, we recall the discussion of basis expansion from Sect. 7.5. The cosines satisfy an orthogonality condition with respect to the L^2 inner product on $[0, \pi]$,

$$\int_0^{\pi} \cos(mx)\cos(nx)\,dx = \begin{cases} 0 & m \neq n, \\ \pi & m = n = 0, \\ \pi/2 & m = n \geq 1, \end{cases}. \tag{8.7}$$

This could be checked with trigonometric identities, but it is perhaps easier is to use the complex form $\cos(kx) = \frac{1}{2}(e^{ikx} + e^{-ikx})$.

Since the sequence $\{\phi_n\}$ is not normalized, the coefficient formula (7.18) must be interpreted as

$$a_n = \frac{1}{\|\phi_n\|^2} \langle h, \phi_n \rangle .$$

By (8.7) the Fourier coefficients are thus given by

$$\begin{aligned} a_0 &= \frac{1}{\pi}\int_0^{\pi} h(x)\,dx, \\ a_n &= \frac{2}{\pi}\int_0^{\pi} h(x)\cos(nx)\,dx, \quad n \geq 1. \end{aligned} \tag{8.8}$$

After substituting (8.5) into (8.8), integration by parts yields

$$a_n = \begin{cases} \frac{\pi^3}{2}, & n = 0, \\ -\frac{48}{\pi n^4}, & n \geq 1, \text{ odd}, \\ 0, & n \geq 2, \text{ even}. \end{cases} \tag{8.9}$$

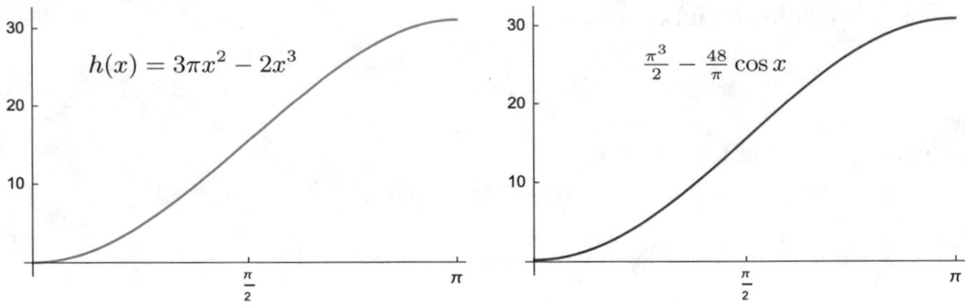

Fig. 8.1 Comparison of the initial condition $h(x)$ with the first two terms of its cosine series

Figure 8.1 shows a comparison between h and the partial sum $S_1[h]$. The close match between these functions is clearly evident. And since the higher coefficients decay by a factor of n^{-4}, convergence of this series seems quite plausible. The resulting solution would be given by

$$u(t,x) = \frac{\pi^3}{2} - \sum_{n \in \mathbb{N}_{\text{odd}}} \frac{48}{\pi n^4} e^{-n^2 t} \cos(nx).$$

Note that the convergence rate improves dramatically as t increases. ◊

8.2 Periodic Fourier Series

We saw examples of Fourier series based on sines in Theorem 5.2 and cosines in Example 8.1. To account for both cases, it is convenient to consider periodic functions on $\mathbb{R}$. We define

$$\mathbb{T} := \mathbb{R}/(2\pi\mathbb{Z}), \tag{8.10}$$

where the quotient notation means that points separated by an integer multiple of 2π are considered equivalent. The space $C^m(\mathbb{T})$ consists of the functions in $C^m(\mathbb{R})$ which are 2π-periodic.

Integrals of functions on $\mathbb{T}$ are defined by restricting the range of integration to an arbitrary interval of length 2π in $\mathbb{R}$. We will write the inner product on $L^2(\mathbb{T})$ as

$$\langle f, g \rangle = \int_{-\pi}^{\pi} f \overline{g} \, dx,$$

but the range of integration could be shifted if needed.

The Helmholtz equation on $\mathbb{T}$ is

$$-\frac{\partial^2 \phi}{\partial x^2} = \lambda \phi,$$

for $\phi \in C^2(\mathbb{T})$, with no need for additional boundary conditions because of the periodicity. The eigenfunctions are the complex exponentials

$$\phi_k(x) := e^{ikx}, \tag{8.11}$$

for $k \in \mathbb{Z}$, with $\lambda_k = k^2$.

It is possible to recover cosine and sine Fourier series from the periodic case, by restricting our attention to even or odd functions on $\mathbb{T}$. We will demonstrate this specialization in the examples and exercises.

The complex exponentials satisfy a simple orthogonality relation,

$$\begin{aligned}\langle \phi_k, \phi_l \rangle &= \int_{-\pi}^{\pi} e^{i(k-l)x}\, dx \\ &= \begin{cases} 2\pi, & k = l, \\ 0, & k \neq l. \end{cases}\end{aligned}$$

We did not include a normalizing factor in (8.11), so $\|\phi_k\|^2 = 2\pi$ and the Fourier coefficients of an integrable function $f \in L^1(\mathbb{T})$ are defined by

$$\begin{aligned} c_k[f] &:= \frac{1}{2\pi} \langle f, \phi_k \rangle \\ &= \frac{1}{2\pi} \int_{-\pi}^{\pi} f(x) e^{-ikx}\, dx. \end{aligned} \tag{8.12}$$

Because the index set is $\mathbb{Z}$ rather than $\mathbb{N}$, we define the partial sums of the periodic Fourier series by truncating on both sides,

$$S_n[f](x) := \sum_{k=-n}^{n} c_k[f] e^{ikx}. \tag{8.13}$$

For the sequence $\{\phi_k\}$, Bessel's inequality (Theorem 7.9) takes the form

$$\sum_{k \in \mathbb{Z}} \left| c_k[f] \right|^2 \leq \frac{1}{2\pi} \|f\|_2^2, \tag{8.14}$$

with equality if and only if $S_n[f] \to f$ in $L^2(\mathbb{T})$.

In the specific example considered in Example 8.1, the Fourier series appeared to converge very quickly. To illustrate potential complications with the convergence, let us consider a function with a jump discontinuity.

Example 8.2 On the interval $[0, \pi]$, define the function

$$h(x) = \begin{cases} 0, & x \in [0, \frac{\pi}{2}], \\ 1, & x \in (\frac{\pi}{2}, \pi], \end{cases}$$

as pictured on the left in Fig. 8.2. As noted above, in order to represent h as a cosine series using the periodic eigenfunctions, we first extend h to $\mathbb{T}$ as an even function, i.e.,

$$h(x) := \begin{cases} 0, & x \in [-\frac{\pi}{2}, \frac{\pi}{2}] + 2\pi\mathbb{Z}, \\ 1, & x \in (\frac{\pi}{2}, \frac{3\pi}{2}] + 2\pi\mathbb{Z}. \end{cases}$$

By (8.12), with a shift to the more convenient interval $[0, 2\pi]$, the Fourier coefficients of h are

$$\begin{aligned} c_k[h] &= \frac{1}{2\pi} \int_{\frac{\pi}{2}}^{\frac{3\pi}{2}} e^{-ikx}\, dx \\ &= \begin{cases} \frac{1}{2}, & k = 0, \\ \frac{(-1)^k}{\pi k} \sin\left(\frac{\pi k}{2}\right), & k \neq 0. \end{cases} \end{aligned}$$

Since $c_{-k}[h] = c_k[h]$, we can combine terms in the partial sums (8.13) to give

$$S_n[h](x) = \frac{1}{2} + 2 \sum_{k=1}^{n} \frac{(-1)^k}{\pi k} \sin\left(\frac{\pi k}{2}\right) \cos(kx).$$

Figure 8.2 shows a sample of these partial sums. In contrast to the case of Example 8.1, where 2 terms of the Fourier series were enough to give a very convincing approximation, we can see significant issues with convergence in the vicinity of the jump, even with 40 terms. ◊

The Fourier series computed in Example 8.2 makes for a good illustration of some different notions of convergence. Consider the sequence of differences $h - S_n[h]$, as

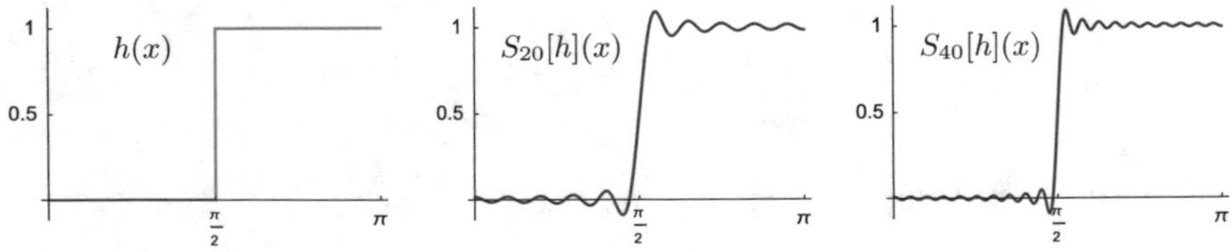

Fig. 8.2 Fourier series expansion for a step function

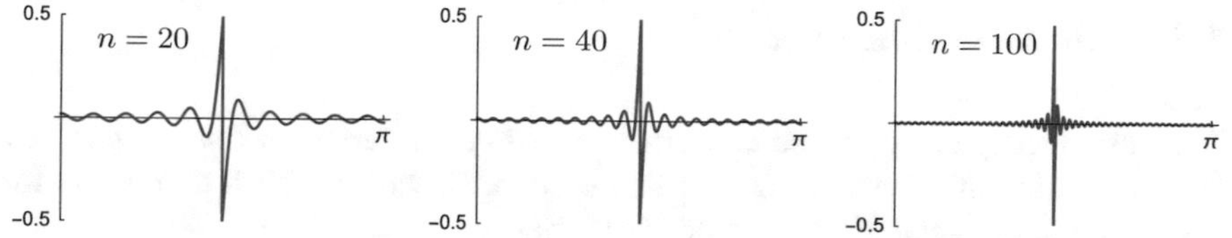

Fig. 8.3 Plots of the differences $h - S_n[h]$

illustrated in Fig. 8.3. If $\{\phi_k\}$ forms an $L^2(\mathbb{T})$ basis, then we would have

$$\lim_{n\to\infty} \int_{-\pi}^{\pi} \left|h - S_n[h]\right|^2 dx \stackrel{?}{=} 0.$$

It is not easy to judge such a limit visually, but this claim is true, as we will prove in Sect. 8.6.

We could instead focus our attention the values of $S_n[h](x)$ for some fixed x. A sequence of functions f_n is said to converge *pointwise* to f (assuming these functions have a common domain) if for each fixed x in the domain,

$$\lim_{n\to\infty} f_n(x) = f(x).$$

In Fig. 8.3, if we focus our attention on some point x away from the center, then the bumps at this point do seem to be decreasing in size as n gets larger. We will verify in Sect. 8.3 that this Fourier series converges pointwise except at $x = \frac{\pi}{2}$.

Another feature that is quite apparent in Fig. 8.3 is the spike near the center. It is possible to prove that such a spike persists, with height essentially constant, for all values of n. The historical term for this effect, which is caused by the jump discontinuity, is the *Gibbs phenomenon*. It was actually first observed in 1848 by Henry Wilbraham, but remained generally unknown until it was rediscovered independently by J. Willard Gibbs in 1899.

The Gibbs phenomenon relates to yet a third definition of convergence. A sequence of bounded functions f_n is said to converge *uniformly* to a function f on a set W if

$$\lim_{n\to\infty} \sup_{x\in W} \left|f_n(x) - f(x)\right| = 0. \tag{8.15}$$

In the cases plotted in Fig. 8.3 we can see that

$$\sup_{x\in\mathbb{T}} \left|h(x) - S_n[h](x)\right| \approx \frac{1}{2}.$$

Since this does not decrease, uniform convergence fails for this series. However, the sequence does converge uniformly on domains that exclude a neighborhood of the jump point, for example on the interval $[0, \frac{\pi}{2} - \varepsilon]$ for $\varepsilon > 0$.

8.3 Pointwise Convergence

The basic theory of pointwise convergence of Fourier series was worked out by Dirichlet in the mid-19th century. In this section we will establish a criterion for pointwise convergence of periodic Fourier series.

Theorem 8.3 *Suppose $f \in L^2(\mathbb{T})$, and that for $x \in \mathbb{T}$ the estimate,*

$$\operatorname*{ess-sup}_{y\in[-\varepsilon,\varepsilon]} \left| \frac{f(x) - f(x-y)}{y} \right| < \infty, \tag{8.16}$$

holds for some $\varepsilon > 0$. Then

$$\lim_{n\to\infty} S_n[f](x) = f(x).$$

The essential supremum was defined in (7.10). The inequality (8.16) means that, after possibly replacing f by an equivalent function in the sense of (7.6), we can assume that

$$\sup_{0<|y|\le\varepsilon} \left| \frac{f(x) - f(x-y)}{y} \right| < \infty. \tag{8.17}$$

This bound holds automatically for $f \in C^1(\mathbb{T})$, by the estimate

$$\begin{aligned} \left| \frac{f(x) - f(x-y)}{y} \right| &= \left| \frac{1}{y} \int_{x-y}^{x} f'(t)\, dt \right| \\ &\le \sup_{t\in\mathbb{T}} \left| f'(t) \right|. \end{aligned}$$

Thus Theorem 8.3 shows that the Fourier series a C^1 function converges pointwise on all of $\mathbb{T}$. The same argument can be extended to functions on $\mathbb{T}$ which are merely piecewise C^1.

It is possible to prove pointwise convergence with a weaker hypothesis than that of Theorem 8.3. However, there are counterexamples, discovered by Fejér and Lebesgue, that show that pointwise convergence of the Fourier series may fail for $f \in C^0(\mathbb{T})$.

Before getting into the proof of Theorem 8.3, let us consider the structure of the partial sums in more detail. Plugging the coefficient formula (8.14) into (8.13) gives

$$\begin{aligned} S_n[f](x) &= \sum_{k=-n}^{n} e^{ikx} \frac{1}{2\pi} \int_0^{2\pi} f(y) e^{-iky}\, dy \\ &= \int_0^{2\pi} f(y) \left(\frac{1}{2\pi} \sum_{k=-n}^{n} e^{ik(x-y)} \right) dy. \end{aligned} \tag{8.18}$$

The function that appears in parentheses is called the *Dirichlet kernel*,

$$D_n(t) := \frac{1}{2\pi}\sum_{k=-n}^{n} e^{ikt}. \tag{8.19}$$

With this definition the formula for the partial sum becomes

$$S_n[f](x) = \int_0^{2\pi} f(y) D_n(x-y)\, dy. \tag{8.20}$$

This could be written as a convolution,

$$S_n[f] = f * D_n.$$

Because the sum (8.19) is finite, it is clear that the Dirichlet kernel is a smooth function on $\mathbb{T}$. It is also easy to compute that

$$\int_0^{2\pi} D_n(t)\, dt = 1 \tag{8.21}$$

for $n \in \mathbb{N}$, since only the $k = 0$ term in (8.19) contributes to the integral. Applying the polynomial identity

$$1 + z + z^2 + \cdots + z^m = \frac{z^{m+1} - 1}{z - 1}.$$

to (8.19) with $z = e^{it}$ gives the explicit formula

$$D_n(t) = \frac{1}{2\pi}\frac{e^{i(n+1)t} - e^{-int}}{e^{it} - 1}. \tag{8.22}$$

Factoring $e^{it/2}$ out of the numerator and denominator reduces this to

$$D_n(t) = \frac{1}{2\pi}\frac{\sin((n+\frac{1}{2})t)}{\sin(\frac{1}{2}t)},$$

which makes it clear that D_n is real-valued.

A plot of $D_n(y)$ for various values of n, as shown in Fig. 8.4, gives some intuition as to why we might expect (8.23) to converge to $f(x)$ as $n \to \infty$. The function $D_n(y)$ concentrates at $y = 0$, and oscillates with increasing frequency away from this point. These oscillations will cause cancellation as $n \to \infty$, except at $y = 0$.

Proof of Theorem 8.3 Because both f and D_n are periodic, a change of variables $y \to x - y$ allows us to rewrite the convolution in the opposite order:

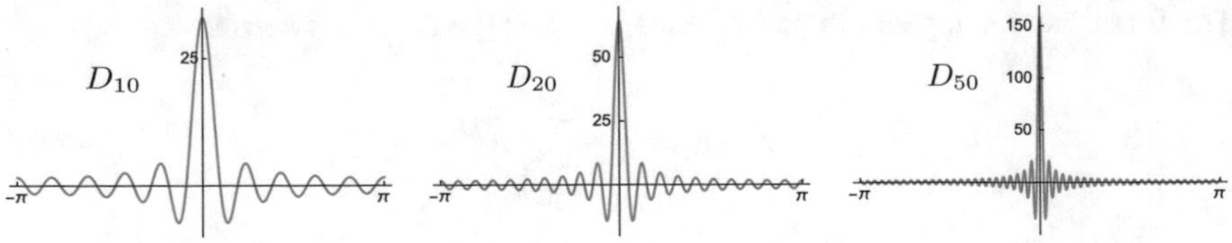

Fig. 8.4 The Dirichlet kernel for increasing values of n

$$S_n[f](x) = \int_{-\pi}^{\pi} D_n(y) f(x-y)\, dy. \tag{8.23}$$

Thus, by (8.21) and (8.23),

$$f(x) - S_n[f](x) = \int_{-\pi}^{\pi} [f(x) - f(x-y)] D_n(y)\, dy.$$

Substituting in with the explicit formula (8.22) for $D_n(t)$ gives

$$f(x) - S_n[f](x) = \frac{1}{2\pi} \int_{-\pi}^{\pi} \frac{f(x) - f(x-y)}{e^{iy} - 1} \left[e^{i(n+1)y} - e^{-iny}\right] dy. \tag{8.24}$$

The crucial observation here is that if we separate the terms inside the brackets, then this looks like a formula for Fourier coefficients.

Assuming that the hypothesis of the theorem is satisfied at $x \in \mathbb{T}$, consider the function

$$h(y) := \frac{f(x) - f(x-y)}{e^{iy} - 1}, \tag{8.25}$$

defined for $y \in \mathbb{T}$ with $y \neq 0$. We can split this into factors as

$$h(y) = \frac{f(x) - f(x-y)}{y} \frac{y}{e^{iy} - 1},$$

note that the first factor is essentially bounded near $y = 0$ by the assumption (8.16). Since $e^{iy} - 1 \sim iy$ as $y \to 0$ by Taylor's approximation, the second factor is also bounded near $y = 0$. The hypothesis (8.16) thus guarantees that $h(y)$ is equivalent to a function that is bounded on the interval $[-\varepsilon, \varepsilon]$. Since $f \in L^2(\mathbb{T})$ and $(e^{iy} - 1)^{-1}$ is bounded for $y \in \pm[\varepsilon, \pi]$, we conclude from this that $h \in L^2(\mathbb{T})$.

We can thus interpret (8.24) in terms of Fourier coefficients,

$$f(x) - S_n[f](x) = c_{-n-1}[h] - c_n[h]. \tag{8.26}$$

Bessel's inequality, which takes the form (8.14) here, implies that $c_k[h] \to 0$ as $k \to \pm\infty$. By (8.26) this establishes pointwise convergence at x. □

8.4 Uniform Convergence

According to the definition (8.15) of uniform convergence, $f_n \to f$ uniformly on $\mathbb{T}$ if

$$\sup_{x\in\mathbb{T}} \left|f_n(x) - f(x)\right| \to 0$$

as $n \to \infty$. This is closely related to the convergence with respect the L^∞ norm, as introduced in Sect. 7.3. If a sequence converges in the L^∞ sense, then after possibly modifying the functions on a set of measure zero we can assume that the convergence is uniform.

Continuity is not necessarily preserved under pointwise limits. For example the sequence e^{-nx^2} converges pointwise on $\mathbb{R}$ but the limit function is discontinuous at $x = 0$. On the other hand, uniform convergence of continuous functions does guarantee continuity.

Lemma 8.4 *Suppose* $\{f_n\} \subset C^0(\Omega)$ *for a domain* $\Omega \subset \mathbb{R}^n$. *If* $\{f_n\}$ *converges uniformly to a function* $f : \Omega \to \mathbb{R}$, *then* f *is also continuous.*

Proof The goal is to show that $f(\boldsymbol{y})$ can be made close to $f(\boldsymbol{x})$ by taking $\boldsymbol{y}$ close to $\boldsymbol{x}$. To make use of the uniform convergence, we note that the triangle inequality implies

$$|f(\boldsymbol{x}) - f(\boldsymbol{y})| \le \left|f(\boldsymbol{x}) - f_n(\boldsymbol{x})\right| + \left|f_n(\boldsymbol{x}) - f_n(\boldsymbol{y})\right| + \left|f_n(\boldsymbol{y}) - f(\boldsymbol{y})\right|. \quad (8.27)$$

For n large we can control the first and third terms on the right by the assumption of uniform convergence. To control the middle term we can use the continuity of f_n.

Fix $\boldsymbol{x} \in \Omega$ and $\varepsilon > 0$. By uniform convergence there exists n so that

$$\sup_{y\in\Omega} \left|f_n(\boldsymbol{y}) - f(\boldsymbol{y})\right| < \varepsilon. \quad (8.28)$$

The fact that f_n is continuous at $\boldsymbol{x}$ means that we can find $\delta > 0$ (depending on $\boldsymbol{x}$) such that for $\boldsymbol{y} \in \Omega$ satisfying $|\boldsymbol{x} - \boldsymbol{y}| < \delta$,

$$\left|f_n(\boldsymbol{x}) - f_n(\boldsymbol{y})\right| < \varepsilon. \quad (8.29)$$

Combining (8.28) and (8.29) with (8.27) shows that for $\boldsymbol{y} \in \Omega$ satisfying $|\boldsymbol{x} - \boldsymbol{y}| < \delta$,

$$|f(\boldsymbol{x}) - f(\boldsymbol{y})| < 3\varepsilon.$$

Thus f is continuous at $\boldsymbol{x}$. □

Uniform convergence is particularly easy to check for periodic Fourier series, because the eigenfunctions ϕ_k satisfy

$$\left|e^{ikx}\right| = 1.$$

Theorem 8.5 *For $f \in C^1(\mathbb{T})$, the sequence of partial sums $S_n[f]$ convergences uniformly to f.*

Proof The assumption that $f \in C^1(\mathbb{T})$ implies $f' \in C^0(\mathbb{T})$. By integration by parts,

$$\begin{aligned} c_k[f'] &= \frac{1}{2\pi}\int_{-\pi}^{\pi} f'(y)e^{-iky}\,dy \\ &= \frac{1}{2\pi} f(y)e^{-iky}\Big|_{-\pi}^{\pi} + \frac{ik}{2\pi}\int_{-\pi}^{\pi} f(y)e^{-iky}\,dy. \end{aligned}$$

The boundary term cancels by periodicity, leaving

$$c_k[f'] = ikc_k[f]. \tag{8.30}$$

Since $f' \in L^2(\mathbb{T})$ also, applying Bessel's inequality in the form (8.14) to the coefficients (8.30) implies that

$$\sum_{k\in\mathbb{Z}} \left|kc_k[f]\right|^2 < \infty. \tag{8.31}$$

Let $\ell^2(\mathbb{Z}\backslash\{0\})$ denote the discrete L^2 space on the set consisting of functions $\mathbb{Z}\backslash\{0\} \to \mathbb{C}$. (The ℓ^p spaces were introduced in Sect. 7.4.) The sequence

$$a_k := |kc_k[f]|$$

defines an element of $\ell^2(\mathbb{Z}\backslash\{0\})$ by (8.31). If we define $b \in \ell^2(\mathbb{Z}\backslash\{0\})$ by $b_k := k^{-1}$, then the sum of the coefficients $c_k[f]$ with $k \neq 0$ can be expressed as an ℓ^2 pairing,

$$\sum_{k\neq 0} |c_k[f]| = \langle a, b\rangle_{\ell^2}.$$

By the Cauchy-Schwarz inequality on $\ell^2(\mathbb{Z}\backslash\{0\})$,

$$\langle a, b\rangle_{\ell^2} \leq \|a\|_{\ell^2}\,\|b\|_{\ell^2} < \infty.$$

Since the norms of a and b are finite, we conclude that

$$\sum_{k\in\mathbb{Z}} |c_k[f]| < \infty. \tag{8.32}$$

Note that we already know that $S_n[f] \to f$ pointwise by Theorem 8.10. This implies that

$$S_n[f](x) - f(x) = \sum_{|k|>n} c_k[f]\phi_k(x)$$

for each $x \in \mathbb{T}$. Because $|\phi_k(x)| = 1$,

$$\left|S_n[f](x) - f(x)\right| \leq \sum_{|k|>n} \left|c_k[f]\right|. \tag{8.33}$$

The right-hand side of (8.33) is independent of x and tends to zero as $n \to \infty$ by (8.32), proving that $S_n[f] \to f$ uniformly. □

8.5 Convergence in L^2

The uniform convergence provided by Theorem 8.5 proves to be very helpful in resolving the L^2 basis question. This is because uniform convergence on $\mathbb{T}$ implies L^2 convergence also, by the integral estimate

$$\begin{aligned}\|f_n - f\|_2^2 &= \int_{-\pi}^{\pi} |f_n - f|^2 \, dx \\ &\leq 2\pi \sup_{x\in\mathbb{T}} \left|f_n(x) - f(x)\right|^2.\end{aligned}$$

Hence Theorem 8.5 gives convergence of Fourier series in L^2 for C^1 functions. In this section we will extend this result to all of L^2.

Theorem 8.6 *The normalized periodic Fourier eigenfunctions*

$$\frac{1}{\sqrt{2\pi}} e^{ikx}, \quad k \in \mathbb{Z},$$

form an orthonormal basis for $L^2(\mathbb{T})$.

Proof Suppose that $u \in L^2(\mathbb{T})$ satisfies

$$\langle u, \phi_k \rangle = 0 \tag{8.34}$$

for each $k \in \mathbb{Z}$. By Theorem 7.10 the conclusion will follow if we can deduce that this implies $u = 0$.

As noted above, for $\psi \in C^1(\mathbb{T})$ Theorem 8.5 implies that $S_n[\psi] \to \psi$ in $L^2(\mathbb{T})$. In particular, this gives

$$\lim_{n\to\infty} \langle u, S_n[\psi] \rangle = \langle u, \psi \rangle \,. \tag{8.35}$$

However, since $S_n[\psi]$ is a finite linear combination of the ϕ_k, the assumption (8.34) implies that

$$\langle u, S_n[\psi] \rangle = 0.$$

Hence from (8.35) we deduce that

$$\langle u, \psi \rangle = 0.$$

Now recall Theorem 7.5, which says that $C^\infty_{\text{cpt}}(0, 2\pi)$ forms a dense subset of $L^2(0, 2\pi)$. This implies also that $C^1(\mathbb{T})$ is dense in $L^2(\mathbb{T})$. Therefore we can choose a sequence $\{\psi_l\}$ in $C^1(\mathbb{T})$ such that $\psi_l \to u$ in $L^2(\mathbb{T})$. Thus

$$\|u\|_2^2 = \lim_{l\to\infty} \langle u, \psi_l \rangle ,$$

and we just showed that all terms on the right are zero under the assumption (8.34). Therefore $u = 0$. □

The combination of Theorems 8.6 and 7.9 immediately yields the following:

Corollary 8.7 (Parseval's identity) *For $f \in L^2(\mathbb{T})$, the periodic Fourier coefficients satisfy*

$$\sum_{k\in\mathbb{Z}} |c_k[f]|^2 = \frac{1}{2\pi} \|f\|_{L^2} .$$

Applying Parseval's identity to $f + g$, where $f, g \in L^2(\mathbb{T})$, and separating out the cross-term yields the corresponding result for the inner product,

$$\langle f, g \rangle = 2\pi \sum_{k\in\mathbb{Z}} c_k[f]\overline{c_k[g]}. \tag{8.36}$$

Example 8.8 In Example 8.2, we found for the step function h that $c_k[h] = \pm\frac{1}{\pi k}$ for k odd, $c_0[h] = \frac{1}{2}$, and otherwise $c_k[h] = 0$. So for this case,

$$\sum_{k\in\mathbb{Z}} |c_k[h]|^2 = \frac{1}{4} + 2 \sum_{k\in\mathbb{N}_{\text{odd}}} \frac{1}{\pi^2 k^2}.$$

On the other hand, $\|h\|_2^2 = \pi$, so Parseval's identity implies

$$\frac{1}{4} + \frac{2}{\pi^2} \sum_{k\in\mathbb{N}_{\text{odd}}} \frac{1}{k^2} = \frac{1}{2}.$$

Thus we obtain the summation formula

$$\sum_{k\in\mathbb{N}_{\text{odd}}} \frac{1}{k^2} = \frac{\pi^2}{8}. \tag{8.37}$$

◇

The space $L^2(0, 2\pi)$ can be identified with $L^2(\mathbb{T})$ by extending functions periodically. Hence Theorem 8.6 also implies that $\left\{\frac{1}{2\pi}e^{ikx}\right\}$ is an orthonormal basis for $L^2(0, 2\pi)$. We can also specialize the periodic results to show that cosine or sine series give orthonormal bases for $L^2(0, \ell)$ with basis functions that satisfy Dirichlet or Neumann boundary conditions, respectively. We will discuss these cases in the exercises.

8.6 Regularity and Fourier Coefficients

In the preceding sections we have made some progress in understanding the representation of a function by Fourier series. However, we still have not addressed one of the primary questions raised in Sect. 8.1: when does a Fourier series yield a classical solution to the original PDE? In this section we will resolve this issue by studying the relationship between the regularity of a function and the decay of its Fourier coefficients.

The starting point for this discussion is the computation used in the proof of Theorem 8.5,

$$c_k[f'] = ikc_k[f]$$

for $f \in C^1(\mathbb{T})$. Repeating this computation inductively gives the following:

Lemma 8.9 *Suppose that $f \in C^m(\mathbb{T})$. Then*

$$c_k[f^{(m)}] = (ik)^m c_k[f]. \tag{8.38}$$

To describe the decay rates of coefficients, we introduce some convenient *order notation*. For $\alpha \in \mathbb{R}$,

$$a_k = o(k^\alpha) \quad \text{means} \quad \lim_{|k|\to\infty} \frac{|a_k|}{|k|^\alpha} = 0.$$

This is commonly referred to as the "little-o" notion of order. There is a corresponding "big-O" definition,

$$a_k = O(k^\alpha) \quad \text{means} \quad |a_k| \le C\,|k|^\alpha\,,$$

for all sufficiently large $|k|$, with C independent of k. Note that the little-o condition is stronger. The content of the statement $a_k = o(k^\alpha)$ is that the ratio a_k/k^α tends to zero, while $a_k = O(k^\alpha)$ says only that the ratio is bounded.

Theorem 8.10 *For $f \in C^m(\mathbb{T})$ with $m \in \mathbb{N}_0$,*

$$\sum_{k\in\mathbb{Z}} k^{2m} \left|c_k[f]\right|^2 < \infty, \tag{8.39}$$

and

$$c_k[f] = o(k^{-m}).$$

Proof The inequality (8.39) follows immediately from a combination of Lemma 8.9 and and Bessel's inequality in the form (8.14). Since the terms in a convergent series must approach zero,

$$\lim_{|k|\to\infty} k^m c_k[f] = 0,$$

which gives the claimed decay estimate. □

Example 8.11 Consider the cosine series (8.6) computed in Example 8.1 for the function $h(x) = 3\pi x^2 - 2x^3$ on $(0, \pi)$. Although $h \in C^\infty(0, \pi)$, the extension of h to $\mathbb{T}$ as an even function is merely C^2. Theorem 8.10 thus implies that $\sum k^4 |c_k[h]|^2 < \infty$.

The periodic Fourier coefficients corresponding to (8.9) are

$$c_k[h] = \begin{cases} \frac{\pi^3}{2}, & k = 0, \\ -\frac{24}{\pi k^4}, & k \text{ odd}, \\ 0, & k \neq 0, \text{ even}. \end{cases}$$

This shows much faster decay than predicted, but not the rapid decay we would have seen if the even periodic extension had been smooth. ◊

Our next goal is to develop a converse to Theorem 8.10 that says that a certain level of decay rate of Fourier coefficients guarantees a corresponding level of differentiability for the function. In fact, the first stage of this result has already been worked out. Suppose $f \in L^2(\mathbb{T})$ and its coefficients satisfy

$$\sum_{k\in\mathbb{Z}} |c_k[f]| < \infty. \tag{8.40}$$

We know that $S_n[f] \to f$ in the L^2 sense by Theorem 8.6. By Theorem 8.5 we also know that $\{S_n[f]\}$ converges uniformly, and so the limit is continuous by Lemma 8.4. Hence we can conclude that (8.40) implies $f \in C^0(\mathbb{T})$. Recall from Sect. 7.3 that when we say an L^2 function is C^m we mean this only up to equivalence, i.e., the original function might require modification on a set of measure zero to make it C^m.

Theorem 8.12 *Suppose $f \in L^2(\mathbb{T})$ has Fourier coefficients satisfying*

$$\sum_{k\in\mathbb{Z}} \left|k^m c_k[f]\right| < \infty, \tag{8.41}$$

for $m \in \mathbb{N}_0$. Then $f \in C^m(\mathbb{T})$.

Proof As remarked above, the $m = 0$ case is already taken care of by Theorems 8.5 and 8.6.

Assume that (8.41) is satisfied for $m = 1$. For convenience, let $c_k := c_k[f]$ and $f_n = S_n[f]$. Since f_n is a (finite) linear combination of smooth functions, the derivatives are given by

$$f_n'(x) = \sum_{k=-n}^{n} ikc_k e^{ikx}.$$

By the $m = 0$ result, the sequence $\{f_n'\}$ converges uniformly to some $g \in C^0(\mathbb{T})$.

Our goal is to show that $g = f'$, which means

$$g(x) = \lim_{y \to 0} \frac{f(x+y) - f(x)}{t},$$

for every $x \in \mathbb{T}$. To argue this we will decompose the difference quotient as

$$\begin{aligned} \frac{f(x+t) - f(x)}{t} - g(x) &= \left[\frac{f_n(x+y) - f_n(x)}{y} - f_n'(x) \right] \\ &\quad + \left(f_n'(x) - g(x)\right) + R_n(x, y). \end{aligned} \tag{8.42}$$

The first term on the right approaches zero by the definition of f_n', and the second term approaches zero as $n \to \infty$ by the construction of g. The remainder term is

$$R_n(x, y) := \sum_{|k|>n} c_k \frac{(e^{ik(x+y)} - e^{ikx})}{y},$$

which converges absolutely for each $y \neq 0$ by (8.41).

We can estimate the remainder by

$$|R_n(x, y)| \le \sum_{|k|>n} |c_k| \left| \frac{e^{iky} - 1}{y} \right|.$$

By noting that $\left|e^{iky} - 1\right| = 2\sin(ky/2)$, a simple calculus estimate gives

$$\left| \frac{e^{iky} - 1}{y} \right| \le |k|$$

for all $y \neq 0$. This implies a uniform estimate on the remainder term,

$$|R_n(x, y)| \le \sum_{|k|>n} |kc_k|. \tag{8.43}$$

In particular, by the assumption (8.41) the remainder term is arbitrarily small for n large.

Fix $x \in \mathbb{T}$ and $\varepsilon > 0$. By (8.43) and the fact that $f_n'(x) \to g(x)$, we can pick n so that

$$\left|f_n'(x) - g(x)\right| < \varepsilon \quad \text{and} \quad |R_n(x, y)| < \varepsilon.$$

for all $y \neq 0$. For this n and x, the definition of $f_n'(x)$ says that we can choose δ such that $0 < |y| < \delta$ implies

$$\left|\frac{f_n(x+y) - f_n(x)}{y} - f_n'(x)\right| < \varepsilon.$$

Applying these estimates to (8.42) shows that for $0 < |y| < \delta$,

$$\left|\frac{f(x+y) - f(x)}{y} - g(x)\right| \le 3\varepsilon.$$

Since ε was arbitrarily small, this shows that $f'(x) = g(x)$. And since g is continuous, we conclude that $f \in C^1(\mathbb{T})$.

The same argument can now be repeated for higher derivatives, assuming (8.41) holds for larger m. □

The hypothesis (8.41) can be reformulated in terms of a decay condition on the coefficients, although this gives a slightly weaker result. If $c_k[f] = O(k^{-\alpha})$, then

$$\sum_{k\in\mathbb{Z}} \left|k^m c_k[f]\right| \le C \sum_{k\in\mathbb{Z}} |k|^{-\alpha+m},$$

and this series converges provided $\alpha > m + 1$. Hence Theorem 8.12 implies that $f \in C^m(\mathbb{T})$ under the condition that

$$c_k[f] = O(k^{-m-1-\epsilon})$$

for some $\varepsilon > 0$.

Let us finally return to the one-dimensional heat equation that motivated this discussion, first considering the periodic case.

Theorem 8.13 *For $h \in C^0(\mathbb{T})$, the heat equation on $[0, \infty) \times \mathbb{T}$,*

$$\frac{\partial u}{\partial t} - \frac{\partial^2 u}{\partial x^2} = 0,$$

admits a solution $u \in C^\infty((0, \infty) \times \mathbb{T})$, defined for $t > 0$ by

$$u(t, x) := \sum_{k\in\mathbb{Z}} c_k[h] e^{-k^2 t} e^{ikx}, \tag{8.44}$$

and satisfying

$$\lim_{t\to 0} u(t, x) = h(x) \tag{8.45}$$

for each $x \in \mathbb{T}$.

Proof For $t > 0$, the Fourier coefficients of $u(t, \cdot)$ decay exponentially, and Theorem 8.12 shows that $u(t, \cdot) \in C^\infty(\mathbb{T})$ for each t. The same arguments used in the proof of that theorem apply to the t derivatives. To see this, let u_n denote the partial sum of (8.44),

$$u_n(t, x) := \sum_{k=-n}^{n} c_k e^{-k^2 t} e^{ikx}.$$

where $c_k := c_k[h]$. As a finite sum, this can be differentiated directly,

$$\frac{\partial u_n}{\partial t}(t, x) = \sum_{k=-n}^{n} (-k^2) c_k e^{-k^2 t} e^{ikx}. \tag{8.46}$$

By Theorem 8.10 the Fourier coefficients of h satisfy $c_k = o(k^{-1})$, so that

$$\left| -k^2 c_k e^{-k^2 t} e^{ikx} \right| \le C k e^{-k^2 t}. \tag{8.47}$$

As $n \to \infty$ the series (8.46) thus converges absolutely for $t > 0$, allowing us to define a function

$$g := \lim_{n\to\infty} \frac{\partial u_n}{\partial t}.$$

For $\varepsilon > 0$, the estimate (8.47) shows that the convergence is uniform for $t \ge \varepsilon$. Lemma 8.4 shows that the limit is continuous for $t \ge \varepsilon$. Since $\varepsilon > 0$ is arbitrary, this implies $g \in C^0((0, \infty) \times \mathbb{T})$.

We can argue that $g = \partial u/\partial t$ by considering

$$\begin{aligned} \frac{u(t+s, x) - u(t, x)}{s} - h(t, x) &= \left[\frac{u_n(t+s, x) - u_n(t, x)}{s} - \frac{\partial u_n}{\partial t}(t, x) \right] \\ &\quad + \left(\frac{\partial u_n}{\partial t}(t, x) - h(t, x) \right) + R_n(t, s, x), \end{aligned}$$

where

$$R_n(t, s, x) := \sum_{|k|>n} c_k e^{-k^2 t} e^{ikx} \left(\frac{e^{-k^2 s} - 1}{s} \right).$$

At this point the argument becomes essentially parallel to the analysis of (8.42), so we will omit the details. The conclusion is that $\partial u/\partial t$ is continuous on $(0, \infty) \times \mathbb{T}$.

The argument can be repeated for higher t derivatives, allowing us to conclude that $u \in C^\infty((0, \infty) \times \mathbb{T})$. Moreover, the partial derivatives of u_n converge to the corresponding derivatives of u, pointwise on $(0, \infty) \times \mathbb{T}$, and uniformly if we restrict to $t \geq \varepsilon$ for some $\varepsilon > 0$. Since u_n satisfies the wave equation for each n by construction, this shows that u satisfies the wave equation also.

At the moment we only have the tools to prove (8.45) under the stronger assumption that $h \in C^1(\mathbb{T})$. In this case we can argue exactly in the proof of Theorem 8.5 that (8.44) converges uniformly for $(t, x) \in [0, \infty) \times \mathbb{T}$. By Lemma 8.4 this shows $u \in C^0([0, \infty) \times \mathbb{T})$ and we can just set $t = 0$ to obtain (8.45).

If h is merely continuous, then this approach breaks down, because the series (8.44) may actually diverge for $t = 0$. We will cover the C^0 case in Chap. 13, after developing an alternate formula for (8.44). □

The one-dimensional heat equation derived in Sect. 6.1 involved Dirichlet or Neumann boundary conditions on an interval $[0, \ell]$. By rescaling the interval to $[0, \pi]$ and then extending functions to $\mathbb{T}$ with either even or odd symmetry, we apply the results for periodic Fourier series results to these cases. In particular, from Theorem 8.13 we deduce the following:

Corollary 8.14 *Suppose $h \in C[0, \ell]$ and satisfies Dirichlet or Neumann boundary conditions. The heat equation on $[0, \infty) \times [0, \ell]$ admits a solution $u \in C^\infty((0, \infty) \times [0, \ell])$, under the same boundary condition, such that*

$$\lim_{t \to 0} u(t, x) = h(x)$$

for each $x \in [0, \ell]$.

The solutions obtained in Theorem 8.13 and Corollary 8.14 are uniquely determined by the initial condition h. (See Exercise 6.4.) The fact that solutions are smooth for $t > 0$, even when h is merely continuous, is a characteristic property of diffusion equations. In fact the smoothing phenomenon carries over to cases where h is L^2 but not even continuous. This is illustrated in Fig. 8.5 for the case considered in Example 8.2.

As these applications show, the Fourier series approach (and spectral analysis in general) is well suited to analyzing the regularity of solutions. However, other qualitative features are perhaps obscured from this viewpoint. For example, we would expect solutions to reflect the physical principle that heat flows from hot to cold. We can see this behavior quite clearly in the plots of Fig. 8.5, but it is not at all apparent in the series formula (8.44).

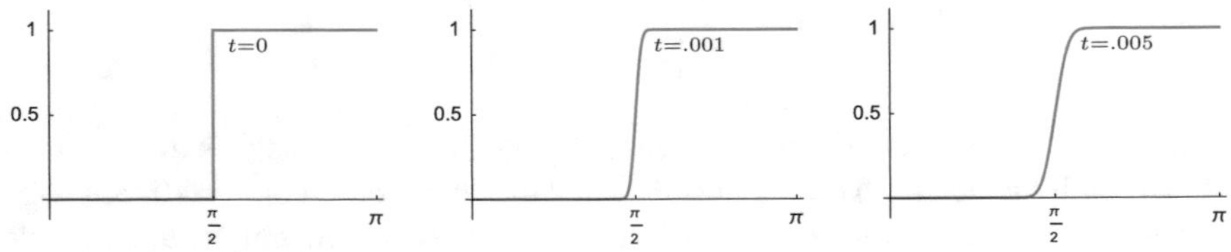

Fig. 8.5 Solutions of the heat equation become smooth for $t > 0$

8.7 Exercises

8.1 For $x \in (0, \pi)$, let

$$f(x) = x.$$

(a) Extend f to an odd function on $\mathbb{T}$ and compute the periodic Fourier coefficients $c_k[f]$ according to (8.12). (Note that the case $k = 0$ needs to be treated separately.) Show that the periodic series reduces to a sine series in this case.

(b) Show that the convergence of the Fourier series at $x = \frac{\pi}{2}$, which is guaranteed by Theorem 8.3, yields the summation formula

$$\frac{\pi}{4} = 1 - \frac{1}{3} + \frac{1}{5} - \frac{1}{7} + \cdots.$$

(c) Show the Parseval identity (Corollary 8.7) leads to the formula

$$\sum_{k=1}^{\infty} \frac{1}{k^2} = \frac{\pi^2}{6}.$$

8.2 For $x \in (0, \pi)$, let

$$g(x) = x.$$

(a) Extend g to an even function on $\mathbb{T}$ and compute the periodic Fourier coefficients $c_k[g]$ according to (8.12). (Note that the case $k = 0$ needs to be treated separately.) Show that the periodic series reduces to a cosine series in this case.

(b) Show that the convergence of the Fourier series at $x = 0$, which is guaranteed by Theorem 8.3, reproduces the formula (8.37).

(c) Show the Parseval identity (Corollary 8.7) implies the formula

$$\sum_{k \in \mathbb{N}_{\text{odd}}} \frac{1}{k^4} = \frac{\pi^4}{96}.$$

8.3 Consider the periodic wave equation

$$\frac{\partial^2 u}{\partial t^2} - \frac{\partial^2 u}{\partial x^2} = 0$$

for $t \in \mathbb{R}$ and $x \in \mathbb{T}$. Suppose the initial conditions are

$$u(0, x) = g(x), \qquad \frac{\partial u}{\partial t}(0, x) = h(x),$$

for $g \in C^{m+1}(\mathbb{T})$ and $h \in C^m(\mathbb{T})$, for $m \in \mathbb{N}$.

(a) Assuming that $u(t,x)$ can be represented as a Fourier series

$$u(t,x) = \sum_{k\in\mathbb{Z}} a_k(t)e^{ikx}, \tag{8.48}$$

find an expression for $a_k(t)$ in terms of the Fourier coefficients of g and h.

(b) Using the assumptions on g and h, together with Theorem 8.10, show that the coefficients $a_k(t)$ satisfy an estimate

$$\sum_{k\in\mathbb{Z}} k^{2m}\,|a_k(t)|^2 \le M < \infty,$$

uniformly for $t \in \mathbb{R}$.

(c) By the arguments used in Theorem 8.13, (b) implies that the series (8.48) converges to a solution u satisfying the initial conditions. What could you conclude about the differentiability of u?

8.4 In $L^2(0,\pi)$ consider the sequence

$$\psi_k(x) := \sqrt{\frac{2}{\pi}}\sin kx,$$

for $k \in \mathbb{N}$.

(a) Show that $\{\psi_k\}$ is an orthonormal sequence.
(b) Suppose that $f \in L^2(0,\pi)$ and $\langle f, \psi_k\rangle = 0$ for all $k \in \mathbb{N}$. Show that $f \equiv 0$. (Hint: extend f to an odd 2π-periodic function on $\mathbb{R}$, which can be regarded as an element of $L^2(\mathbb{T})$. Then apply Theorem 8.6.)
(c) Conclude that $\{\psi_k\}$ is an orthonormal basis for $L^2(0,\pi)$.

8.5 Suppose that $f \in L^2(-\pi,\pi)$ satisfies

$$\int_{-\pi}^{\pi} x^l f(x)\,dx = 0$$

for all $l \in \mathbb{N}_0$.

(a) Show that $\langle q_{m,k}, f\rangle = 0$ for all $m \in \mathbb{N}$ and $k \in \mathbb{Z}$, where

$$q_{m,k}(x) := \sum_{l=0}^{m} \frac{(-ikx)^l}{l!}.$$

(b) Note that

$$\lim_{m\to\infty} q_{m,k}(x) = e^{-ikx}$$

by the definition of the complex exponential. Show that this convergence is uniform for $x \in [-\pi,\pi]$ (with k fixed).

(c) Use (a) and (b) to show that for $f \in L^2(-\pi, \pi)$,

$$\int_{-\pi}^{\pi} e^{-ikx} f(x)\, dx = 0$$

for all $k \in \mathbb{Z}$.

(d) Conclude from Theorem 8.6 that $f \equiv 0$. (In other words, the monomials $1, x, x^2, \ldots$ form a basis for $L^2(-\pi, \pi)$, although not an orthonormal one.)

8.6 The *Legendre polynomials* are functions of $z \in [-1, 1]$ defined by

$$P_k(z) := \frac{1}{2^k k!} \frac{d^k}{dz^k}(z^2 - 1)^k,$$

for $k \in \mathbb{N}_0$. (This corresponds to the case $m = 0$ in (5.31).) These are solutions of the eigenvalue equation

$$LP_k = k(k+1)P_k,$$

where

$$L := \frac{d}{dz}\left[(z^2 - 1)\frac{d}{dz}\right].$$

(a) For $u, v \in C^2[-1, 1]$, check that L satisfies a formal self-adjointness condition,

$$\langle u, Lv \rangle_{L^2} = \langle Lu, v \rangle_{L^2}.$$

Conclude that the P_k's with distinct values of k are orthogonal in $L^2(-1, 1)$.

(b) Use the result of Exercise 8.5 to show that $\{P_k\}$ forms an orthogonal basis for $L^2(-1, 1)$. (The P_k are normalized by the condition $P_k(1) = 0$, rather than by unit L^2 norm.)

Chapter 9
Maximum Principles

We saw in Sect. 4.7 that conservation of energy can be used to derive uniqueness for solutions of the wave equation. In this chapter we will consider another approach to issues of uniqueness and stability, based on maximum values. This method applies generally to elliptic equations, which describe equilibrium states, and to parabolic equations, which are generally used to model diffusion.

9.1 Model Problem: The Laplace Equation

As noted in Sect. 5.2, the classical evolution equations such as the heat or wave equation have the form

$$P_t u - \Delta u = 0,$$

where P_t denotes some combination of time derivatives. In an equilibrium state, for which the solution is independent of time, these equations all reduce to

$$\Delta u = 0,$$

which is called the *Laplace equation*. A solution of the Laplace equation is also called a *harmonic function*. The Laplace equation on a bounded domain Ω is generally formulated with an inhomogeneous Dirichlet boundary condition,

$$u|_{\partial\Omega} = f$$

for $f : \partial\Omega \to \mathbb{R}$.

D. Borthwick, *Introduction to Partial Differential Equations*,
Universitext, DOI 10.1007/978-3-319-48936-0_9

The Laplace equation frequently appears in applications involving vector fields. A conservative vector field $\boldsymbol{v} \in C^0(\Omega; \mathbb{R}^n)$ can be represented as the gradient of a potential function $\phi \in C^1(\Omega; \mathbb{R})$,

$$\boldsymbol{v} = \nabla\phi,$$

If the vector field $\boldsymbol{v}$ is also solenoidal ($\nabla \cdot \boldsymbol{v} = 0$), then the potential satisfies the Laplace equation

$$\Delta\phi = 0.$$

In fluid dynamics in $\mathbb{R}^3$, for example, the velocity field is solenoidal for an incompressible fluid, such as water, and conservative precisely when the flow is *irrotational* ($\nabla \times \boldsymbol{v} = 0$).

Electrostatics provides another important source of Laplace problems. In the absence of charges, the electric field $\boldsymbol{E}$ is conservative and is commonly written as

$$\boldsymbol{E} = -\nabla\phi$$

where ϕ is the electric potential. On the other hand, Gauss's law of electrostatics says that $\nabla \cdot \boldsymbol{E}$ is proportional to the electric charge density. Hence, the electric potential for a charge-free region satisfies the Laplace equation.

In the remainder of this section we will consider a particular classical case, the Laplace problem on the unit disk. Circular symmetry allows us to solve the equation explicitly using Fourier series, and the resulting formula gives some insight into the general behavior of harmonic functions.

Let $\mathbb{D}$ denote the open unit disk in $\mathbb{R}^2$. Given $g \in C^0(\partial\mathbb{D})$, our goal is to solve

$$\Delta u = 0, \quad u|_{\partial\mathbb{D}} = g. \tag{9.1}$$

In Sect. 5.3 we used separation of variables in polar coordinates to find the family of harmonic functions,

$$\phi_k(r, \theta) := r^{|k|} e^{ik\theta},$$

for $k \in \mathbb{Z}$. The boundary $\partial\mathbb{D}$ is naturally identified with the space $\mathbb{T} := \mathbb{R}/2\pi\mathbb{Z}$ introduced in Sect. 8.2, parametrized by θ.

Consider the periodic Fourier series expansion,

$$g(\theta) = \sum_{k\in\mathbb{Z}} c_k[g] e^{ik\theta}, \tag{9.2}$$

where

$$c_k[g] := \frac{1}{2\pi} \int_0^{2\pi} e^{-ik\theta} g(\theta)\, d\theta.$$

Given that

$$e^{ik\theta} = \phi_k(1, \theta),$$

we might hope to construct a solution of (9.1) by setting

$$u(r, \theta) = \sum_{k \in \mathbb{Z}} c_k[g]\phi_k(r, \theta). \tag{9.3}$$

Theorem 8.10 shows that the sequence $\{c_k[g]\}$ is bounded for g continuous. Note also that

$$|\phi_k(r, \theta)| = r^{|k|}$$

and

$$\sum_{k \in \mathbb{Z}} r^{|k|} < \infty$$

for $r < 1$ by geometric series. This implies that (9.3) converges absolutely for $r < 1$. In fact the convergence is uniform on $\{r \leq R\}$ for $R < 1$.

We can write $u(r, \theta)$ more explicitly by substituting the definition of $c_k[g]$ into the integral,

$$u(r, \theta) = \sum_{k \in \mathbb{Z}} \frac{1}{2\pi} \int_0^{2\pi} r^{|k|} e^{ik(\theta - \eta)} g(\eta)\, d\eta.$$

Uniform convergence in θ for $r < 1$ allows us to move the sum inside the integral, yielding the formula

$$u(r, \theta) = \frac{1}{2\pi} \int_0^{2\pi} P_r(\theta - \eta) g(\eta)\, d\eta, \tag{9.4}$$

where

$$P_r(\theta) := \sum_{k \in \mathbb{Z}} r^{|k|} e^{ik\theta}. \tag{9.5}$$

This function is called the *Poisson kernel*. Its behavior as $r \to 1$ is illustrated in Fig. 9.1.

Summing by geometric series gives the formula

$$\begin{aligned} P_r(\theta) &= 1 + \sum_{k=1}^{\infty} (re^{i\theta})^k + \sum_{k=1}^{\infty} (re^{-i\theta})^k \\ &= 1 + \frac{re^{i\theta}}{1 - re^{i\theta}} + \frac{re^{-i\theta}}{1 - re^{-i\theta}} \\ &= \frac{1 - r^2}{1 - 2r\cos\theta + r^2}. \end{aligned} \tag{9.6}$$

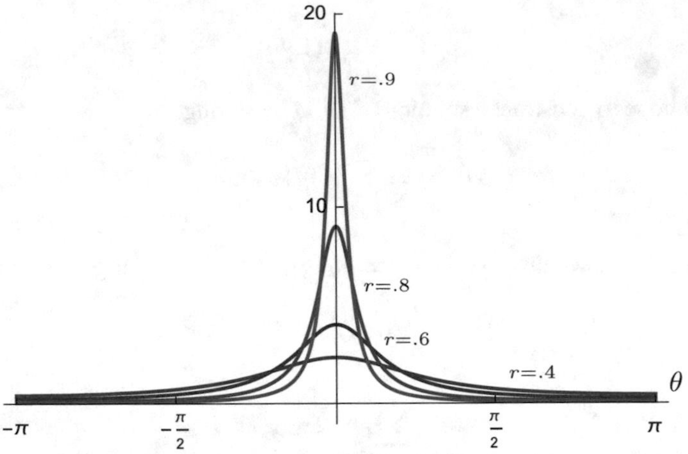

Fig. 9.1 The Poisson kernel $P_r(\theta)$ for a succession of radii

From the series formula (9.5) we can also deduce directly that

$$\frac{1}{2\pi}\int_0^{2\pi} P_r(\theta)\, d\theta = 1, \tag{9.7}$$

since the only nonzero contribution comes from the term $k = 0$.

By periodicity, a change of variables $\eta \to \theta - \eta$ in (9.4) gives the alternate form

$$u(r,\theta) = \frac{1}{2\pi}\int_0^{2\pi} P_r(\eta) g(\theta - \eta)\, d\eta. \tag{9.8}$$

In view of (9.7), this could be interpreted as a weighted average of f with a weight function that depends on r. As $r \to 1^-$ this weight function becomes concentrated at 0, as Fig. 9.1 demonstrates. This is the mechanism by which we expect to have $u(r,\theta) \to g(\theta)$ as $r \to 1^-$.

Theorem 9.1 *For $f \in C^0(\partial\mathbb{D})$, the Laplace equation,*

$$\Delta u = 0 \text{ in } \mathbb{D}, \qquad u|_{\partial\mathbb{D}} = g,$$

admits a classical solution $u \in C^\infty(\mathbb{D}) \cap C^0(\bar{\mathbb{D}})$ given by the Poisson integral (9.4).

Proof The function $P_r(\theta)$ is smooth for $r < 1$, and it follows from (9.5) that

$$\Delta P_r(\theta) = 0,$$

where $P_r(\theta)$ is interpreted as a function on $\mathbb{D}$ written in polar coordinates. By passing derivatives inside the integral, we can deduce from (9.4) that $u \in C^\infty(\mathbb{D})$ and

$$\Delta u = 0.$$

To complete the proof we need to check that

$$\lim_{r \to 1^-} u(r, \theta) = g(\theta) \tag{9.9}$$

for every $\theta \in \partial\mathbb{D}$, which will also show that $u \in C^0(\bar{\mathbb{D}})$. Note that (9.9) is not the same as claiming that the Fourier series for g converges, which is not necessarily true. The difference lies in the order of the limits. In (9.9) we take the limit of the Fourier series first for $r < 1$, and then the limit $r \to 1^-$. This limit exists, as we will see below, but if we first set $r = 1$ in (9.3) then the sum over k may diverge.

By (9.7) and (9.8) we can write

$$u(r, \theta) - g(\theta) = \frac{1}{2\pi} \int_{-\pi}^{\pi} P_r(\eta) \Big[g(\theta - \eta) - g(\theta) \Big] d\eta. \tag{9.10}$$

The goal is to estimate the left-hand side for r close to 1. Fix $\theta \in \mathbb{D}$ and let $\varepsilon > 0$. Since g is continuous, there exists $\delta > 0$ so that

$$|g(\theta - \eta) - g(\theta)| < \varepsilon \tag{9.11}$$

for $|\eta| < \delta$. For $|\eta| \geq \delta$ we can estimate

$$\max_{\delta \leq |\eta| \leq \pi} P_r(\eta) = P_r(\delta). \tag{9.12}$$

Thus, splitting the integral (9.10) at $|\eta| = \delta$ gives

$$\begin{aligned} |u(r, \theta) - g(\theta)| &\leq \frac{1}{2\pi} \int_{-\delta}^{\delta} P_r(\eta) \Big| g(\theta - \eta) - g(\theta) \Big| \, d\eta \\ &\quad + \frac{1}{2\pi} \int_{\delta < |\eta| < \pi} P_r(\eta) \Big| g(\theta - \eta) - g(\theta) \Big| \, d\eta \\ &\leq \frac{\varepsilon}{2\pi} \int_{-\delta}^{\delta} P_r(\eta) d\eta + \frac{P_r(\delta)}{2\pi} \int_{\delta < |\eta| < \pi} \Big| g(\theta - \eta) - g(\theta) \Big| \, d\eta. \end{aligned}$$

By (9.7) and the fact that $P_r > 0$,

$$\frac{1}{2\pi} \int_{-\delta}^{\delta} P_r(\eta) \, d\eta \leq 1.$$

Furthermore, since g is continuous, $|g(\theta - \eta) - g(\theta)|$ is bounded by some constant M for all θ and η. This reduces the bound to

$$|u(r, \theta) - g(\theta)| \leq \varepsilon + M P_r(\delta).$$

We can now use the fact that

$$\lim_{r \to 1^-} P_r(\delta) = 0$$

to choose $R < 1$ so that

$$M P_r(\delta) \leq \varepsilon$$

for $R < r < 1$. We conclude that

$$|u(r, \theta) - g(\theta)| \leq 2\varepsilon$$

for $R < r < 1$. Since ε was arbitrary, this shows

$$\lim_{r \to 1^-} |u(r, \theta) - g(\theta)| = 0.$$

□

For students who know some complex analysis, we note that the formula (9.4) could be deduced from the Cauchy integral formula, because any harmonic function on $\mathbb{D}$ is the real part of a holomorphic function.

Example 9.2 For $0 < a < \pi$, suppose the boundary function is given by

$$g(\theta) = \begin{cases} 1 - \frac{|\theta|}{a}, & |\theta| \leq a, \\ 0, & a < |\theta| \leq \pi, \end{cases}$$

as shown in Fig. 9.2.

This boundary condition could represent, for example, a hot spot at one point on the edge of a metal plate. The corresponding equilibrium temperature distribution within the plate is given by calculating the Fourier coefficients of g and substituting into (9.3). The resulting solution,

$$u(r, \theta) = \frac{a}{2\pi} + \frac{2}{\pi a} \sum_{k=1}^{\infty} \frac{1 - \cos(ka)}{k^2} \cos(k\theta),$$

is illustrated in Fig. 9.3. ◊

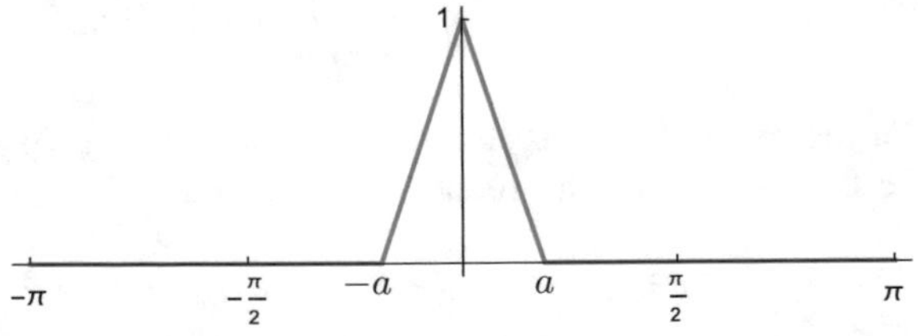

Fig. 9.2 Boundary function with a triangular peak

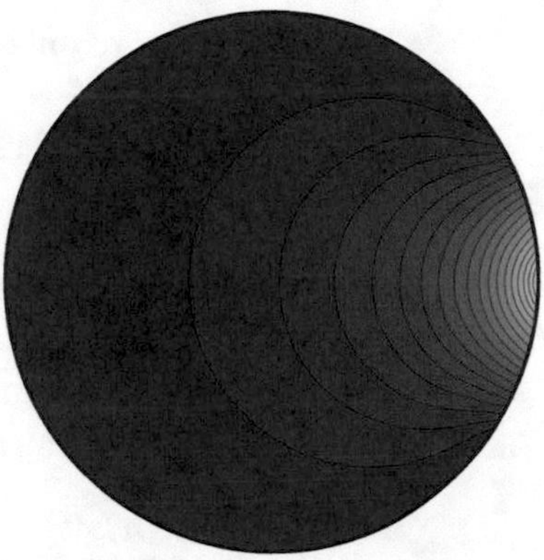

Fig. 9.3 Contour plot of the harmonic function from Example 9.2

9.2 Mean Value Formula

Setting $r = 0$ in the Poisson formula (9.4) gives

$$u(0) = \frac{1}{2\pi}\int_0^{2\pi} g(\theta)\, d\theta, \tag{9.13}$$

because $P_0(\theta) = 1$. In other words, the value of a harmonic function at the center of the disk is equal to its average value on the boundary. This phenomenon is illustrated in Fig. 9.4. In this section we will extend (9.13) to an averaging formula that works in any dimension.

The $(n-1)$-dimensional volume of a sphere of radius r is

$$\mathrm{vol}[\partial B(\boldsymbol{x}_0; r)] = A_n r^{n-1}, \tag{9.14}$$

where A_n denotes the volume of the unit sphere in $\mathbb{R}^n$, as defined in (2.13). It follows from the radial integral formula (2.10) that

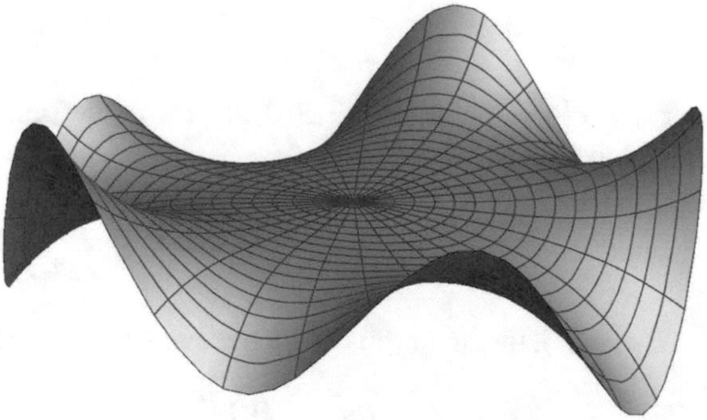

Fig. 9.4 Mean value property of a harmonic function

$$\text{vol}[B(\boldsymbol{x}_0; r)] = \frac{A_n r^n}{n}. \tag{9.15}$$

To state the mean value formula for a ball of radius R, we introduce the family of radial functions,

$$G_R(\boldsymbol{x}) := \begin{cases} \frac{1}{2\pi} \ln(\frac{r}{R}), & n = 2, \\ \frac{1}{(n-2)A_n}\left[\frac{1}{R^{n-2}} - \frac{1}{r^{n-2}}\right], & n \geq 3. \end{cases} \tag{9.16}$$

The function G_R is the unique solution of the equations

$$\frac{\partial G_R}{\partial r} = \frac{1}{A_n r^{n-1}}, \quad G_R\Big|_{r=R} = 0. \tag{9.17}$$

Note that G_R is integrable on $B(0; R)$, despite the singularity at the origin, because the radial volume element is $A_n r^{n-1}\, dr$ by (2.10).

Theorem 9.3 (Mean value formula) *Assume that $u \in C^2(\Omega)$ on a domain $\Omega \subset \mathbb{R}^n$ with $n \geq 2$. For $R > 0$ such that $\overline{B(\boldsymbol{x}_0; R)} \subset \Omega$,*

$$u(\boldsymbol{x}_0) = \frac{1}{A_n R^{n-1}} \int_{\partial B(\boldsymbol{x}_0; R)} u(\boldsymbol{x}) dS + \int_{B(\boldsymbol{x}_0; R)} G_R(\boldsymbol{x} - \boldsymbol{x}_0) \Delta u(\boldsymbol{x}) d^n\boldsymbol{x}.$$

Proof By a change of variables, it suffices to consider the case $\boldsymbol{x}_0 = 0$. The formula (2.15) for the radial component of the Laplacian implies that

$$\Delta G_R(\boldsymbol{x}) = 0$$

for $\boldsymbol{x} \neq 0$. For $\varepsilon > 0$, we can therefore apply Green's second identity (Theorem 2.11) on the domain $\{\varepsilon < r < R\}$ to obtain

$$\begin{aligned} \int_{\{\varepsilon<r<R\}} G_R \Delta u\, d^n\boldsymbol{x} = &\int_{\{r=R\}} \left(G_R \frac{\partial u}{\partial r} - u \frac{\partial G_R}{\partial r}\right) dS \\ &- \int_{\{r=\varepsilon\}} \left(G_R \frac{\partial u}{\partial r} - u \frac{\partial G_R}{\partial r}\right) dS. \end{aligned} \tag{9.18}$$

Because G_R is integrable on $B(0; R)$, on the left-hand side of (9.18) we can take $\varepsilon \to 0$ to obtain

$$\lim_{\varepsilon \to 0} \int_{\{\varepsilon<r<R\}} G_R \Delta u\, d^n\boldsymbol{x} = \int_{B(0;R)} G_R \Delta u\, d^n\boldsymbol{x}. \tag{9.19}$$

By (9.17), the first term on the right in (9.18) reduces to

$$\int_{\{r=R\}} \left(G_R \frac{\partial u}{\partial r} - u \frac{\partial G_R}{\partial r}\right) dS = -\frac{1}{A_n R^{n-1}} \int_{\{r=R\}} u\, dS. \tag{9.20}$$

The second term on the right in (9.18) is

$$\int_{\{r=\varepsilon\}} \left(G_R \frac{\partial u}{\partial r} - u \frac{\partial G_R}{\partial r} \right) dS$$
$$= G_R(\varepsilon) \int_{\{r=\varepsilon\}} \frac{\partial u}{\partial r}\, dS + \frac{1}{A_n \varepsilon^{n-1}} \int_{\{r=\varepsilon\}} u\, dS.$$

The first of these integrals can be estimated by noting that $\partial u/\partial r$ is a directional derivative and thus bounded by the magnitude of $|\nabla u|$. By the assumption that $u \in C^2(\Omega)$, $|\partial u/\partial r|$ is therefore bounded by a constant C for $r \le R$, yielding the estimate

$$\left| \int_{\{r=\varepsilon\}} \frac{\partial u}{\partial r}\, dS \right| \le C A_n \varepsilon^{n-1}.$$

Since the divergent term in $G_R(\varepsilon)$ as $\varepsilon \to 0$ is proportional to ε^{2-n} for $n \ge 3$ and $\log \varepsilon$ for $n = 2$, this implies

$$\lim_{\varepsilon \to 0} \left[G_R(\varepsilon) \int_{\{r=\varepsilon\}} \frac{\partial u}{\partial r}\, dS \right] = 0.$$

Hence

$$\lim_{\varepsilon \to 0} \int_{\{r=\varepsilon\}} \left(G_R \frac{\partial u}{\partial r} - u \frac{\partial G_R}{\partial r} \right) dS = \lim_{\varepsilon \to 0} \left[\frac{1}{A_n \varepsilon^{n-1}} \int_{\{r=\varepsilon\}} u\, dS \right].$$

The term in brackets is the average of u over a sphere of radius ε. Since u is continuous, this average approaches $u(0)$ as $\varepsilon \to 0$, so that

$$\lim_{\varepsilon \to 0} \int_{\{r=\varepsilon\}} \left(G_R \frac{\partial u}{\partial r} - u \frac{\partial G_R}{\partial r} \right) dS = u(0). \tag{9.21}$$

Applying (9.19), (9.20), and (9.21) to (9.18) gives

$$\int_{B(0;R)} G_R \Delta u\, d^n\mathbf{x} = u(0) - \frac{1}{A_n R^{n-1}} \int_{\partial B(0;R)} u\, dS,$$

which completes the proof. □

For harmonic functions, Theorem 9.3 gives a generalization of the circle formula (Theorem 9.3) to spherical averages in higher dimensions. As we will now show, the mean value property can be stated in a equivalent form in terms of averages over a ball.

Corollary 9.4 (Mean value for harmonic functions) *Suppose $\Omega \subset \mathbb{R}^n$ for $n \ge 2$. For $u \in C^2(\Omega)$ the following properties are equivalent:*

(A) The function u is harmonic on Ω.
(B) For $\overline{B(\boldsymbol{x}_0;\, R)} \subset \Omega$,

$$u(\boldsymbol{x}_0) = \frac{1}{A_n R^{n-1}} \int_{\partial B(\boldsymbol{x}_0;\, R)} u\, dS.$$

(C) For $\overline{B(\boldsymbol{x}_0;\, R)} \subset \Omega$,

$$u(\boldsymbol{x}_0) = \frac{n}{A_n R^n} \int_{B(\boldsymbol{x}_0;\, R)} u\, d^n\boldsymbol{x}.$$

Proof The fact that (A) implies (B) follows immediately by setting $\Delta u = 0$ in the formula of Theorem 9.3.

To see that (B) and (C) are equivalent, fix some $\boldsymbol{x}_0 \in \Omega$ and define

$$h(r) := \int_{B(\boldsymbol{x}_0; r)} u\, d^n\boldsymbol{x},$$

for $r \geq 0$ such that $\overline{B(\boldsymbol{x}_0;\, r)} \subset \Omega$. As we saw in Exercise 2.4, the derivative of $h(r)$ is given by a surface integral

$$h'(r) = \int_{\partial B(\boldsymbol{x}_0; r)} u\, dS.$$

Hence property (B) says that

$$h'(r) = A_n r^{n-1} u(\boldsymbol{x}_0),$$

while property (C) says that

$$h(r) = \frac{A_n r^n}{n} u(\boldsymbol{x}_0).$$

Since $h(0) = 0$ by definition, these are two statements are equivalent.

Finally, we need to show that (B) implies (A). Assuming that (B) holds, Theorem 9.3 gives

$$\int_{B(\boldsymbol{x}_0;\, R)} G_R(\boldsymbol{x} - \boldsymbol{x}_0) \Delta u(\boldsymbol{x}) d^n\boldsymbol{x} = 0 \tag{9.22}$$

provided $\overline{B(\boldsymbol{x}_0;\, R)} \subset \Omega$. Suppose $\Delta u(\boldsymbol{x}_0) < 0$ for some $\boldsymbol{x}_0 \in \Omega$. Then by continuity there exists some $\varepsilon > 0$ and $\delta > 0$ such that $\Delta u \leq -\varepsilon$ on $B(\boldsymbol{x}_0;\, \delta)$. Since G_R is strictly negative and decreasing as $r \to 0$, this implies

$$\int_{B(\boldsymbol{x}_0; \delta)} G_R(\boldsymbol{x} - \boldsymbol{x}_0) \Delta u(\boldsymbol{x}) d^n\boldsymbol{x} > -\varepsilon G_R\big|_{r=\delta} > 0,$$

which contradicts (9.22). The same argument applies if $\Delta u(\boldsymbol{x}_0) > 0$. We thus conclude that (B) implies $\Delta u \equiv 0$. □

9.3 Strong Principle for Subharmonic Functions

A real-valued C^2 function that satisfies

$$-\Delta u \le 0$$

is called *subharmonic*. The case $-\Delta u \ge 0$ is similarly called *superharmonic*. We will focus on the subharmonic case. The results can easily be translated to the superharmonic case by replacing u by $-u$.

The "weak" maximum principle says that for a subharmonic function the maximum value occurs at a boundary point. We will prove here a "strong" version of this principle, which says furthermore that if that the global maximum occurs at an interior point then the function is constant.

Theorem 9.5 (Strong maximum principle) *Let $\Omega \subset \mathbb{R}^n$ be a bounded domain. If $u \in C^2(\Omega; \mathbb{R}) \cap C^0(\overline{\Omega})$ is subharmonic then*

$$\max_{\overline{\Omega}} u = \max_{\partial\Omega} u.$$

The maximum is attained at an interior point only if u is a constant function.

Proof By the extreme value theorem (Theorem A.2), u achieves a global maximum at some point $\boldsymbol{x}_0 \in \overline{\Omega}$. If $\boldsymbol{x}_0 \in \partial\Omega$ then the claimed equality clearly holds. The goal is thus to show that $\boldsymbol{x}_0 \in \Omega$ implies that u is constant.

Because Ω is open, an interior point $\boldsymbol{x}_0$ has a neighborhood contained in Ω. We may thus assume that $\overline{B(\boldsymbol{x}_0; R)} \subset \Omega$ for some $R > 0$. Applying Theorem 9.3 to this ball gives

$$u(\boldsymbol{x}_0) = \frac{1}{A_n R^{n-1}} \int_{\partial B(\boldsymbol{x}_0; R)} u(\boldsymbol{x}) dS + \int_{B(\boldsymbol{x}_0; R)} G_R(\boldsymbol{x} - \boldsymbol{x}_0) \Delta u(\boldsymbol{x}) d^n\boldsymbol{x}.$$

By the definition (9.16), $G_R \le 0$ for $0 < r \le R$. Therefore, since $\Delta u \ge 0$ by assumption,

$$u(\boldsymbol{x}_0) \le \frac{1}{A_n R^{n-1}} \int_{\partial B(\boldsymbol{x}_0; R)} u(\boldsymbol{x}) dS. \tag{9.23}$$

Using (9.14), we can subtract $u(\boldsymbol{x}_0)$ from both sides to obtain

$$\frac{1}{A_n R^{n-1}} \int_{\partial B(\boldsymbol{x}_0; R)} [u(\boldsymbol{x}) - u(\boldsymbol{x}_0)]\, dS \ge 0. \tag{9.24}$$

By assumption $u(\boldsymbol{x}_0)$ is the global maximum of u, implying that the integrand of (9.24) is nonpositive. The inequality therefore shows that the integrand vanishes, and we conclude that $u(\boldsymbol{x}) = u(\boldsymbol{x}_0)$ on $\partial B(\boldsymbol{x}_0; R)$.

Note that the same argument works for every radius $r < R$, so this argument shows that $u \equiv u(\boldsymbol{x}_0)$ on all of $B(\boldsymbol{x}_0; R)$.

To extend the conclusion to the full domain, let M denote the maximum value of u on $\overline{\Omega}$. We can write Ω as a disjoint union $E \cup F$, where

$$\begin{aligned} E &:= \{\boldsymbol{x} \in \Omega;\ u(\boldsymbol{x}) < M\}, \\ F &:= \{\boldsymbol{x} \in \Omega;\ u(\boldsymbol{x}) = M\}. \end{aligned}$$

By the argument given above, a point $\boldsymbol{x} \in F$ has a neighborhood $B(\boldsymbol{x}; R) \subset \Omega$ on which u is equal to M. Hence F is open.

On the other hand, for $\boldsymbol{x} \in E$ we can set $\varepsilon = M - u(\boldsymbol{x})$ and use the continuity of u to find a $\delta > 0$ such that

$$|u(\boldsymbol{x}) - u(\boldsymbol{y})| < \varepsilon$$

for $\boldsymbol{y} \in B(\boldsymbol{x}; \delta)$. This implies in particular that $u(\boldsymbol{y}) < M$, so that $B(\boldsymbol{x}; \delta) \in E$. Thus E is open also.

Recall from Sect. 2.3 that the fact that Ω is connected means that the domain cannot be written as a disjoint union of nonempty open sets. Since $\Omega = E \cup F$ with E and F both open, one of the two sets is empty. If E is empty then u is constant on Ω, while if F is empty then the maximum of u is not attained in the interior. □

For a superharmonic function $u \in C^2(\Omega; \mathbb{R}) \cap C^0(\overline{\Omega})$, reversing the sign yields a minimum principle,

$$\min_{\overline{\Omega}} u = \min_{\partial\Omega} u.$$

Both principles apply to a harmonic function u, which therefore satisfies

$$\min_{\partial\Omega} u \le u(\boldsymbol{x}) \le \max_{\partial\Omega} u$$

for all $\boldsymbol{x} \in \Omega$.

The maximum principle implies the following stability result for the Laplace equation.

Corollary 9.6 *Suppose that $u_1, u_2 \in C^2(\Omega) \cap C^0(\overline{\Omega})$ are solutions of the Laplace equation $\Delta u = 0$ with boundary values*

$$u_1|_{\partial\Omega} = g_1, \qquad u_2|_{\partial\Omega} = g_2,$$

for $g_1, g_2 \in C^0(\partial\Omega)$. Then

$$\max_{\overline{\Omega}} |u_2 - u_1| \le \max_{\partial\Omega} |g_2 - g_1|. \tag{9.25}$$

In particular, a solution the Laplace equation is uniquely determined by its boundary data.

Proof By superposition, $u_2 - u_1$ is a harmonic function with boundary data $g_2 - g_1$. Theorem 9.5 applies to $\pm \operatorname{Re}(u_2 - u_1)$ as well as $\pm \operatorname{Im}(u_2 - u_1)$. Combining these estimates yields the inequality (9.25). □

Note that uniqueness of solutions of the Laplace equation also follows directly from Green's first identity (Theorem 2.10), in the case where Ω has piecewise C^1 boundary and $u \in C^2(\Omega; \mathbb{R})$. If $\Delta u = 0$, then setting $v = u$ in Green's formula gives

$$\int_{\Omega} \|\nabla u\|^2 \, d^n x = \int_{\partial\Omega} u \frac{\partial u}{\partial \nu} \, dS.$$

Thus if $u = 0$ on $\partial\Omega$, then

$$\int_{\Omega} \|\nabla u\|^2 \, d^n x = 0.$$

Since the integrand is positive, this implies $\nabla u \equiv 0$, so that u is constant. The assumption $u|_{\partial\Omega} = 0$ then gives $u \equiv 0$ on the full domain. (This is the "energy method" argument, as introduced in Sect. 4.7.)

One advantage that maximum principle has over the energy method is the explicit stability formula (9.25). In terms of the L^∞ norm introduced in Sect. 7.3, this inequality could be written

$$\|u_2 - u_1\|_\infty \le \|g_2 - g_1\|_\infty \, .$$

This is an explicit formulation of the continuity requirement for well-posedness: a small change in boundary data results in a correspondingly small change in the solution.

9.4 Weak Principle for Elliptic Equations

Although the mean value formula gives a direct proof of the strong maximum principle, this approach applies only to the Laplacian itself. In this section we will present an alternative approach that generalizes quite easily to operators with variable coefficients.

On a domain $\Omega \subset \mathbb{R}^n$ let us consider a second order elliptic operator of the form

$$L = -\sum_{i,j=1}^{n} a_{ij}(\boldsymbol{x}) \frac{\partial^2}{\partial x_i \partial x_j} + \sum_{j=1}^{n} b_j(\boldsymbol{x}) \frac{\partial}{\partial x_j}, \tag{9.26}$$

where the coefficients a_{ij} and b_j are continuous functions on Ω. As defined in Sect. 1.3, ellipticity means that the symmetric matrix $[a_{ij}]$ is positive definite at each point.

For the maximum principle we need a stronger assumption, called *uniform ellipticity*, that says that for some fixed constant $\kappa > 0$,

$$\sum_{i,j=1}^{n} a_{ij}(\boldsymbol{x})v_i v_j \geq \kappa \|\boldsymbol{v}\|^2 \tag{9.27}$$

for all $\boldsymbol{x} \in \Omega$ and $\boldsymbol{v} \in \mathbb{R}^n$. An equivalent way to say this is that the smallest eigenvalue of $[a_{ij}]$ is bounded below by κ at each point $\boldsymbol{x}$.

Theorem 9.7 (Weak maximum principle) *Suppose $\Omega \subset \mathbb{R}^n$ is bounded, and L is an operator of the form* (9.26) *satisfying the uniform ellipticity condition* (9.27). *If $u \in C^2(\Omega; \mathbb{R}) \cap C^0(\overline{\Omega})$ satisfies*

$$Lu \leq 0$$

in Ω, then

$$\max_{\overline{\Omega}} u = \max_{\partial\Omega} u.$$

Proof For the moment let u be a general function in $C^2(\Omega; \mathbb{R})$. Suppose that u has a local maximum at $\boldsymbol{x}_0 \in \Omega$. The first partial derivatives of u vanish at a local maximum, so that

$$Lu(\boldsymbol{x}_0) = -\sum_{i,j=1}^{n} a_{ij}(\boldsymbol{x}_0)\frac{\partial^2 u}{\partial x_i \partial x_j}(\boldsymbol{x}_0). \tag{9.28}$$

Furthermore, we claim that the matrix of second partials of u is negative definite at $\boldsymbol{x}_0$, meaning

$$\sum_{i,j=1}^{n} \frac{\partial^2 u}{\partial x_i \partial x_j}(\boldsymbol{x}_0)v_i v_j \leq 0$$

for $\boldsymbol{v} \in \mathbb{R}^n$. To see this, set $h(t) := u(\boldsymbol{x}_0 + t\boldsymbol{v})$ and note that h has a local maximum at $t = 0$, implying $h''(0) \leq 0$. Evaluating $h''(0)$ yields the inequality stated above.

The right-hand side of (9.28) could be written as $\operatorname{tr}(AB)$ where A and B are the positive symmetric matrices

$$A = \left[a_{ij}(\boldsymbol{x}_0)\right], \quad B = \left[-\frac{\partial^2 u}{\partial x_i \partial x_j}(\boldsymbol{x}_0)\right].$$

By switching to a basis in which A is diagonal, $\operatorname{tr}(AB)$ can be written in terms of the eigenvalues $\{\lambda_j\}$ of A as

$$\operatorname{tr}(AB) = \sum_{j=1}^{n} \lambda_j b_{jj}.$$

If the eigenvalues are ordered $\lambda_1 \leq \cdots \leq \lambda_n$, then

$$\mathrm{tr}(AB) \geq \lambda_1 \,\mathrm{tr}\, B.$$

The positivity of B implies $\mathrm{tr}\, B \geq 0$, so we conclude that

$$Lu(\boldsymbol{x}_0) \geq 0.$$

Thus argument shows that $Lu(\boldsymbol{x}_0) \geq 0$ for $\boldsymbol{x}_0$ a local interior maximum. Therefore, the strict inequality $Lu < 0$ implies that u cannot have a local interior maximum and that

$$\max_{\overline{\Omega}} u = \max_{\partial\Omega} u. \tag{9.29}$$

To complete the proof, we must relax the hypothesis to $Lu \leq 0$. The strategy is to perturb u slightly to reduce to the previous case. For $M > 0$, let

$$h(\boldsymbol{x}) := e^{Mx_1}.$$

By the definition of L,

$$Lh = \left[-a_{11}M^2 + b_1 M\right] h.$$

The ellipticity condition (9.27) implies that $a_{11} \geq \kappa$, so by choosing

$$M > \frac{1}{\kappa} \max_{\overline{\Omega}} b_1,$$

we can guarantee that

$$Lh < 0.$$

If we now assume now that u satisfies the hypothesis $Lu \leq 0$, then

$$L(u + \varepsilon h) < 0$$

for $\varepsilon > 0$. Applying (9.29) to $u + \varepsilon h$ gives

$$\max_{\overline{\Omega}}(u + \varepsilon h) = \max_{\partial\Omega}(u + \varepsilon h). \tag{9.30}$$

Since $h \geq 0$, clearly

$$\max_{\overline{\Omega}} u \leq \max_{\overline{\Omega}}(u + \varepsilon h).$$

On the other hand, since Ω is bounded, we may assume $x_1 < R$ in $\overline{\Omega}$, for some R sufficiently large. This implies

$$h \leq e^{MR},$$

so that

$$\max_{\partial\Omega}(u + \varepsilon h) \leq \max_{\partial\Omega} u + \varepsilon e^{MR}.$$

From (9.30) we therefore conclude that

$$\max_{\overline{\Omega}} u \leq \max_{\partial\Omega} u + \varepsilon e^{MR}$$

for all $\varepsilon > 0$. Since M and R are independent of ε, we can take $\varepsilon \to 0$ to conclude that

$$\max_{\overline{\Omega}} u \leq \max_{\partial\Omega} u,$$

and the result follows. □

The maximum principle implies in particular the only solution of $Lu = 0$ with $u|_{\partial\Omega} = 0$ is $u \equiv 0$. Hence a solution of the equation

$$Lu = f, \quad u|_{\partial\Omega} = g,$$

is uniquely determined by f and g if it exists.

9.5 Application to the Heat Equation

Fourier's law of heat conduction, as introduced in Sect. 6.1, suggests a maximum principle for solutions of the heat equation. Because heat flows away from a spatial maximum of the temperature, a local spatial maximum of the temperature should be impossible at time $t > 0$. The global maximum of the temperature therefore must occur either at $t = 0$ or on the boundary.

Although it is possible to prove the maximum principle via a mean value formula as in the proof of Theorem 9.5, in this section we will follow the more direct approach from Sect. 9.4, which has the advantage of generalizing to operators with variable coefficients.

Because the heat equation is second order with respect to spatial variables and first order in the time variable, it makes sense to define a domain for classical solutions that takes this structure into account. For $\Omega \subset \mathbb{R}^n$, define

$$C^{\text{heat}}(\Omega) := \left\{u \in C^0\left([0,\infty) \times \overline{\Omega}; \mathbb{R}\right);\ u(\cdot, \boldsymbol{x}) \in C^1(0,\infty), u(t,\cdot) \in C^2(\Omega)\right\}.$$

Note that this definition includes only real-valued functions.

Theorem 9.8 *Suppose $\Omega \subset \mathbb{R}^n$ is a bounded domain and $u \in C^{\text{heat}}(\Omega)$ satisfies*

$$\frac{\partial u}{\partial t} - \Delta u \leq 0, \tag{9.31}$$

on $(0, T) \times \Omega$. *Then the maximum value of* u *within* $[0, T] \times \overline{\Omega}$ *occurs at a point* $(t_0, \boldsymbol{x}_0)$ *with either* $t_0 = 0$ *or* $\boldsymbol{x}_0 \in \partial\Omega$.

Proof Suppose that u attains a maximum at $(t_0, \boldsymbol{x}_0) \subset (0, T) \times \Omega$. By the same calculus argument used in Sect. 9.4, this implies

$$\frac{\partial u}{\partial t}(t_0, \boldsymbol{x}_0) = 0, \tag{9.32}$$

as well as

$$\frac{\partial u}{\partial x_j}(t_0, \boldsymbol{x}_0) = 0, \quad \frac{\partial^2 u}{\partial x_j^2}(t_0, \boldsymbol{x}_0) \leq 0. \tag{9.33}$$

In particular,

$$\left[\frac{\partial u}{\partial t} - \Delta u\right](t_0, \boldsymbol{x}_0) \geq 0. \tag{9.34}$$

If (9.31) were a strict inequality this would complete the proof.

To proceed we use a perturbation strategy as in the proof of Theorem 9.7. For $\varepsilon > 0$, set

$$u_\varepsilon := u + \varepsilon\,|\boldsymbol{x}|^2.$$

Because $\Delta\,|\boldsymbol{x}|^2 = 2n$, the hypothesis on u gives

$$\frac{\partial u_\varepsilon}{\partial t} - \Delta u_\varepsilon = \frac{\partial u}{\partial t} - \Delta u - 2n\varepsilon < 0. \tag{9.35}$$

The existence of a local maximum for u_ε within $(0, T) \times \Omega$ is ruled out by (9.34). We conclude that u_ε attains a global maximum at a boundary point of $[0, T] \times \overline{\Omega}$.

Let us label this point $(t_\varepsilon, \boldsymbol{x}_\varepsilon)$, so that

$$\max_{[0,T]\times\overline{\Omega}} u_\varepsilon = u_\varepsilon(t_\varepsilon, \boldsymbol{x}_\varepsilon). \tag{9.36}$$

Since $(t_\varepsilon, \boldsymbol{x}_\varepsilon)$ is on the boundary, either $t_\varepsilon = 0$, $t_\varepsilon = T$, or $\boldsymbol{x}_\varepsilon \in \partial\Omega$.

Suppose that $t_\varepsilon = T$ and $\boldsymbol{x}_\varepsilon \in \Omega$. Then $u_\varepsilon(t, x_\varepsilon) \leq u_\varepsilon(T, x_\varepsilon)$ for $t \in [0, T]$, implying that

$$\frac{\partial u_\varepsilon}{\partial t}(T, \boldsymbol{x}_\varepsilon) \geq 0.$$

By (9.35), this implies also that $\Delta u_\varepsilon(T, \boldsymbol{x}_\varepsilon) > 0$, which is ruled out by (9.33). Hence $t_\varepsilon \neq T$ if $\boldsymbol{x}_\varepsilon \in \Omega$.

Therefore $(t_\varepsilon, \boldsymbol{x}_\varepsilon)$ lies in the set

$$\Gamma := (\{0\} \times \Omega) \cap ([0, T] \times \partial\Omega)\,. \tag{9.37}$$

Let R be sufficiently large so that $\Omega \subset B(0; R)$. This means that $|\boldsymbol{x}| \le R$ on $\overline{\Omega}$, so the inequality

$$u \le u_\varepsilon \le u + \varepsilon R^2$$

holds at every point in $[0, T] \times \overline{\Omega}$. From (9.36) we can thus conclude that

$$\begin{aligned} \max_{[0,T]\times\overline{\Omega}} u &\le u_\varepsilon(t_\varepsilon, \boldsymbol{x}_\varepsilon) \\ &\le u(t_\varepsilon, \boldsymbol{x}_\varepsilon) + \varepsilon R^2. \end{aligned} \tag{9.38}$$

This implies that

$$\max_{[0,T]\times\overline{\Omega}} u \le \max_{\Gamma} u + \varepsilon R^2,$$

because $(t_\varepsilon, \boldsymbol{x}_\varepsilon) \in \Gamma$. Since this inequality holds for every $\varepsilon > 0$, this proves

$$\max_{[0,T]\times\overline{\Omega}} u \le \max_{\Gamma} u.$$

□

For a solution of the heat equation, both $\pm u$ satisfy the hypothesis of Theorem 9.8, which implies that

$$\min_{\Gamma} u \le u(t, \boldsymbol{x}) \le \max_{\Gamma} u,$$

for $(t, \boldsymbol{x}) \in (0, T) \times \Omega$, where Γ is defined by (9.37). In particular this yields the following:

Corollary 9.9 *Let $\Omega \in \mathbb{R}^n$ be a bounded domain. A solution of the heat equation $u \in C^{\mathrm{heat}}(\Omega)$ is uniquely determined by $u|_{\partial\Omega}$ and $u|_{t=0}$.*

The same arguments could be applied to the more general parabolic equation

$$\frac{\partial u}{\partial t} - Lu = 0,$$

where L is a uniformly elliptic operator as defined in (9.27).

In Sect. 6.3, we stated without proof a uniqueness result for solutions of the heat equation on $\mathbb{R}^n$. We now have the means to prove this, by establishing a maximum principle for $\mathbb{R}^n$ as a corollary of Theorem 9.8.

Corollary 9.10 *Suppose that u is a classical solution of the heat equation*

$$\frac{\partial u}{\partial t} - \Delta u = 0, \qquad u|_{t=0} = g, \tag{9.39}$$

on $[0, \infty) \times \mathbb{R}^n$, and that u is bounded on $[0, T] \times \mathbb{R}^n$ for $T > 0$. Then

$$\max_{[0,\infty)\times\mathbb{R}^n} u \le \max_{\mathbb{R}^n} g. \tag{9.40}$$

Proof Assume that u satisfies (9.39) and also

$$u(t,\boldsymbol{x}) \le M$$

for $t \in [0,T]$ and $\boldsymbol{x} \in \mathbb{R}^n$. For $\boldsymbol{y} \in \mathbb{R}^n$ and $\varepsilon > 0$ set

$$v(t,\boldsymbol{x}) := u(t,\boldsymbol{x}) - \varepsilon(T-t)^{-\frac{n}{2}} e^{\frac{|x-y|^2}{4(T-t)}}.$$

The ε term resembles the heat kernel defined by (6.16), except that the sign in the exponential is reversed. Direct differentiation shows that this expression satisfies the heat equation on $(0,T)\times\mathbb{R}^n$, and hence v does also.

For $R > 0$, let us apply the maximum principle of Theorem 9.8 to v on the domain $(0,T)\times B(\boldsymbol{y};R)$. By construction,

$$v(0,\boldsymbol{x}) \le g(\boldsymbol{x}),$$

and for $\boldsymbol{x} \in \partial B(\boldsymbol{y};R)$,

$$\begin{aligned} v(t,\boldsymbol{x}) &\le M - \varepsilon(T-t)^{-\frac{n}{2}} e^{\frac{R^2}{4(T-t)}} \\ &\le M - \varepsilon T^{-\frac{n}{2}} e^{R^2/4T}. \end{aligned}$$

With T fixed, the right-hand side of this second inequality is arbitrarily negative for large R. Therefore, for sufficiently large R, Theorem 9.8 implies that

$$\max_{[0,T]\times B(\boldsymbol{y};R)} v \le \max_{B(\boldsymbol{y};R)} g.$$

In particular, setting $\boldsymbol{x} = \boldsymbol{y}$ in this inequality gives

$$v(t,\boldsymbol{y}) \le \max_{\mathbb{R}^n} g.$$

for $t \in [0,T]$ and $\boldsymbol{y} \in \mathbb{R}^n$. By the definition of v, this implies that

$$u(t,\boldsymbol{y}) \le \max_{\mathbb{R}^n} g + \varepsilon(T-t)^{-\frac{n}{2}}.$$

We can now take $\varepsilon \to 0$ and $T \to \infty$ to conclude that

$$u(t,\boldsymbol{y}) \le \max_{\mathbb{R}^n} g$$

for all $t \in [0,\infty)$ and $\boldsymbol{y} \in \mathbb{R}^n$. □

The argument given here can be refined to show that conclusion (9.40) holds under the weaker growth condition

$$u(t,\boldsymbol{x}) \le Me^{c|\boldsymbol{x}|^2}$$

for $t \in [0,T]$.

Corollary 9.10 implies Theorem 6.3 by the argument used in Corollary 9.9. That is, if u_1 and u_2 are bounded solutions of (6.19), then $\pm(u_1 - u_2)$ solves (6.19) with $g = 0$. It then follows from (9.40) that $u_1 = u_2$.

9.6 Exercises

9.1 Suppose that $u, \phi \in C^2(\Omega;\mathbb{R}) \cap C^0(\overline{\Omega})$ on a bounded domain $\Omega \subset \mathbb{R}^n$. Assume that u subharmonic and ϕ harmonic, with matching boundary values:

$$u|_{\partial\Omega} = \phi|_{\partial\Omega}.$$

Show that

$$u \le \phi$$

at all points of Ω. (This is the motivation for the term "subharmonic".)

9.2 *Liouville's theorem* says that a bounded harmonic function on $\mathbb{R}^n$ is constant. To show this, assume $u \in C^2(\mathbb{R}^n)$ is harmonic and satisfies

$$|u(\boldsymbol{x})| \le M$$

for all $\boldsymbol{x} \in \mathbb{R}^n$.

(a) For $\boldsymbol{x}_0 \in \mathbb{R}^n$, set $r_0 = |\boldsymbol{x}_0|$. Use Corollary 9.4 at the centers 0 and $\boldsymbol{x}_0$ to show that

$$u(0) - u(\boldsymbol{x}_0) = \frac{n}{A_n R^n}\left[\int_{B(0;R)} u\, d^n\boldsymbol{x} - \int_{B(\boldsymbol{x}_0;R)} u\, d^n\boldsymbol{x}\right] \tag{9.41}$$

for $R > 0$. Note that the integrals cancel on the intersection of the two balls.

(b) Show that

$$\begin{aligned}\operatorname{vol}\left[B(0;R)\backslash B(\boldsymbol{x}_0;R)\right] &\le \operatorname{vol}\left[B(0;R)\backslash B(\tfrac{\boldsymbol{x}_0}{2}; R - \tfrac{r_0}{2})\right]\\ &= \frac{A_n}{n}\left[R^n - (R - \tfrac{r_0}{2})^n\right],\end{aligned}$$

and the same for $B(\boldsymbol{x}_0;R)\backslash B(0;R)$.

(c) Apply the volume estimates and the fact that $|u| \leq M$ to (9.41) to estimate

$$|u(0) - u(\boldsymbol{x}_0)| \leq 2M \left[\frac{R^n - (R - \frac{r_0}{2})^n}{R^n} \right].$$

Take the limit $R \to \infty$ to show that $u(\boldsymbol{x}_0) = u(0)$.

9.3 Suppose that $\Omega \subset \mathbb{R}^n$ is bounded, with $\Omega \subset B(0; R)$, and assume that $u \in C^2(\Omega; \mathbb{R}) \cap C^0(\overline{\Omega})$ satisfies

$$-\Delta u = f, \quad u|_{\partial\Omega} = 0,$$

for $f \in C^0(\overline{\Omega})$.

(a) Find a constant $c > 0$ (depending on f and R), such that $u + c\,|\boldsymbol{x}|^2$ is subharmonic on Ω.
(b) For this value of c, apply the maximum principle to $u + c\,|\boldsymbol{x}|^2$ to deduce that

$$\max_{\overline{\Omega}} |u| \leq C \max_{\overline{\Omega}} |f|,$$

where C depends only on R.

9.4 Suppose u is a harmonic function on a domain that includes $\overline{B(0; 4R)}$ for some $R > 0$, and assume $u \geq 0$. Show that

$$\max_{B(0;R)} u \leq 3^n \min_{B(0;R)} u.$$

Hint: For $\boldsymbol{x} \in B(0; R)$, apply the maximum principle to write $u(\boldsymbol{x})$ as an integral over the balls $B(\boldsymbol{x}; R)$ and $B(\boldsymbol{x}; 3R)$. Then show that

$$B(\boldsymbol{x}; R) \subset B(0; 2R) \subset B(\boldsymbol{x}; 3R),$$

and use this to estimate the integrals.

9.5 Suppose $u \in C^2(B; R) \cap C^0(\overline{B})$ is a nonconstant subharmonic function and assume that the maximum of u on B is attained at the point $\boldsymbol{x}_0 \in \partial B$. This automatically implies that $\frac{\partial u}{\partial r}(\boldsymbol{x}_0) \geq 0$. *Hopf's lemma* says that this inequality is strict,

$$\frac{\partial u}{\partial r}(\boldsymbol{x}_0) > 0.$$

To show this, let $B := B(0; R) \subset \mathbb{R}^n$ for some $R > 0$, and set

$$A := \{R/2 < |\boldsymbol{x}| < R\}.$$

(a) Consider the function

$$h(\boldsymbol{x}) := e^{-2n|\boldsymbol{x}|^2/R^2} - e^{-2n}.$$

Compute Δh and show that h is subharmonic on A.

(b) Set

$$m = \max_{\{r=R/2\}} u, \quad M = \max_{\{r=R\}} u,$$

and show that $m < M$.

(c) For $\varepsilon > 0$ set

$$u_\varepsilon := u + \varepsilon h,$$

and show that by taking ε sufficiently small we may assume that

$$\max_{\partial A} u_\varepsilon \leq M.$$

(d) Show that $u_\varepsilon(\boldsymbol{x}) \leq M$ for $\boldsymbol{x} \in A$, and hence that

$$\frac{\partial u_\varepsilon}{\partial r}(\boldsymbol{x}_0) \geq 0.$$

(e) By computing $\partial u_\varepsilon/\partial r$ and taking $\varepsilon \to 0$, conclude that

$$\frac{\partial u}{\partial r}(\boldsymbol{x}_0) > 0.$$

Chapter 10
Weak Solutions

In Sect. 1.2 we observed that d'Alembert's formula for a solution of the wave equation makes sense even when the initial data are not differentiable. This concept of a *weak solution* that is not actually required to solve the equation literally has come up in other contexts as well, for example in the discussion of the traffic equation in Sect. 3.4. In this chapter we will discuss the mathematical formulation of this generalized notion of solution.

Weak solutions first appeared in physical applications as idealized, limiting cases of true solutions. For example, one might replace a smooth density function by a simpler piecewise linear approximation, as illustrated in Fig. 10.1, in order to simplify computations. (We used this idea in Example 3.9.)

Up until the late 19th century, the limiting process by which weak solutions were obtained was understood rather loosely, and justified mainly by physical intuition. Weak solutions proved to be extremely useful, and eventually a consistent mathematical framework was developed.

10.1 Test Functions and Weak Derivatives

Consider a linear equation of the form $Lu = f$, where L is a differential operator on a domain $\Omega \subset \mathbb{R}^n$. Suppose that u represents a physical quantity such as temperature or density. Direct observation of such quantities at a single point is a practical impossibility. Even the most sensitive instrument will only be able to measure the weighted average over some small region.

To formalize this notion of a local average, we use the concept of a *test function* $\psi \in C^\infty_{\text{cpt}}(\Omega)$. The test function defines a local measurement of a quantity u through the integral

$$\int_\Omega u\psi \, d^n\mathbf{x}. \tag{10.1}$$

D. Borthwick, *Introduction to Partial Differential Equations*, Universitext, DOI 10.1007/978-3-319-48936-0_10

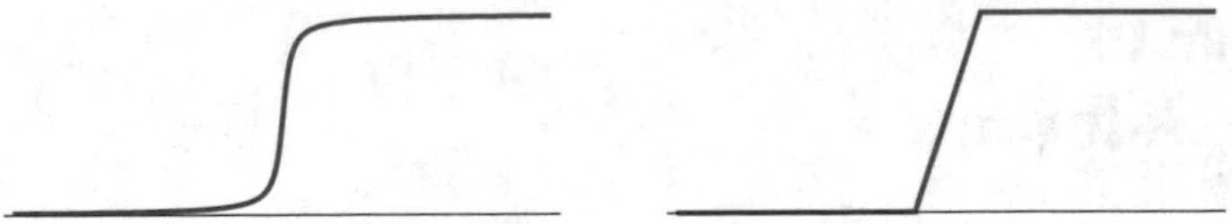

Fig. 10.1 A smooth function and its piecewise linear approximation

The function ψ plays the role of a experimental probe that takes a particular sample of the values of u.

Let us consider how we would "detect" a derivative using test functions. Suppose for the moment that $u \in C^1(\mathbb{R})$, with $u' = f$. If we measure this derivative associated using the test function $\psi \in C^\infty_{\text{cpt}}(\mathbb{R})$, the result is

$$\int_{\mathbb{R}} u'\psi \, dx = \int_{\mathbb{R}} f\psi \, dx. \tag{10.2}$$

The fact that $u' = f$ is equivalent to the statement that (10.2) holds for all $\psi \in C^\infty_{\text{cpt}}(\mathbb{R})$.

Note that the left-hand side could be integrated by parts, since ψ has compact support, yielding

$$-\int_{\mathbb{R}} u\psi' \, dx = \int_{\mathbb{R}} f\psi \, dx. \tag{10.3}$$

This condition now makes sense even when u fails to be differentiable. The only requirement is that u and f be integrable on compact sets, a property we refer to as *local integrability*. We can say that locally integrable functions satisfy $u' = f$ in the weak sense provided (10.3) holds for all $\psi \in C^\infty_{\text{cpt}}(\mathbb{R})$.

To generalize this definition to a domain $\Omega \subset \mathbb{R}^n$, let us define the space of locally integrable functions,

$$L^1_{\text{loc}}(\Omega) := \left\{ f : \Omega \to \mathbb{C};\ f|_K \in L^1(K) \text{ for all compact } K \subset \Omega \right\}.$$

The same equivalence relation (7.6) used for L^p spaces applies to L^1_{loc}, i.e., functions that differ on a set of measure zero are considered to be the same.

Inspired by (10.3), for u and $f \in L^1_{\text{loc}}(\Omega)$ we say that

$$\frac{\partial u}{\partial x_j} = f$$

as a *weak derivative* if

$$\int_\Omega u \frac{\partial \psi}{\partial x_j} \, dx = -\int_\Omega f\psi \, dx \tag{10.4}$$

for all $\psi \in C^\infty_{\text{cpt}}(\mathbb{R})$. The condition (10.4) determines f uniquely as an element of $L^1_{\text{loc}}(\Omega)$, by the following:

Lemma 10.1 *If $f \in L^1_{\text{loc}}(\Omega)$ satisfies*

$$\int_\Omega f\psi\, d^n\mathbf{x} = 0$$

for all $\psi \in C^\infty_{\text{cpt}}(\Omega)$, then $f \equiv 0$.

Proof It suffices to consider the case when Ω is bounded, since a larger domain could be subdivided into bounded pieces. For bounded Ω the local integrability of f implies that $f \in L^2(\Omega)$. By Theorem 7.5 we can choose a sequence ψ_k in $C^\infty_{\text{cpt}}(\Omega)$ such that $\psi_k \to f$ in $L^2(\Omega)$. This implies that

$$\lim_{k\to\infty} \langle f, \psi_k\rangle = \|f\|_2.$$

The inner products $\langle f, \psi_k\rangle$ are zero by hypothesis, so we conclude that $f \equiv 0$. □

Example 10.2 In $\mathbb{R}$, consider the piecewise linear function

$$g(x) := \begin{cases} 0, & x < 0, \\ x, & 0 \le x \le 1, \\ 1, & x > 0. \end{cases}$$

If we ignore the points where g is not differentiable, then we would expect that the derivative of g is given by

$$f(x) = \begin{cases} 1, & 0 \le x \le 1, \\ 0, & |x| > 1. \end{cases}$$

These functions are illustrated in Fig. 10.2.

Let us check that this works in the sense of weak derivatives. We take $\psi \in C^\infty_{\text{cpt}}(\mathbb{R})$ and compute

$$\int_{-\infty}^\infty g\psi'\, dx = \int_0^1 x\psi'(x)\, dx + \int_1^\infty \psi'\, dx.$$

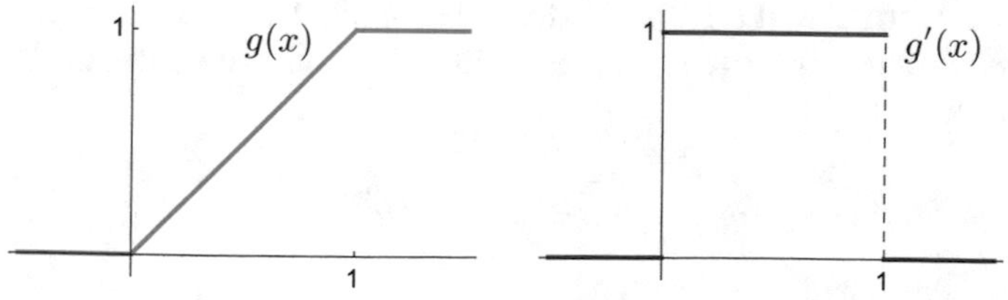

Fig. 10.2 A piecewise linear function and its weak derivative

Using integration by parts on the first term and evaluating the second gives

$$\begin{aligned}\int_{-\infty}^{\infty} g\psi'\,dx &= \psi(1) - \int_0^1 \psi\,dx - \psi(1)\\ &= -\int_{\infty}^{\infty} f\psi\,dx.\end{aligned}$$

This verifies that $g' = f$ in the weak sense. ◊

Example 10.3 For $t \in \mathbb{R}$, define $w \in L^1_{\mathrm{loc}}(\Omega)$ by

$$w(t) = \begin{cases} w_-(t), & t < 0\\ w_+(t), & t \geq 0, \end{cases} \tag{10.5}$$

where $w_\pm \in C^1(\mathbb{R})$. For $\psi \in C^\infty_{\mathrm{cpt}}(\mathbb{R})$,

$$\begin{aligned}-\int_{-\infty}^{\infty} w\psi'\,dt &= -\int_{-\infty}^{0} w_-\psi'\,dt - \int_0^{\infty} w_+\psi'\,dt\\ &= [w_+(0) - w_-(0)]\psi(0) + \int_{-\infty}^{0} w_-'\psi\,dt + \int_0^{\infty} w_+'\psi\,dt.\end{aligned}$$

The term proportional to $\psi(0)$ could not possibly come from the integral of ψ against a locally integrable function, because the value of the integrand at a single point does not affect the integral. Hence w admits a weak derivative only under the matching condition

$$w_-(0) = w_+(0).$$

If this is satisfied, then the derivative is

$$w'(t) = \begin{cases} w_-'(t), & t < 0,\\ w_+'(t), & t > 0. \end{cases}$$

◊

Weak derivatives of higher order are defined by an extension of (10.4). To write the corresponding formulas, it is helpful to have a simplified notation for higher partials. For each *multi-index* $\alpha = (\alpha_1, \ldots, \alpha_n)$ with $\alpha_j \in \mathbb{N}_0$, we define the differential operator on $\mathbb{R}^n$,

$$D^\alpha := \frac{\partial^{\alpha_1}}{\partial x_1^{\alpha_1}} \cdots \frac{\partial^{\alpha_n}}{\partial x_n^{\alpha_n}}. \tag{10.6}$$

The order of this operator is denoted by

$$|\alpha| := \alpha_1 + \cdots + \alpha_n.$$

Repeated integration by parts introduces a minus sign for each derivative. Therefore, a function $u \in L^1_{\text{loc}}(\Omega)$ admits a weak derivative $D^\alpha u \in L^1_{\text{loc}}(\Omega)$ if

$$\int_\Omega (D^\alpha u)\psi \, d^n\mathbf{x} = (-1)^{|\alpha|} \int_\Omega u D^\alpha \psi \, d^n\mathbf{x} \tag{10.7}$$

for all $\psi \in C^\infty_{\text{cpt}}(\Omega)$.

It might seem that we should distinguish between classical and weak derivatives in the notation. This is made unnecessary by the following:

Theorem 10.4 (Consistency of weak derivatives) *If $u \in C^m(\Omega)$ then u is weakly differentiable to order k and the weak derivatives equal the classical derivatives.*

Conversely, if $u \in L^1_{\text{loc}}(\Omega)$ admits weak derivatives $D^\alpha u$ for $|\alpha| \le m$, and each $D^\alpha u$ can be represented by a continuous function, then u is equivalent to a function in $C^m(\Omega)$ whose classical derivatives match the weak derivatives.

In one direction the argument is straightfoward. Classical derivatives satisfy the criterion (10.7) by integration by parts, so they automatically qualify as weak derivatives. Lemma 10.1 shows that weak derivatives are uniquely defined.

The argument for the converse statement, that continuity of the weak derivatives $D^\alpha u$ implies classical differentiability, is much more technical and we will not be able to give the details. The basic idea is to show that one can approximate u by a sequence $\psi_k \in C^\infty_{\text{cpt}}(\Omega)$ such that $D^\alpha \psi_k \to D^\alpha u$ uniformly on every compact subset of Ω for $|\alpha| \le m$. The fact that the weak derivatives $D^\alpha u$ are continuous makes this possible. Uniform convergence then allows the classical derivatives of u to be computed as limits of the functions $D^\alpha \psi_k$.

10.2 Weak Solutions of Continuity Equations

Consider the continuity equation on $\mathbb{R}$ introduced in Sect. 3.1,

$$\frac{\partial u}{\partial t} + \frac{\partial q}{\partial x} = 0, \quad u|_{t=0} = g. \tag{10.8}$$

The flux q could depend on u as well as t and x. To allow for the nonlinear case, we will assume that u is real-valued here.

Suppose for the moment that q is differentiable and u is a classical solution of (10.8). Let ψ be a test function in $C^\infty_{\text{cpt}}([0,\infty) \times \mathbb{R})$. Use of the closed interval $[0, \infty)$ means that ψ and its derivatives are not necessarily zero at $t = 0$. Pairing $\frac{\partial u}{\partial t}$ with ψ and integrating by parts thus generates a boundary term,

$$\int_0^\infty \frac{\partial u}{\partial t}\psi \, dt = -u\psi\Big|_{t=0} - \int_0^\infty u \frac{\partial \psi}{\partial t} \, dt.$$

On the other hand, the spatial integration parts has no boundary term,

$$\int_{-\infty}^{\infty} \frac{\partial q}{\partial x}\psi\, dx = -\int_{-\infty}^{\infty} q\frac{\partial \psi}{\partial x}\, dx.$$

When the left-hand side of (10.8) is paired with ψ and integrated over both t and x, the result is thus

$$\int_0^{\infty}\int_{-\infty}^{\infty}\Big[\frac{\partial u}{\partial t} + \frac{\partial q}{\partial x}\Big]\psi\, dx\, dt = -\int_0^{\infty}\int_{-\infty}^{\infty}\Big[u\frac{\partial \psi}{\partial t} + q\frac{\partial \psi}{\partial x}\Big]\psi\, dx\, dt$$
$$- \int_{-\infty}^{\infty} u\psi|_{t=0}\, dx.$$

If u is a classical solution of (10.8), then

$$\int_0^{\infty}\int_{-\infty}^{\infty}\Big[u\frac{\partial \psi}{\partial t} + q\frac{\partial \psi}{\partial x}\Big]dx\, dt + \int_{-\infty}^{\infty} g\psi|_{t=0}\, dx = 0 \tag{10.9}$$

for all $\psi \in C^{\infty}_{\text{cpt}}([0,\infty) \times \mathbb{R})$.

The $t = 0$ integral in (10.9) makes sense for $g \in L^1_{\text{loc}}(\mathbb{R})$. Under this assumption, we define $u \in L^1_{\text{loc}}((0,\infty) \times \mathbb{R}; \mathbb{R})$ to be a weak solution of (10.8) provided $q \in L^1_{\text{loc}}((0,\infty) \times \mathbb{R}; \mathbb{R})$ and (10.9) holds for all test functions.

Example 10.5 Consider the linear conservation equation with constant velocity, which means $q = cu$ in (10.8). By the method of characteristics (Theorem 3.2), the solution is

$$u(t,x) = g(x - ct).$$

Let us check that this defines a weak solution for $g \in L^1_{\text{loc}}(\mathbb{R})$.

For $\psi \in C^{\infty}_{\text{cpt}}(\mathbb{R})$ the first term in (10.9) is

$$\int_0^{\infty}\int_{-\infty}^{\infty} g(x - ct)\Big[\frac{\partial \psi}{\partial t}(t,x) - c\frac{\partial \psi}{\partial x}(t,x)\Big]dx\, dt. \tag{10.10}$$

To evaluate the integral, introduce the variables

$$\tau = t, \quad y = x - ct,$$

and define

$$\tilde{\psi}(\tau, y) := \psi(\tau, y + c\tau).$$

By the chain rule,

$$\frac{\partial \tilde{\psi}}{\partial \tau} = \frac{\partial \psi}{\partial t} - c\frac{\partial \psi}{\partial x}. \tag{10.11}$$

The Jacobian determinant of the transformation $(\tau, y) \mapsto (t, x)$ is 1, so that

$$\int_0^\infty \int_{-\infty}^\infty u\left[\frac{\partial \psi}{\partial t} - c\frac{\partial \psi}{\partial x}\right] dx\, dt = \int_0^\infty \int_{-\infty}^\infty g(y)\frac{\partial \tilde{\psi}}{\partial \tau}\, dy\, d\tau.$$

The τ integration can now be done directly,

$$\int_0^\infty \frac{\partial \tilde{\psi}}{\partial \tau}(\tau, y)\, d\tau = -\tilde{\psi}(0, y).$$

This gives

$$\begin{aligned}\int_0^\infty \int_{-\infty}^\infty u\left[\frac{\partial \psi}{\partial t} - c\frac{\partial \psi}{\partial x}\right] dx\, dt &= -\int_{-\infty}^\infty g(y)\tilde{\psi}(0, y)\, dy \\ &= -\int_{-\infty}^\infty g(y)\psi(0, y)\, dy,\end{aligned}$$

which verifies (10.9). ◇

We saw in Example 10.3 that in one dimension a jump discontinuity precludes the existence of a weak derivative. Example 10.5 shows that this is not the case in higher dimension. For $g \in L^1_{\text{loc}}(\mathbb{R})$, the solution $u(t, x) = g(x - ct)$ could be highly discontinuous. The direction of the derivative is crucial here; regularity is required only along the characteristics.

As an application of the weak formulation (10.9), let us return to an issue that arose in the traffic model in Sect. 3.4. For certain initial conditions the characteristic lines crossed each other, ruling out a classical solution of the PDE. We will see that weak solutions can still exist in this case.

Consider a one-dimensional quasilinear equation of the form

$$\frac{\partial u}{\partial t} - \frac{\partial}{\partial x}q(u) = 0 \tag{10.12}$$

with the flux $q(u)$ a smooth function of u which is independent of t and x. As we saw in Sect. 3.4, the characteristics are straight lines whose slope depends on the initial conditions. Let us study the situation pictured in Fig. 3.11, where a shock forms as characteristic lines cross at some point. For simplicity we assume that the initial crossing occurs at the origin.

One possible way to resolve the issue of crossing characteristics is to subdivide the (t, x) plane into two regions by drawing a *shock curve* C, as illustrated in Fig. 10.3. Suppose that classical solutions $u_\pm$ are derived by the method of characteristics above and below this curve. We will show that this combination yields a weak solution provided a certain jump condition is satisfied along C. The jump condition was discovered in the 19th century by engineers William Rankine and Pierre Hugoniot, who developed the first theories of shock waves in the context of gas dynamics.

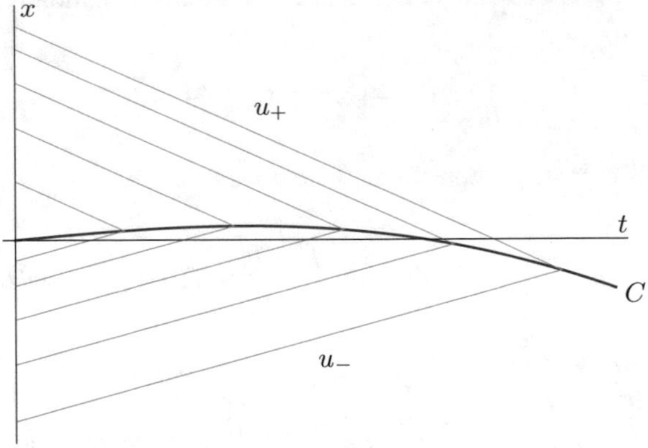

Fig. 10.3 Shock curve with solutions $u_\pm$ on either side

Theorem 10.6 (Rankine-Hugoniot condition) *Let C be a curve parametrized as $x = \sigma(t)$ with $\sigma \in C^1[0,\infty)$. Suppose that u is a weak solution of* (10.12) *given by*

$$u(t,x) = \begin{cases} u_-(t,x), & x < \sigma(t), \\ u_+(t,x), & x > \sigma(t), \end{cases}$$

where $u_\pm$ are classical solutions. Then, at each point of C,

$$q(u_+) - q(u_-) = (u_+ - u_-)\sigma'. \tag{10.13}$$

Proof Since we are not concerned with the boundary conditions, we consider a test function $\psi \in C^\infty_{\mathrm{cpt}}((0,\infty)\times\mathbb{R})$, for which (10.9) specializes to

$$\int_0^\infty \int_{-\infty}^\infty \left[u\frac{\partial\psi}{\partial t} + q(u)\frac{\partial\psi}{\partial x}\right] dx\, dt = 0. \tag{10.14}$$

Since the solutions $u_\pm$ are classical and σ is C^1, we can separate the integral (10.14) at the shock curve and integrate by parts on either side.

Consider first the u_- side. For the term involving the x derivative the integration by parts is straightforward:

$$\begin{aligned}\int_0^\infty \int_{-\infty}^{\sigma(t)} q(u_-)\frac{\partial\psi}{\partial x}\, dx\, dt = &-\int_0^\infty \int_{-\infty}^{\sigma(t)} \psi\frac{\partial}{\partial x}q(u_-)\, dx\, dt \\ &+ \int_0^\infty \psi q(u_-)\big|_{x=\sigma(t)}\, dt.\end{aligned}$$

For the t-derivative term we start by using the fundamental theorem of calculus to derive

$$\frac{d}{dt}\int_{-\infty}^{\sigma(t)} \psi u_- \, dx = \int_{-\infty}^{\sigma(t)} \left[\frac{\partial \psi}{\partial t} u_- + \psi \frac{\partial u_-}{\partial t}\right] dx + \sigma'(t)\psi u_-\big|_{x=\sigma(t)}.$$

By the compact support of ψ, the integral over t of the left-hand side vanishes, yielding

$$\begin{aligned}\int_0^\infty \int_{-\infty}^{\sigma(t)} \frac{\partial \psi}{\partial t} u_- \, dx\, dt = &-\int_0^\infty \int_{-\infty}^{\sigma(t)} \psi \frac{\partial u_-}{\partial t}\, dx\, dt \\ &- \int_0^\infty \sigma'(t)\psi u_-\big|_{x=\sigma(t)}\, dt.\end{aligned}$$

Combining these integration by parts formulas gives

$$\begin{aligned}&\int_0^\infty \int_{-\infty}^{\sigma(t)} \left[u_- \frac{\partial \psi}{\partial t} + q(u_-)\frac{\partial \psi}{\partial x}\right] dx\, dt \\ &= -\int_0^\infty \int_{-\infty}^{\sigma(t)} \psi \left[\frac{\partial u_-}{\partial t} - \frac{\partial}{\partial x} q(u_-)\right] dx\, dt \\ &\quad - \int_0^\infty \left[\sigma'(t) u_- - q(u_-)\right]\psi\Big|_{x=\sigma(t)}\, dt.\end{aligned}$$

The first term on the right vanishes by the assumption that u_- is a classical solution, leaving

$$\int_0^\infty \int_{-\infty}^{\sigma(t)} \left[u_- \frac{\partial \psi}{\partial t} + q(u_-)\frac{\partial \psi}{\partial x}\right] dx\, dt = -\int_0^\infty \left[\sigma' u_- - q(u_-)\right]\psi\Big|_{x=\sigma(t)}\, dt.$$

The corresponding calculation on the u_+ side yields

$$\int_0^\infty \int_{\sigma(t)}^{\infty} \left[u_+ \frac{\partial \psi}{\partial t} + q(u_+)\frac{\partial \psi}{\partial x}\right] dx\, dt = \int_0^\infty \left[\sigma' u_+ - q(u_+)\right]\psi\Big|_{x=\sigma(t)}\, dt.$$

By (10.14) the sum of the u_- and u_+ integrals is zero, which implies

$$\int_0^\infty \left[(u_+ - u_-)\sigma' - (q(u_+) - q(u_-))\right]\psi\Big|_{x=\sigma(t)}\, dt.$$

Since this holds for all $\psi \in C^\infty((0,\infty) \times \mathbb{R})$, we conclude that

$$\left[(u_+ - u_-)\sigma' - (q(u_+) - q(u_-))\right]\Big|_{x=\sigma(t)} = 0$$

for all $t > 0$. □

Example 10.7 Consider the traffic equation introduced in Sect. 3.4,

$$\frac{\partial u}{\partial t} + (1 - 2u)\frac{\partial u}{\partial x} = 0,$$

for which $q(u) = u - u^2$. For the initial condition take a step function,

$$g(x) := \begin{cases} a, & x < 0, \\ b, & x > 0. \end{cases} \tag{10.15}$$

From 3.28 the characteristic lines are given by

$$x(t) = \begin{cases} x_0 + (1 - 2a)t, & x_0 < 0, \\ x_0 + (1 - 2b)t, & x_0 > 0. \end{cases}$$

These intersect to form a shock provided $a < b$.

The solutions above and below the shock line are given by constants,

$$u_-(t, x) = a, \quad u_+(t, x) = b.$$

The Rankine-Hugoniot condition (10.13) thus reduces to

$$q(b) - q(a) = (b - a)\sigma'.$$

Substituting with $q(u) = u - u^2$ reduces this condition to

$$\sigma' = 1 - b - a.$$

Since the discontinuity starts at the origin, the shock curve is thus given by

$$\sigma(t) = (1 - b - a)t.$$

Hence the weak solution is

$$u(t, x) = \begin{cases} a, & x < (1 - b - a)t, \\ b, & x > (1 - b - a)t. \end{cases}$$

Some cases are illustrated in Fig. 10.4. In the plot on the left, the shock wave propagates backwards. ◊

For certain initial conditions, the definition (10.9) of a weak solution is not sufficient to determine the solution uniquely. For example, if we had taken $a > b$ in (10.15), then instead of overlapping the characteristics originating from $t = 0$ would separate, leaving a triangular region with no characteristic lines. An additional physical condition is required to specify the solution uniquely in this case. We will discuss this further in the exercises.

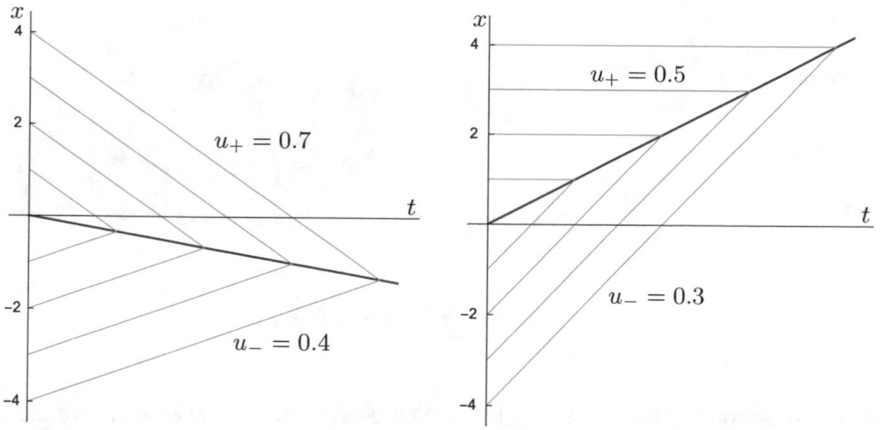

Fig. 10.4 Characteristic lines meeting at the shock wave

10.3 Sobolev Spaces

Boundary values are not well defined for locally integrable functions. We were able to avoid this issue in the discussion of the continuity equation in Sect. 10.2, because solutions were required to be constant along characteristics. In general, the formulation of boundary or initial conditions requires a class of functions with greater regularity.

The most obvious class to consider consists of functions that admit weak higher partial derivatives. However, it proves to be very helpful to strengthen the integrability requirements as well. Such function spaces were introduced by Sergei Sobolev in the mid 20th century and have since become fundamental tools of analysis.

The Sobolev spaces based on L^2 are defined by

$$H^m(\Omega) := \left\{u \in L^2(\Omega);\ D^\alpha u \in L^2(\Omega) \text{ for all } |\alpha| \leq m\right\},$$

for $m \in \mathbb{N}_0$, with derivatives interpreted in the weak sense. An extended family of Sobolev spaces $W^{m,p}$ is given by replacing L^2 with L^p in the definition. The extended family is important in the analysis of nonlinear PDE, but our focus will be limited to linear applications involving H^m.

Sobolev spaces are useful as theoretical tools, but they also have a practical side. For a bounded domain Ω, the space $H^1(\Omega)$ includes the continuous *piecewise linear* functions. A function is called piecewise linear if the domain can be decomposed into a finite number of polygonal subdomains, on which the function is linear. Figure 10.5 shows a two-dimensional example. Sobolev spaces provide a natural framework for the approximation of solutions by computationally simple classes of functions.

The space $H^m(\Omega)$ carries a natural inner product,

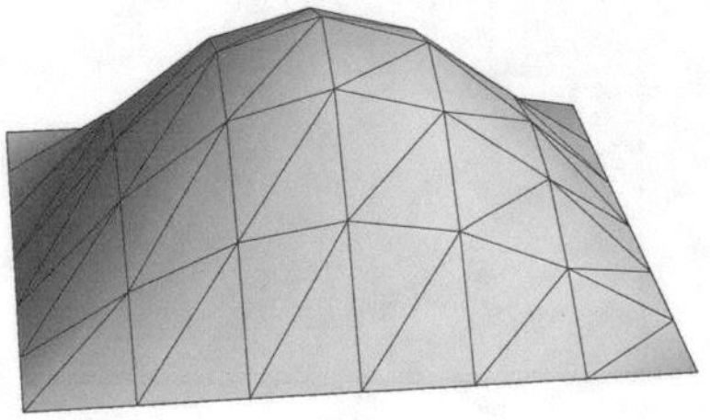

Fig. 10.5 Graph of a piecewise linear H^1 function on the unit square

$$\langle u, v\rangle_{H^m} := \sum_{|\alpha|\le m} \langle D^\alpha u, D^\alpha v\rangle. \tag{10.16}$$

(Our convention will be that a bracket without subscript denotes the L^2 inner product.) The corresponding norm is

$$\|u\|_{H^m} := \Big(\sum_{|\alpha|\le m} \|D^\alpha u\|_2^2\Big)^{\frac{1}{2}}. \tag{10.17}$$

Lebesgue integration theory gives us the following completeness result, analogous to Theorem 7.7.

Theorem 10.8 *For $\Omega \subset \mathbb{R}^n$ and $m \in \mathbb{N}_0$, $H^m(\Omega)$ is a Hilbert space.*

Recall that Theorem 7.5 says that $C^\infty_{\mathrm{cpt}}(\Omega)$ is a dense subspace of $L^2(\Omega)$. This means that the closure of $C^\infty_{\mathrm{cpt}}(\Omega)$ with respect to the L^2 norm is $L^2(\Omega)$. This result no longer holds for the Sobolev space $H^m(\Omega)$ with $m \ge 1$. In particular, the closure of $C^\infty_{\mathrm{cpt}}(\Omega)$ with respect to the H^1 norm defines a subspace

$$H^1_0(\Omega) = \left\{ u \in H^m(\Omega);\ \lim_{k\to\infty} \|u - \psi_k\|_{H^1} = 0 \text{ for } \psi_k \in C^\infty_{\mathrm{cpt}}(\Omega) \right\}. \tag{10.18}$$

By Lemma 7.8 $H^1_0(\Omega)$ is also a Hilbert space with respect to the H^1 norm.

If $\partial\Omega$ is piecewise C^1, then for functions in $H^1(\Omega)$ it is possible to define boundary restrictions in $L^2(\partial\Omega)$ that generalize the boundary restriction of a continuous function. In this case, $H^1_0(\Omega)$ consists precisely of the functions whose boundary restriction vanishes. Thus the space $H^1_0(\Omega)$ can be interpreted as the class of H^1 functions satisfying Dirichlet boundary conditions on $\partial\Omega$.

The theory of boundary restrictions is too technical for us to cover here, but we can at least show how this works in the one-dimensional case.

Theorem 10.9 *If $u \in H^1_0(a, b)$ then u is continuous on $[a, b]$ and equal to zero at the endpoints.*

Proof Suppose $u \in H^1_0(a, b)$. By definition, there exists a sequence of $C^\infty_{\mathrm{cpt}}(a, b)$ such that

$$\lim_{k\to\infty} \|\psi_k - u\|_{H^1} = 0.$$

For $x \in [a, b]$,

$$\psi_j(x) - \psi_k(x) = \int_a^x [\psi_j'(t) - \psi_k'(t)]\, dt$$

The integral on the right could be expressed as an inner product on $\mathbb{R}$,

$$\int_a^x [\psi_j'(t) - \psi_k'(t)]\, dt = \langle \psi_j' - \psi_k', \chi_{[a,x]} \rangle,$$

where χ_I denotes the characteristic function of the interval I. Thus, by the Cauchy-Schwarz inequality (Theorem 7.1),

$$|\psi_j(x) - \psi_k(x)| \le \sqrt{x-a}\, \|\psi_j' - \psi_k'\|_2.$$

In view of the definition of the H^1 norm, this implies the uniform bound

$$\|\psi_j - \psi_k\|_\infty \le \sqrt{b-a}\, \|\psi_j - \psi_k\|_{H^1} \tag{10.19}$$

Since $\{\psi_k\}$ converges and is therefore Cauchy with respect to the H^1 norm, it follows from (10.19) implies that the sequence $\{\psi_k\}$ is also Cauchy in the uniform sense. By the completeness of $L^\infty(a, b)$ (Theorem 7.7) and Lemma 8.4, this implies that $\psi_k \to g$ uniformly for some $g \in C^0[a, b]$.

At this point we have $\psi_k \to u$ in H^1 and $\psi_k \to g$ uniformly. Uniform convergence on a bounded interval implies convergence in L^2, by a simple integral estimate. Therefore $\psi_k \to g$ in L^2 also, implying that $u = g$ in L^2. Hence $u \in C^0[a, b]$.

To show that u vanishes at the endpoints, note that

$$\max\{|u(a)|, |u(b)|\} \le \sup_{[a,b]} |\psi_k - u|. \tag{10.20}$$

because $\psi_k(a) = \psi_k(b) = 0$. By uniform convergence, the left-hand side of (10.20) approaches zero as $k \to \infty$, showing that

$$u(a) = u(b) = 0.$$

□

In higher dimensions, functions in H^1 are not necessarily continuous. However, H^m does imply continuity if m is sufficiently large relative to the dimension. We will develop this regularity theory in Sect. 10.4.

We conclude this section with an extension property that will prove useful in Chap. 11.

Lemma 10.10 *For $\Omega \subset \widetilde{\Omega} \subset \mathbb{R}^n$, the extension by zero of an element of $H_0^1(\Omega)$ gives an element of $H_0^1(\widetilde{\Omega})$.*

Proof For $u \in H_0^1(\Omega)$, let $\tilde{u}$ denote the extension by zero to $\widetilde{\Omega}$. The weak gradient $\nabla u \in L^2(\Omega; \mathbb{R}^n)$ can also be extended by zero to $\widetilde{\nabla u} \in L^2(\widetilde{\Omega}; \mathbb{R}^n)$. We need to show that $\widetilde{\nabla u}$ is the weak gradient of $\tilde{u}$. This is the condition that

$$\int_{\widetilde{\Omega}} \phi \widetilde{\nabla u}\, d^n\boldsymbol{x} = -\int_{\widetilde{\Omega}} \tilde{u} \nabla \phi\, d^n\boldsymbol{x}, \tag{10.21}$$

for all $\phi \in C^\infty_{\text{cpt}}(\widetilde{\Omega})$.

By the definition of $H_0^1(\Omega)$, there exists a sequence of $\psi_k \in C^\infty_{\text{cpt}}(\Omega)$ such that $\psi_k \to u$ in the H^1 norm. Since ψ_k has compact support within Ω, integration by parts gives

$$\int_{\Omega} \phi \nabla \psi_k\, d^n\boldsymbol{x} = -\int_{\Omega} \psi_k \nabla \phi\, d^n\boldsymbol{x}. \tag{10.22}$$

By the H^1 convergence $\psi_k \to u$, we can take the limit $k \to \infty$ on both sides of (10.22) to obtain

$$\int_{\Omega} \phi \nabla u\, d^n\boldsymbol{x} = -\int_{\Omega} u \nabla \phi\, d^n\boldsymbol{x}.$$

Since $\tilde{u}$ and $\widetilde{\nabla u}$ are equal to u and ∇u on Ω and vanish on $\widetilde{\Omega} - \Omega$, this is equivalent to (10.21). □

10.4 Sobolev Regularity

In this section we will consider the relationship between weak regularity, defined in terms of Sobolev spaces, and regularity in the classical sense. This connection plays a central role in the application of Sobolev spaces to PDE.

Theorem 10.11 (Sobolev embedding theorem) *Suppose $\Omega \subset \mathbb{R}^n$ is a bounded domain. If $m > k + \frac{n}{2}$, then*

$$H^m(\Omega) \subset C^k(\Omega).$$

This result can be sharpened and extended in various ways. One important variant includes differentiability up to the boundary under certain conditions on $\partial\Omega$. For example, if the boundary $\partial\Omega$ is piecewise C^1 then it is possible to show that

$$H^m(\Omega) \subset C^k(\overline{\Omega}).$$

These boundary results are quite important but too technically difficult for us to include here.

The strategy we will use for Theorem 10.11 is based on the connection established in Sect. 8.6 between regularity and the decay of Fourier coefficients. Recall the definition $\mathbb{T} := \mathbb{R}/2\pi\mathbb{Z}$ introduced in Sect. 8.2. To extend Fourier series to higher dimensions we introduce the corresponding space

$$\mathbb{T}^n := \mathbb{R}^n/(2\pi\mathbb{Z})^n.$$

A function on $\mathbb{T}^n$ is a function on $\mathbb{R}^n$ which is 2π-periodic in each coordinate.

The periodic Fourier series theory from can be carried over to $\mathbb{T}^n$ directly. For $f \in L^2(\mathbb{T}^n)$ and $\boldsymbol{k} \in \mathbb{Z}^n$ we define the coefficients

$$c_{\boldsymbol{k}}[f] := \frac{1}{(2\pi)^n} \int_{\mathbb{T}^n} e^{-i\boldsymbol{k}\cdot\boldsymbol{x}} f(\boldsymbol{x})\, d^n\boldsymbol{x}. \tag{10.23}$$

The integral over $\mathbb{T}^n$ can be taken over $[-\pi, \pi]^n$, or any translate of this cube. The argument from Theorem 8.6 can be adapted, with minor notational changes, to prove the following:

Theorem 10.12 *For $f \in L^2(\mathbb{T}^n)$, the series*

$$\sum_{\boldsymbol{k}\in\mathbb{Z}^n} c_{\boldsymbol{k}}[f] e^{i\boldsymbol{k}\cdot\boldsymbol{x}}$$

converges to f in the L^2 norm.

As a corollary, we obtain the generalization of the Parseval identity (8.36),

$$\langle f, g\rangle = (2\pi)^n \sum_{\boldsymbol{k}\in\mathbb{Z}^n} c_{\boldsymbol{k}}[f]\overline{c_{\boldsymbol{k}}[g]} \tag{10.24}$$

for $f, g \in L^2(\mathbb{T}^n)$.

Because of the periodic structure of $\mathbb{T}^n$, it is not necessary to assume that test functions have compact support. For $f \in L^1_{loc}(\mathbb{T})$ the weak derivative $D^\alpha f \in L^1_{loc}(\mathbb{T})$ is defined by the condition that

$$\int_0^{2\pi} \psi D^\alpha f\, dx = (-1)^{|\alpha|} \int_0^{2\pi} f D^\alpha \psi\, dx \tag{10.25}$$

for all $\psi \in C^\infty(\mathbb{T})$. The space $H^m(\mathbb{T}^n)$ consists of functions in $L^2(\mathbb{T}^n)$ which have weak partial derivatives up to order m contained in $L^2(\mathbb{T}^n)$.

It is convenient notate powers of the components of $\boldsymbol{k}$ by analogy with D^α,

$$\boldsymbol{k}^\alpha := k_1^{\alpha_1} \cdots k_n^{\alpha_n},$$

for $\alpha = (\alpha_1, \ldots, \alpha_n)$ with $\alpha_j \in \mathbb{N}_0$. A simple computation shows that

$$D^\alpha e^{i\boldsymbol{k}\cdot\boldsymbol{x}} = (i\boldsymbol{k})^\alpha e^{i\boldsymbol{k}\cdot\boldsymbol{x}}.$$

Thus, for $f \in H^m(\mathbb{T})$, substituting $e^{i\boldsymbol{k}\cdot\boldsymbol{x}}$ into (10.25) gives

$$c_{\boldsymbol{k}}[D^\alpha f] = (i\boldsymbol{k})^\alpha c_{\boldsymbol{k}}[f] \tag{10.26}$$

for $|\alpha| \leq m$. This generalizes the integration by parts formula (8.30).

Theorem 10.13 *A function $f \in L^2(\mathbb{T})$ lies in $H^m(\mathbb{T})$ for $m \in \mathbb{N}$ if and only if*

$$\sum_{\boldsymbol{k}\in\mathbb{Z}^n} |\boldsymbol{k}|^{2m} |c_{\boldsymbol{k}}[f]|^2 < \infty. \tag{10.27}$$

Proof By (10.26) and Bessel's inequality (Proposition 7.9), the condition that $D^\alpha f \in L^2(\mathbb{T}^n)$ implies that

$$\sum_{\boldsymbol{k}\in\mathbb{Z}^n} \left|\boldsymbol{k}^\alpha c_{\boldsymbol{k}}[f]\right|^2 < \infty. \tag{10.28}$$

This holds for all $|\alpha| \leq m$, implying (10.27).

Conversely, if $f \in L^2(\mathbb{T}^n)$ satisfies (10.27), then (10.28) holds for $|\alpha| \leq m$. We can therefore define functions $g_\alpha \in L^2(\mathbb{T}^n)$ by the Fourier series

$$g_\alpha(\boldsymbol{x}) := \sum_{\boldsymbol{k}\in\mathbb{Z}^n} (i\boldsymbol{k})^\alpha c_{\boldsymbol{k}}[f] e^{i\boldsymbol{k}\cdot\boldsymbol{x}}.$$

By Parseval's identity (10.24), the inner product of g_α with $\psi \in C^\infty_{\text{cpt}}(\mathbb{T}^n)$ gives

$$\begin{aligned}
\langle g_\alpha, \psi\rangle &= (2\pi)^n \sum_{\boldsymbol{k}\in\mathbb{Z}^n} (i\boldsymbol{k})^\alpha c_{\boldsymbol{k}}[f]\overline{c_{\boldsymbol{k}}[\psi]} \\
&= (-1)^{|\alpha|} 2\pi \sum_{k\in\mathbb{Z}} c_{\boldsymbol{k}}[f]\overline{c_{\boldsymbol{k}}[D^\alpha\psi]} \\
&= (-1)^{|\alpha|}\langle f, D^\alpha\psi\rangle.
\end{aligned}$$

This shows that the weak derivative $D^\alpha f$ exists and is equal to g_α. □

Theorem 10.13 makes the connection between Sobolev regularity and decay of Fourier coefficients. Our task is now to translate this back into classical regularity.

Theorem 10.14 (Periodic Sobolev embedding) *If $m > q + \frac{n}{2}$, then*

$$H^m(\mathbb{T}^n) \subset C^q(\mathbb{T}^n).$$

Proof Using the notation for discrete spaces introduced in Sect. 7.4, the space $\ell^2(\mathbb{Z}^n)$ is defined as the Hilbert space of functions $\mathbb{Z}^n \to \mathbb{C}$, equipped with the inner product

$$\langle \beta, \gamma \rangle_{\ell^2} := \sum_{\boldsymbol{k} \in \mathbb{Z}^n} \beta(\boldsymbol{k}) \overline{\gamma(\boldsymbol{k})}.$$

Consider the function

$$\beta(\boldsymbol{k}) := (1 + |\boldsymbol{k}|)^{-m}.$$

The ℓ^2 norm of β can be estimated with an integral,

$$\begin{aligned} \|\beta\|_{\ell^2}^2 &:= \sum_{\boldsymbol{k} \in \mathbb{Z}^n} (1 + |\boldsymbol{k}|)^{-2m} \\ &\le \int_{\mathbb{R}^n} (1 + |\boldsymbol{x}|)^{-2m} \, d^n \boldsymbol{x} \\ &= A_n \int_0^\infty (1 + r)^{-2m} r^{n-1} \, dr. \end{aligned}$$

The integral is finite if $2m > n$, implying that $\beta \in \ell^2(\mathbb{Z}^n)$ for $m > \frac{n}{2}$.

By Theorem 10.13, for $f \in H^m(\mathbb{T}^n)$ we can also define an element of $\ell^2(\mathbb{Z}^n)$ by

$$\gamma(\boldsymbol{k}) := (1 + |\boldsymbol{k}|)^m |c_{\boldsymbol{k}}[f]|,$$

so that

$$\langle \beta, \gamma \rangle_{\ell^2} = \sum_{\boldsymbol{k} \in \mathbb{Z}^n} |c_{\boldsymbol{k}}[f]|.$$

It then follows from the Cauchy-Schwarz inequality on ℓ^2 that

$$\sum_{\boldsymbol{k} \in \mathbb{Z}^n} |c_{\boldsymbol{k}}[f]| \le \|\beta\|_{\ell^2} \|\gamma\|_{\ell^2}, \tag{10.29}$$

which is finite for $m > \frac{n}{2}$.

Since $|e^{i\boldsymbol{k}\cdot\boldsymbol{x}}| = 1$, the estimate (10.29) implies that the Fourier series for f converges uniformly. By Lemma 8.4 the limit of this series is continuous. Thus, after possible replacement by an equivalent function in L^2, f is continuous.

This argument shows that

$$H^m(\mathbb{T}^n) \subset C^0(\mathbb{T}^n) \tag{10.30}$$

for $m > \frac{n}{2}$. To apply it to higher derivatives we note that if $f \in H^m(\mathbb{T}^n)$ for $m > q + \frac{n}{2}$ then for $|\alpha| \le q$ the weak derivatives $D^\alpha f$ will lie in $H^{m-q}(\mathbb{T}^N)$. For $m > q + \frac{n}{2}$ it follows from (10.30) that these derivatives are continuous. By Theorem 10.4, this shows that $u \in C^q(\mathbb{T}^n)$. □

We are now prepared to derive the Sobolev embedding result for a bounded domain as a consequence of Theorem 10.14.

Proof of Theorem 10.11 Suppose $u \in H^m(\Omega)$ for $\Omega \in \mathbb{R}^n$, let $\boldsymbol{x}_0 \in \Omega$. Because Ω is open, we can choose $\varepsilon > 0$ small enough that

$$B(\boldsymbol{x}_0; \varepsilon) \subset \Omega.$$

Suppose that $\psi \in C^\infty_{\text{cpt}}(\Omega)$ has support contained in $B(\boldsymbol{x}_0; \varepsilon)$ and is equal to 1 inside $B(\boldsymbol{x}_0; \varepsilon/2)$. (Such a function can be constructed as in Example 2.2.) Since ψ is smooth, $u\psi \in H^m(\Omega)$ also. Thus, assuming $\varepsilon < 2\pi$, we can extend $u\psi$ by periodicity to a function in $H^m(\mathbb{T}^n)$. Theorem 10.14 then shows that $u\psi \in C^k(\mathbb{T}^n)$ if $m > k + n/2$. Since $u\psi$ and u agree in a neighborhood of $\boldsymbol{x}_0$, this shows that u is k-times continuously differentiable at $\boldsymbol{x}_0$. This argument applies at every interior point of Ω, so we conclude that $u \in C^k(\Omega)$. □

10.5 Weak Formulation of Elliptic Equations

The Laplace equation introduced in Sect. 9.1 is the prototypical elliptic equation. Another classic example is the Poisson equation $-\Delta u = f$, which we will discuss in more detail in Sect. 11.1.

If $\Omega \subset \mathbb{R}^n$ is a bounded domain, then for $u, \psi \in C^\infty_{\text{cpt}}(\Omega)$, Green's first identity (Theorem 2.10) gives

$$\int_\Omega \psi \Delta u \, d^n\boldsymbol{x} = -\int_\Omega \nabla u \cdot \nabla \psi \, d^n\boldsymbol{x}. \tag{10.31}$$

On the other hand, the H^1 inner product on Ω is given by

$$\langle u, \psi \rangle_{H^1} := \langle u, \psi \rangle + \int_\Omega \nabla u \cdot \nabla \psi \, d^n\boldsymbol{x}. \tag{10.32}$$

The right-hand side of (10.31) is thus well-defined for $u \in H^1_0(\Omega)$.

To account for applications to the Helmholtz equation as well as the Laplace equation, let us consider the PDE

$$-\Delta u = \lambda u + f, \qquad u|_{\partial\Omega} = 0. \tag{10.33}$$

For $f \in L^2(\Omega)$, we say that $u \in H^1_0(\Omega)$ constitutes a weak solution of (10.33) if

$$\int_\Omega \Big[\nabla u \cdot \nabla \psi - \lambda u \psi - f\psi\Big] d^n\boldsymbol{x} = 0 \tag{10.34}$$

for every $\psi \in C^\infty_{\text{cpt}}(\Omega)$. This definition could be extended to more general elliptic equations of the form $Lu = f$, but for simplicity we restrict our attention to the case of the Laplacian. We will study the existence and regularity of solutions of (10.34) extensively in Chap. 11. For now, we consider a simple one-dimensional case to illustrate how the definition works.

Example 10.15 On the interval [0, 2], consider the equation

$$u'' = f, \quad u(0) = u(2) = 0,$$

with

$$f(x) = \begin{cases} x, & 0 \le x \le 1, \\ -1, & 1 \le x \le 2. \end{cases}$$

Since f is piecewise linear, it makes sense to try using classical solutions on the two subintervals. Imposing the boundary and continuity requirements gives a family of possible solutions

$$u(x) = \begin{cases} \frac{1}{6}x^3 - ax, & 0 \le x \le 1, \\ -\frac{1}{2}x^2 + (a + \frac{4}{3})x - 2a - \frac{2}{3}, & 1 \le x \le 2. \end{cases}$$

To determine a we apply the weak solution condition,

$$\int_0^2 \Big[u'\psi' + f\psi \Big] dx = 0, \tag{10.35}$$

for $\psi \in C^\infty_{\text{cpt}}(0, 2)$. Using integration by parts, the first term evaluates to

$$\int_0^2 u'\psi'\,dx = \int_0^1 (\tfrac{1}{2}x^2 - a)\psi'(x)\,dx + \int_1^2 (-x + a + \tfrac{4}{3})\psi'(x)\,dx$$

$$\begin{aligned} \int_0^2 u'\psi'\,dx &= (\tfrac{1}{2} - a)\psi(1) - \int_0^1 x\psi(x)\,dx - (a + \tfrac{1}{3})\psi(1) + \int_1^2 \psi(x)\,dx \\ &= (\tfrac{1}{6} - 2a)\psi(1) - \int_0^2 f\psi\,dx. \end{aligned}$$

The weak solution condition (10.35) requires $a = \frac{1}{12}$. This gives

$$u(x) = \begin{cases} \frac{1}{6}x^3 - \frac{1}{12}x, & 0 \le x \le 1, \\ -\frac{1}{2}x^2 + \frac{17}{12}x - \frac{5}{6}, & 1 \le x \le 2. \end{cases}$$

This result is illustrated in Fig. 10.6. Note that the condition on a corresponds to a matching of the first derivatives at $x = 1$, so that $u \in C^1[0, 2]$. ◊

We will show in Sect. 11.3 that solutions of (10.34) are unique, so the function obtained in Example 10.15 is the only possible solution. The matching of derivatives required for this solution is indicative of a more general regularity property for solutions of elliptic equation, which we will discuss in detail in Sect. 11.4.

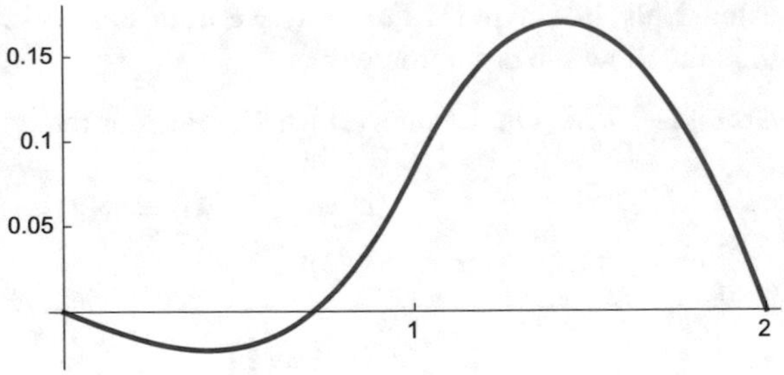

Fig. 10.6 One-dimensional weak solution

10.6 Weak Formulation of Evolution Equations

The heat and wave equations are the primary examples of linear evolution equations. Weak solutions for these equations can be defined by essentially the same strategy used in Sect. 10.5. Starting from a classical solution, we pair with a test function and use integration by parts to find the corresponding integral equation. Unfortunately, the time dependence creates some technicalities in the definition that we are not equipped to fully resolve here, but we can at least illustrate the basic philosophy by working through some examples.

Consider first the wave equation on a bounded domain $\Omega \subset \mathbb{R}^n$ with Dirichlet boundary conditions,

$$\frac{\partial^2 u}{\partial t^2} - \Delta u = 0, \quad u|_{x \in \partial\Omega} = 0, \tag{10.36}$$

subject to the initial conditions

$$u|_{t=0} = g, \quad \left.\frac{\partial u}{\partial t}\right|_{t=0} = h.$$

Assuming u is a classical solution, pairing the wave equation for u with a test function $\psi \in C^\infty_{\mathrm{cpt}}([0,\infty) \times \Omega)$ gives

$$\int_0^\infty \int_\Omega \left[\psi \frac{\partial^2 u}{\partial t^2} - \psi \Delta u\right] d^n\mathbf{x}\, dt = 0.$$

Integration by parts for the Δu term works just as in (10.31), yielding

$$\int_\Omega \psi \Delta u\, d^n\mathbf{x} = -\int_\Omega \nabla u \cdot \nabla \psi\, d^n\mathbf{x}.$$

In the t variable, we integrate by parts twice and pick up a boundary term each time because the test function is not assumed to vanish at $t = 0$. The result is

$$\begin{aligned}\int_0^\infty \psi \frac{\partial^2 u}{\partial t^2}\, dt &= -g \frac{\partial \psi}{\partial t}\bigg|_{t=0} - \int_0^\infty \frac{\partial \psi}{\partial t}\frac{\partial u}{\partial t}\, dt \\ &= -g \frac{\partial \psi}{\partial t}\bigg|_{t=0} + h\psi|_{t=0} + \int_0^\infty u \frac{\partial^2 \psi}{\partial t^2}\, dt.\end{aligned}$$

Combining this with the spatial integral yields

$$\begin{aligned}\int_0^\infty \int_\Omega \left[u \frac{\partial^2 \psi}{\partial t^2} + \nabla u \cdot \nabla \psi\right] d^n \boldsymbol{x}\, dt \\ = -\int_\Omega g \frac{\partial \psi}{\partial t}\bigg|_{t=0} d^n \boldsymbol{x} + \int_\Omega h\psi|_{t=0}\, d^n \boldsymbol{x}.\end{aligned} \tag{10.37}$$

As in Sect. 10.5, the Dirichlet boundary condition is imposed by assuming that $u(t, \cdot) \in H_0^1(\Omega)$ for all t. To make sense of the boundary and initial terms we need to assume at least that $g \in L^1_{\mathrm{loc}}(\Omega)$ and $h \in L^1_{\mathrm{loc}}[0, \infty)$. To interpret (10.37) we also need to require that the spatial pairing of ∇u with $\nabla \psi$ is integrable over t. This condition is more technical. It turns out to be sufficient to assume that $\|u(t, \cdot)\|_{H^1}$ is integrable as a function of t, but we will not attempt to justify this here. Instead we will limit our discussion to examples for which the existence of the integrals in (10.37) is clear.

Example 10.16 Consider the piecewise linear d'Alembert solution for the wave equation introduced in Sect. 1.2. On $[0, 2]$ we take the initial conditions $h = 0$ and

$$g(x) := \begin{cases} x, & 0 \le x \le 1, \\ 2 - x, & 1 \le x \le 2. \end{cases}$$

By Theorem 4.5, d'Alembert's solution is given by extending g to an odd periodic function on $\mathbb{R}$ with period 4, and then setting

$$u(t, x) = \frac{1}{2}\Big[g(x + t) + g(x - t)\Big].$$

The linear components of the resulting solution are shown in Fig. 10.7. Because u is piecewise linear and vanishes at $x = 0$ and 2, it is clear that $u(t, \cdot) \in H_0^1(0, 2)$ for each t.

For this case the weak solution condition (10.37) specializes to

$$\int_0^\infty \int_0^2 \left[u \frac{\partial^2 \psi}{\partial t^2} + \frac{\partial u}{\partial x}\frac{\partial \psi}{\partial x}\right] dx\, dt = -\int_0^2 g \frac{\partial \psi}{\partial t}\bigg|_{t=0} dx. \tag{10.38}$$

Checking this is essentially a matter of integration by parts, but the integrals must be broken into many pieces for large t. As a sample case, let us assume that ψ has support in $[0, 1) \times (0, 2)$.

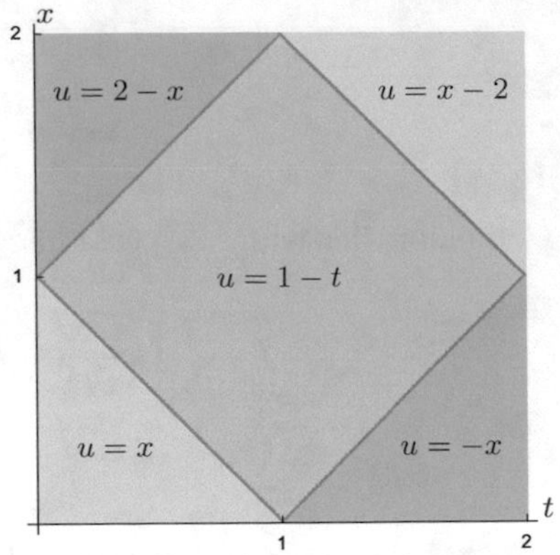

Fig. 10.7 Piecewise linear wave solution $u(t, x)$

The first integral in (10.38) becomes

$$\begin{aligned}\int_0^1\int_0^2 u\frac{\partial^2\psi}{\partial t^2}\,dx\,dt &= \int_0^1\Big[\int_0^{1-x} x\frac{\partial^2\psi}{\partial t^2}\,dt + \int_{1-x}^1 (1-t)\frac{\partial^2\psi}{\partial t^2}\,dt\Big]dx\\ &\quad \int_1^2\Big[\int_{x-1}^1 (1-t)\frac{\partial^2\psi}{\partial t^2}\,dt + \int_0^{x-1} (2-x)\frac{\partial^2\psi}{\partial t^2}\,dt\Big]dx\\ &= \int_0^1\Big[-x\frac{\partial\psi}{\partial t}(0,x) - \psi(1-x,x)\Big]dx\\ &\quad \int_1^2\Big[-(2-x)\frac{\partial\psi}{\partial t}(0,x) - \psi(x-1,x)\Big]dx\end{aligned}$$

Similarly, the second term in (10.38) evaluates to

$$\begin{aligned}\int_0^\infty\int_0^2 \frac{\partial u}{\partial x}\frac{\partial\psi}{\partial x}\,dx\,dt &= \int_0^1\int_0^{1-t}\frac{\partial\psi}{\partial x}\,dx\,dt - \int_0^1\int_{1+t}^2\frac{\partial\psi}{\partial x}\,dx\,dt\\ &= \int_0^1 \psi(t,1-t)\,dt + \int_0^1 \psi(t,1+t)\,dt\\ &= \int_0^1 \psi(1-x,x)\,dx + \int_1^2 \psi(x-1,x)\,dx.\end{aligned}$$

Adding these pieces together gives

$$\begin{aligned}\int_0^\infty\int_0^2\Big[u\frac{\partial^2\psi}{\partial t^2} + \frac{\partial u}{\partial x}\frac{\partial\psi}{\partial x}\Big]dx\,dt &= -\int_0^1 x\frac{\partial\psi}{\partial t}(0,x)\,dx - \int_1^2 (2-x)\frac{\partial\psi}{\partial t}(0,x)\,dx\\ &= -\int_0^2 g(x)\frac{\partial\psi}{\partial t}(0,x)\,dx,\end{aligned}$$

which verifies (10.38) for this case. ◊

Now let us consider the weak formulation of the heat equation with Dirichlet boundary conditions,

$$\frac{\partial u}{\partial t} - \Delta u = 0, \quad u|_{x \in \partial\Omega} = 0, \quad u|_{t=0} = h. \tag{10.39}$$

Derivation of the integral equation works just as for the wave equation, except that there is only a single integration by parts in the time variable. Assuming that $u(t, \cdot) \in H_0^1(\Omega)$ for each $t > 0$ and $h \in L^1_{\mathrm{loc}}(\Omega)$, the weak solution condition is

$$\int_0^\infty \int_\Omega \Big[-u\frac{\partial \psi}{\partial t} + \nabla u \cdot \nabla \psi\Big] d^n\boldsymbol{x}\, dt = \int_\Omega h\psi|_{t=0}\, d^n\boldsymbol{x} \tag{10.40}$$

for all $\psi \in C^\infty_{\mathrm{cpt}}([0, \infty) \times \Omega)$.

Example 10.17 Consider the heat equation on the interval $(0, \pi)$, with initial condition $h \in L^2(0, \pi; \mathbb{R})$. In view of the Dirichlet boundary conditions, we use the orthonormal basis for $L^2(0, \pi)$ developed in Exercise 8.4, given by the sine functions

$$\phi_k(x) := \sqrt{\frac{2}{\pi}} \sin(kx)$$

for $k \in \mathbb{N}$. The coefficients associated to h are

$$a_k := \int_0^\pi h(x)\phi_k(x)\, dx,$$

and $\sum a_k \phi_k$ converges to h in the L^2 sense by Theorem 8.6. The corresponding heat solution is

$$u(t, x) = \sum_{k=1}^\infty a_k e^{-k^2 t} \phi_k(x). \tag{10.41}$$

For example, solution corresponding to a step function h is illustrated in Fig. 10.8.

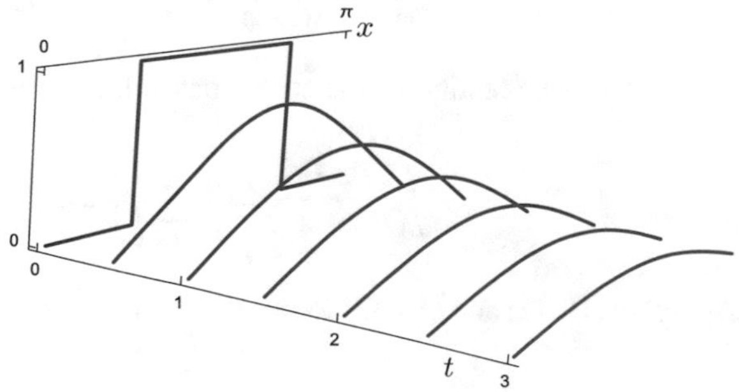

Fig. 10.8 Heat solution with L^2 initial data

As noted in Corollary 8.14, this yields a classical solution if h is continuous. If h is not continuous then u might not have a well-defined limit as $t \to 0$. Nevertheless, we can check that the weak solution condition is satisfied. Given a (real-valued) test function $\psi \in C^\infty_{\mathrm{cpt}}([0,\infty)\times(0,\pi);\mathbb{R})$, define the time-dependent Fourier coefficients

$$b_k(t) := \int_0^\pi \psi(t,x)\phi_k(x)\,dx.$$

By the smoothness of ψ, Theorem 8.10 implies that the coefficients satisfy $b_k(t) = O(k^{-\infty})$, uniformly in t, and so the series

$$\psi(t,x) = \sum_{k=1}^\infty b_k(t)\phi_k(x)$$

converges uniformly as well as in L^2. By the same principle, the series

$$\frac{\partial\psi}{\partial t}(t,x) = \sum_{k=1}^\infty b_k'(t)\phi_k(x)$$

is also uniformly convergent. Since $\{\phi_k\}$ is an orthonormal basis, we deduce from Parseval's identity (8.36) that

$$\int_0^\pi u\frac{\partial\psi}{\partial t}\,dt = \sum_{k=1}^\infty a_k e^{-k^2 t} b_k'(t) \tag{10.42}$$

for $t \ge 0$.

Similarly, for $t \ge 0$ we have L^2 convergent series

$$\begin{aligned}\frac{\partial\psi}{\partial x}(t,x) &= \sum_{k=1}^\infty b_k(t)k\sqrt{\frac{2}{\pi}}\cos(kx),\\ \frac{\partial u}{\partial x}(t,x) &= \sum_{k=1}^\infty a_k(t)e^{-k^2 t}k\sqrt{\frac{2}{\pi}}\cos(kx).\end{aligned}$$

By the Parseval identity for the cosine basis, this gives

$$\int_0^\pi \frac{\partial u}{\partial x}\frac{\partial\psi}{\partial x}\,dx = \sum_{k=1}^\infty k^2 a_k e^{-k^2 t} b_k(t). \tag{10.43}$$

Applying (10.42) and (10.43) to the left-hand side of (10.40) yields

$$\int_0^\infty \int_0^\pi \Big[-u\frac{\partial\psi}{\partial t} + \frac{\partial u}{\partial x}\frac{\partial\psi}{\partial x}\psi\Big] d^n\boldsymbol{x}\, dt$$
$$= \int_0^\infty \Big[\sum_{k=1}^{\infty} a_k e^{-k^2 t}\Big(k^2 b_k(t) - b_k'(t)\Big)\Big] dt. \tag{10.44}$$

Switching the order of the summation and integration is justified if series converges uniformly on the domain of the integral, but that is not necessarily the case here. To check this carefully, we break the sum at some value $k = N$. For the finite sum there is no convergence issue, so that

$$\int_0^\infty \Big[\sum_{k=1}^{N} a_k e^{-k^2 t}\Big(k^2 b_k(t) - b_k'(t)\Big)\Big] dt = \sum_{k=1}^{N} \int_0^\infty a_k \frac{d}{dt}\Big(-e^{-k^2 t} b_k(t)\Big) dt$$
$$= \sum_{k=1}^{N} a_k b_k(0).$$

To estimate the tail of the sum, note that the sequence $\{a_k\}$ is bounded because $\sum |a_k|^2 < \infty$. For $b_k(t)$, we apply repeated integration by parts to deduce

$$\int_0^\pi \phi_k(x)\Big(\frac{\partial}{\partial x}\Big)^{2m} \psi(t,x)\, dx = \int_0^\pi \psi(t,x)\Big(\frac{\partial}{\partial x}\Big)^{2m} \phi_k(x)\, dx$$
$$= (-1)^m k^{2m} \int_0^\pi \psi(t,x)\phi_k(x)\, dx$$
$$= (-1)^m k^{2m} b_k(t).$$

for $m \in \mathbb{N}$. Since $\psi \in C^\infty_{\text{cpt}}([0,\infty) \times (0,\pi))$, this gives an estimate

$$|b_k(t)| \le C_m k^{-2m},$$

where C_m is independent of t. The same reasoning applies to $b_k'(t)$. Combining the $m = 2$ estimate for b_k with the $m = 1$ case for b_k' gives

$$\left| a_k e^{-k^2 t}\Big(k^2 b_k(t) - b_k'(t)\Big)\right| \le C k^{-2}.$$

This shows that

$$\left| \sum_{k=N+1}^{\infty} a_k e^{-k^2 t}\Big(k^2 b_k(t) - b_k'(t)\Big)\right| \le C N^{-1}, \tag{10.45}$$

independently of t.

Now fix $M > 0$ so the support of ψ is contained in $[0, M]$. Applying (10.45) to the integral gives

$$\left| \int_0^M \left[\sum_{k=N+1}^{\infty} a_k e^{-k^2 t} \left(k^2 b_k(t) - b_k'(t) \right) \right] dt \right| \leq CMN^{-1}.$$

Returning to (10.44), our analysis of the sum over k now gives

$$\int_0^\infty \int_0^\pi \left[-u \frac{\partial \psi}{\partial t} + \frac{\partial u}{\partial x} \frac{\partial \psi}{\partial x} \psi \right] d^n \mathbf{x}\, dt = \sum_{k=1}^{N} a_k b_k(0) + O(N^{-1}).$$

By taking $N \to \infty$, we deduce

$$\int_0^\infty \int_0^\pi \left[-u \frac{\partial \psi}{\partial t} + \frac{\partial u}{\partial x} \frac{\partial \psi}{\partial x} \psi \right] d^n \mathbf{x}\, dt = \sum_{k=1}^{\infty} a_k b_k(0).$$

On the other hand Parseval's identity gives

$$\int_0^\pi h(x) \psi(0, x)\, dx = \sum_{k=1}^{\infty} a_k b_k(0),$$

so the weak solution condition (10.40) is satisfied. ♢

10.7 Exercises

10.1 On $\mathbb{R}$ consider the ordinary differential equation

$$x \frac{du}{dx} = 1.$$

(a) Develop a weak formulation of this ODE in terms of pairing with a test function $\psi \in C^\infty_{\text{cpt}}(\mathbb{R})$

(b) Show that $u(x) = \log |x|$ is locally integrable and solves the equation in the weak sense.

10.2 In Exercise 3.6 we studied Burger's equation,

$$\frac{\partial u}{\partial t} + u \frac{\partial u}{\partial x} = 0,$$

with the initial condition

$$u(0,x) = \begin{cases} a, & x \le 0, \\ a(1-x) + bx, & 0 < x < 1, \\ b, & x \ge 1. \end{cases}$$

For $a > b$ a shock forms at some positive time. Assuming u is a weak solution, find the equation of the shock curve starting from this point.

10.3 Consider the traffic equation

$$\frac{\partial u}{\partial t} + (1 - 2u)\frac{\partial u}{\partial x} = 0$$

with initial data

$$g(x) = \begin{cases} 0, & x > 0, \\ 1, & x < 0. \end{cases}$$

(a) Sketch the characteristic lines for this initial condition, and show that they leave a triangular region uncovered.

(b) Show that the constant solution $u(t,x) = g(x)$ satisfies the Rankine-Hugoniot condition for the shock curve $\sigma(t) = 0$ and thus gives a weak-solution of the traffic equation. (This solution is considered unphysical because characteristic lines emerge from the shock line to fill the triangular region.)

(c) The physical solution is specified by an *entropy* condition that says that characteristic lines may only intersect when followed forwards in time. Show that the continuous function

$$u(t,x) = \begin{cases} 0, & x > t, \\ \frac{1}{2} - \frac{x}{2t}, & -t < x < t, \\ 1, & x < -t, \end{cases}$$

satisfies the weak solution condition (10.9) (with $q = u - u^2$), as well as this entropy condition. (This type of solution is called a *rarefaction wave*.)

10.4 Define $f \in L^1_{\mathrm{loc}}(\mathbb{R}^n)$ by

$$f(\boldsymbol{x}) = \begin{cases} f_+(\boldsymbol{x}), & x_n > 0, \\ f_-(\boldsymbol{x}), & x_n < 0. \end{cases}$$

with $f_\pm \in C^1(\mathbb{R}^n)$.

(a) For $j = 1, \ldots, n-1$, show that f has weak partial derivatives given by $\frac{\partial f_\pm}{\partial x_j}$ for $\pm x_n > 0$.

(b) Show that the weak partial $\frac{\partial f}{\partial x_n}$ exists and is given by $\frac{\partial f_{\pm}}{\partial x_n}$ for $\pm x_n > 0$ only if f extends to a continuous function at $x_n = 0$.

10.5 Let $\mathbb{D} \subset \mathbb{R}^2$ be the unit disk $\{r < 1\}$ with $r := |\boldsymbol{x}|$. Consider the function $u(\boldsymbol{x}) = r^{\alpha}$ with $\alpha \in \mathbb{R}$ constant.

(a) Compute the ordinary partial derivatives $\frac{\partial u}{\partial x_j}$, $j = 1, 2$, for $r \neq 0$.

(b) Show that for $\alpha > -1$ these partials lie in $L^1(\mathbb{D})$ and define weak derivatives.

(c) For what values of α is $u \in H^1(\mathbb{D})$?

10.6 In $\mathbb{R}^3$ consider the equation

$$\Delta u = \begin{cases} 1, & r \leq a, \\ 0, & r > a, \end{cases}$$

with $r := |\boldsymbol{x}|$ and $a > 0$. (With appropriate physical constants this is the equation for the gravitational potential of a spherical planet of radius a.)

(a) Assuming that u depends only on r, formulate a weak solution condition in terms of pairing with a test function $\psi(r)$ with $\psi \in C^{\infty}_{\text{cpt}}[0, \infty)$.

(b) Find the unique solution which is smooth at $r = 0$ and for which $u(r) \to 0$ as $r \to \infty$.

10.7 Let $\Omega \subset \mathbb{R}^n$ be bounded with $\partial\Omega$ piecewise C^1. For $u \in C^2(\overline{\Omega})$ and $f \in C^0(\overline{\Omega})$, suppose that

$$\int_{\Omega} \left[\nabla u \cdot \nabla \psi - f\psi \right] d^n\boldsymbol{x} = 0. \tag{10.46}$$

for all $\psi \in C^{\infty}(\overline{\Omega})$. Show that u satisfies the Poisson equation with Neumann boundary condition,

$$-\Delta u = f, \qquad \left.\frac{\partial u}{\partial \nu}\right|_{\partial\Omega} = 0. \tag{10.47}$$

(Thus (10.46) allows a weak formulation of (10.47) for $u \in H^1(\Omega)$.)

Chapter 11
Variational Methods

Recall the formulas for the kinetic and potential energy of a solution of the wave equation derived in Sect. 4.7. At equilibrium the kinetic energy is zero, and by physical reasoning the system should occupy a state of minimum potential energy. This suggests a strategy of reformulating the Laplace equation, which models the equilibrium state, as a minimization problem for the energy.

In this application, the potential energy term from the wave equation is called the *Dirichlet energy*. For a bounded domain $\Omega \subset \mathbb{R}^n$ and $w \in C^2(\overline{\Omega})$, let

$$\mathcal{E}[w] := \frac{1}{2}\int_{\Omega} |\nabla w|^2 \, d^n\boldsymbol{x}. \tag{11.1}$$

The term *functional* is used to describe functions such as $\mathcal{E}[\cdot]$, to indicate that the domain is a function space.

To see how minimization of energy is related to the Laplace equation, let us suppose that $u \in C^2(\overline{\Omega}; \mathbb{R})$ satisfies

$$\mathcal{E}[u] \le \mathcal{E}[u + \varphi]$$

for all $\varphi \in C^{\infty}_{\text{cpt}}(\Omega; \mathbb{R})$. This implies that for $t \in \mathbb{R}$ the function $t \to \mathcal{E}[u + t\varphi]$ achieves a global minimum at $t = 0$. Hence

$$\frac{d}{dt}\mathcal{E}[u + t\varphi]\bigg|_{t=0} = 0. \tag{11.2}$$

Differentiation under the integral in the definition of $\mathcal{E}[u + t\varphi]$ gives

---

The original version of the book was revised: Belated corrections from author have been incorporated. The erratum to the book is available at https://doi.org/10.1007/978-3-319-48936-0_14

D. Borthwick, *Introduction to Partial Differential Equations*,
Universitext, DOI 10.1007/978-3-319-48936-0_11

$$\frac{d}{dt}\mathcal{E}[u+t\varphi]\bigg|_{t=0} = \frac{1}{2}\frac{d}{dt}\int_{\Omega}\left(|\nabla u|^2 + 2t\nabla u\cdot\nabla\varphi + t^2|\nabla\varphi|^2\right)d^n\boldsymbol{x}\bigg|_{t=0}$$
$$= \int_{\Omega}\nabla u\cdot\nabla\varphi\, d^n\boldsymbol{x}.$$

By Green's first identity (Theorem 2.10) and the fact that φ vanishes on $\partial\Omega$,

$$\int_{\Omega}\nabla u\cdot\nabla\varphi\, d^n\boldsymbol{x} = -\int_{\Omega}\varphi\Delta u\, d^n\boldsymbol{x}.$$

Thus (11.2) is equivalent to

$$\int_{\Omega}\varphi\Delta u\, d^n\boldsymbol{x} = 0.$$

This holds for all $\varphi\in C^{\infty}_{\mathrm{cpt}}(\Omega;\mathbb{R})$ if and only if $\Delta u = 0$ on Ω.

11.1 Model Problem: The Poisson Equation

The empirical law describing the electric field in the presence of a charge distribution was formulated by Gauss in the mid-19th century. Gauss's law states that the outward flux of the electric field through a closed surface is proportional to the total electric charge contained within the region bounded by the surface. More specifically, if $\Omega\subset\mathbb{R}^3$ is a bounded domain with piecewise C^1 boundary $\partial\Omega$ and ρ is the charge density within Ω, then

$$\int_{\partial\Omega}\boldsymbol{E}\cdot\boldsymbol{\nu}\, dS = 4\pi k\int_{\Omega}\rho\, d^3\boldsymbol{x},$$

where k is called Coulomb's constant.

Using the divergence theorem (Theorem 2.6), we can rewrite the flux integral as

$$\int_{\partial\Omega}\boldsymbol{E}\cdot\boldsymbol{\nu}\, dS = \int_{\Omega}\nabla\cdot\boldsymbol{E}\, d^3\boldsymbol{x},$$

so that Gauss's law becomes

$$\int_{\Omega}\nabla\cdot\boldsymbol{E}\, d^3\boldsymbol{x} = 4\pi k\int_{\Omega}\rho\, d^3\boldsymbol{x}.$$

This holds for an arbitrary region if and only if

$$\nabla\cdot\boldsymbol{E} = 4\pi k\rho \tag{11.3}$$

(the differential form of Gauss's law).

In Sect. 9.1 we noted that the electric potential ϕ and electric field $\boldsymbol{E}$ are related by

$$\boldsymbol{E} = -\nabla\phi.$$

Substituting this into (11.3) gives the *Poisson equation*

$$-\Delta\phi = 4\pi k\rho.$$

A common electrostatics problem is to find the electric potential caused by a distribution of charges within a region Ω bounded by a conducting material. The electric field must be perpendicular to a conducting surface, implying that the boundary restriction $\phi|_{\partial\Omega}$ is constant. If the boundary is connected then we can set this constant to 0, so the potential satisfies the Poisson equation with Dirichlet boundary conditions.

Other forms of the Poisson equation appear in various contexts. For example in Newtonian gravity the relationship between gravitational potential Φ and the mass density ρ is

$$\Delta\Phi = 4\pi G\rho,$$

where G is Newton's gravitational constant. In this application the domain is $\mathbb{R}^3$, and the physical assumption that $\Phi \to 0$ at infinity plays the role of a boundary condition.

11.2 Dirichlet's Principle

To solve Poisson's equation using a minimization argument, it proves to be very helpful to use the weak formulation of the equation. This is because $H_0^1(\Omega)$ is a Hilbert space. The completeness of $H_0^1(\Omega)$ with respect to the H^1 norm will play an essential role in establishing the existence of a minimizing function.

For $\Omega \subset \mathbb{R}^n$ bounded, and $f \in C^0(\Omega)$, the classical Poisson problem is to find a function $u \in C^2(\Omega) \cap C^0(\overline{\Omega})$ so that

$$-\Delta u = f, \quad u|_{\partial\Omega} = 0. \tag{11.4}$$

The weak formulation of (11.4) is a special case of (10.34), We take $f \in L^2(\Omega)$ and the goal is to find $u \in H_0^1(\Omega)$ such that

$$\int_\Omega \left[\nabla u \cdot \nabla\psi - f\psi\right] d^n\boldsymbol{x} = 0, \tag{11.5}$$

for every $\psi \in C^\infty_{\text{cpt}}(\Omega)$.

For convenience let us consider real-valued functions. (In the complex case we could split the Poisson problem into real and imaginary parts.) In view of (11.5), we define the functional

$$\mathcal{D}_f[w] := \mathcal{E}[w] - \langle f, w \rangle, \tag{11.6}$$

for $f \in L^2(\Omega; \mathbb{R})$ and $w \in H_0^1(\Omega; \mathbb{R})$, where $\mathcal{E}[\cdot]$ is the Dirichlet energy (11.1).

Theorem 11.1 (Dirichlet's principle) *Suppose $\Omega \subset \mathbb{R}^n$ is a bounded domain and $f \in L^2(\Omega; \mathbb{R})$. If $u \in H_0^1(\Omega; \mathbb{R})$ satisfies*

$$\mathcal{D}_f[u] \leq \mathcal{D}_f[w]$$

for all $w \in H_0^1(\Omega; \mathbb{R})$, then u is a weak solution of the Poisson equation, in the sense of (11.5).

Proof Since $C_{\mathrm{cpt}}^\infty(\Omega; \mathbb{R}) \subset H_0^1(\Omega)$, the assumption on u implies

$$\mathcal{D}_f[u] \leq \mathcal{D}_f[u + t\psi]$$

for $\psi \in C_{\mathrm{cpt}}^\infty(\Omega; \mathbb{R})$ and $t \in \mathbb{R}$. Therefore

$$\begin{aligned} 0 &= \frac{d}{dt}\mathcal{D}_f[u + t\psi]\Big|_{t=0} \\ &= \frac{d}{dt}\Big(\mathcal{E}[u + t\psi] - \langle f, u + t\psi \rangle\Big)\Big|_{t=0} \\ &= \int_\Omega \big[\nabla u \cdot \nabla \psi - f\psi\big]\, d^n\boldsymbol{x}. \end{aligned}$$

□

Dirichlet's principle is a classic example of a *variational* method. This terminology refers to the family of variations $u + t\psi$ used to derive the PDE from the minimization problem. In Sect. 11.3 we will show that $\mathcal{D}_f[\cdot]$ attains a minimum within $H_0^1(\Omega)$, guaranteeing the existence of a weak solution. Furthermore, in Sect. 11.4 we will see that the weak solution is actually a classical solution under certain conditions.

11.3 Coercivity and Existence of a Minimum

The functional $\mathcal{D}_f$ defined in (11.6) consists of a quadratic term plus a linear term. The Dirichlet minimization problem is thus analogous to minimizing the polynomial $ax^2 + bx$ for $x \in \mathbb{R}$. This polynomial obviously has a minimum if and only if $a > 0$. For the Dirichlet case the analogous condition is a lower bound on the quadratic term $\mathcal{E}[\cdot]$. The original form of this result was proven by Henri Poincaré.

Theorem 11.2 (Poincaré Inequality) *For a bounded domain $\Omega \subset \mathbb{R}^n$, there is a constant $\kappa > 0$, depending only on Ω, such that*

$$\|u\|_2^2 \le \kappa^2 \mathcal{E}[u]$$

for all $u \in H_0^1(\Omega)$.

Proof Because $C_{\text{cpt}}^\infty(\Omega)$ is dense in $H_0^1(\Omega)$, we can restrict our attention to smooth compactly supported functions at first and extend to the general case later. To illustrate the core argument we start with the $n = 1$ case.

Consider a bounded interval $(a, b) \subset \mathbb{R}$. For $\psi \in C_{\text{cpt}}^\infty(a, b)$, our goal is to compare the size of $\psi(x)$ to values of its derivative ψ'. The obvious connection between them comes from the fundamental theorem of calculus:

$$\psi(x) = \int_a^x \psi'(t)\, dt.$$

The right-hand side can be written as an L^2 pairing with the characteristic function of $[a, x]$,

$$\psi(x) = \left\langle \psi', \chi_{[a,x]} \right\rangle.$$

The Cauchy-Schwarz inequality (Theorem 7.1) then gives

$$\begin{aligned} |\psi(x)|^2 &\le \|\chi_{[a,x]}\|_2^2 \, \|\psi'\|_2^2 \\ &\le (b-a)\mathcal{E}[\psi], \end{aligned}$$

for all $x \in [a, b]$. We can integrate this estimate over x to obtain

$$\|\psi\|_2^2 \le (b-a)^2 \mathcal{E}[\psi]$$

for $\psi \in C_{\text{cpt}}^\infty(a, b)$.

Now let us consider the higher dimensional case $\Omega \subset \mathbb{R}^n$. The domain is assumed to be bounded, so

$$\Omega \subset \mathcal{R} := [-M, M]^n$$

for some large M. Functions in $C_{\text{cpt}}^\infty(\Omega)$ can be extended by zero to smooth functions on $\mathcal{R}$, so it suffices to derive the Poincaré inequality for $\psi \in C_{\text{cpt}}^\infty(\mathcal{R})$. Following the one-dimensional case, we apply the fundamental theorem of calculus in the x_1 variable to write

$$\psi(\boldsymbol{x}) = \int_{-M}^{x_1} \frac{\partial \psi}{\partial x_1}(y, x_2, \ldots, x_n)\, dy.$$

By the Cauchy-Schwarz inequality on $L^2(-M, M)$,

$$|\psi(\boldsymbol{x})|^2 \le 2M \int_{-M}^{M} \left| \frac{\partial \psi}{\partial x_1}(y, x_2, \ldots, x_n) \right|^2 dy,$$

for all $\boldsymbol{x} \in \mathcal{R}$. Integrating this estimate over $\boldsymbol{x} \in \mathcal{R}$ yields

$$\|\psi\|_2^2 \le 4M^2 \left\| \frac{\partial \psi}{\partial x_1} \right\|_2^2. \tag{11.7}$$

By the definition (11.1), the energy is given by

$$\begin{aligned} \mathcal{E}[\psi] &= \frac{1}{2} \int_{\mathcal{R}} \nabla\psi \cdot \nabla\psi \, d^n\boldsymbol{x} \\ &= \frac{1}{2} \int_{\mathcal{R}} \left[\left(\frac{\partial \psi}{\partial x_1} \right)^2 + \cdots + \left(\frac{\partial \psi}{\partial x_n} \right)^2 \right] d^n\boldsymbol{x} \\ &= \frac{1}{2} \sum_{j=1}^n \left\| \frac{\partial \psi}{\partial x_j} \right\|_2^2. \end{aligned}$$

Thus the bound (11.7) implies the estimate,

$$\|\psi\|_2^2 \le 8M^2 \mathcal{E}[\psi], \tag{11.8}$$

for $\psi \in C^\infty_{\mathrm{cpt}}(\Omega)$.

To complete the argument, suppose that $u \in H^1_0(\Omega)$. By the definition of H^1_0 there exists an approximating sequence $\{\psi_k\} \in C^\infty_{\mathrm{cpt}}(\Omega)$ such that $\psi_k \to u$ in the H^1 norm. By (11.8) the inequality

$$\|\psi_k\|_2^2 \le 8M^2 \mathcal{E}[\psi_k] \tag{11.9}$$

holds for each $k \in \mathbb{N}$. Our goal is thus to take the limit $k \to \infty$ on both sides.

For the energy side note that

$$\begin{aligned} \mathcal{E}[u] - \mathcal{E}[\psi_k] &= \sum_{j=1}^n \left(\left\langle \frac{\partial u}{\partial x_j}, \frac{\partial u}{\partial x_j} \right\rangle - \left\langle \frac{\partial \psi_k}{\partial x_j}, \frac{\partial \psi_k}{\partial x_j} \right\rangle \right) \\ &= \sum_{j=1}^n \left(\left\langle \frac{\partial (u - \psi_k)}{\partial x_j}, \frac{\partial u}{\partial x_j} \right\rangle + \left\langle \frac{\partial \psi_k}{\partial x_j}, \frac{\partial (u - \psi_k)}{\partial x_j} \right\rangle \right) \end{aligned}$$

Hence by the Cauchy-Schwarz inequality

$$|\mathcal{E}[u] - \mathcal{E}[\psi_k]| \le \sum_{j=1}^n \left\| \frac{\partial (u - \psi_k)}{\partial x_j} \right\|_2 \left(\left\| \frac{\partial u}{\partial x_j} \right\|_2 + \left\| \frac{\partial \psi_k}{\partial x_j} \right\|_2 \right).$$

By the definition of the H^1 norm this yields

$$|\mathcal{E}[u] - \mathcal{E}[\psi_k]| \le \|u - \psi_k\|_{H^1} \left(\|u\|_{H^1} + \|\psi_k\|_{H^1} \right).$$

In particular, $\psi_k \to u$ in H^1 implies that

$$\lim_{k\to\infty} \mathcal{E}[\psi_k] = \mathcal{E}[u].$$

Thus taking the limit $k \to \infty$ in (11.9) gives

$$\|u\|_2^2 \leq 8M^2\mathcal{E}[u].$$

□

The *Poincaré constant* $\kappa(\Omega)$ for a bounded domain Ω is defined to be the optimal choice of κ in Theorem 11.2. In other words,

$$\kappa(\Omega) := \sup_{u\in H_0^1(\Omega)\setminus\{0\}} \frac{\|u\|_2}{\mathcal{E}[u]^{\frac{1}{2}}}.$$

Our proof gives the rough estimate

$$\kappa(\Omega) \leq \sqrt{8}\ \mathrm{diam}(\Omega),$$

which is rather poor compared to the best known bounds. In Sect. 11.5 we will establish a direct relationship between the Poincaré constant and the lowest eigenvalue of Δ on Ω.

Since our goal is to minimize $\mathcal{D}_f[\cdot]$ over H_0^1, it is useful to express the conclusion of Theorem 11.2 in terms of the H^1 norm. Because

$$\|u\|_{H^1}^2 = \|u\|_2^2 + \mathcal{E}[u], \tag{11.10}$$

(11.8) is equivalent to

$$\|u\|_{H^1}^2 \leq (\kappa^2 + 1)\mathcal{E}[u] \tag{11.11}$$

for all $u \in H_0^1(\Omega)$. A quadratic functional on a Hilbert space is called *coercive* if its ratio to the norm squared is bounded below. The Poincaré inequality thus states that $\mathcal{E}[\cdot]$ is coercive on $H_0^1(\Omega)$.

The identity (11.10) also gives an upper bound,

$$\mathcal{E}[u] \leq \|u\|_{H^1}^2. \tag{11.12}$$

A quadratic functional is called *bounded* if its ratio to the norm squared is bounded above. For the energy this condition is automatic.

We are now prepared to tackle the minimization problem for $\mathcal{D}_f[\cdot]$, by exploiting the fact that $\mathcal{E}[\cdot]$ is both coercive and bounded.

Theorem 11.3 *For a bounded domain $\Omega \subset \mathbb{R}^n$ and $f \in L^2(\Omega)$ there is a unique function $u \in H_0^1(\Omega)$ such that*

$$\mathcal{D}_f[u] \le \mathcal{D}_f[w]$$

for all $w \in H_0^1(\Omega)$, where $\mathcal{D}_f[\cdot]$ is defined by (11.6).

Proof By the triangle inequality,

$$\mathcal{D}_f[w] \ge \mathcal{E}[w] - |\langle f, w\rangle| .$$

Applying (11.11) to the energy and the Cauchy-Schwarz inequality to the inner product gives

$$\begin{aligned} \mathcal{D}_f[w] &\ge \frac{1}{\kappa^2+1}\|w\|_{H^1}^2 - \|f\|_2\|w\|_2 \\ &\ge \frac{1}{\kappa^2+1}\|w\|_{H^1}^2 - \|f\|_2\|w\|_{H^1}. \end{aligned} \tag{11.13}$$

The right-hand side has the form $cx^2 - bx$ where $c = 1/(\kappa^2+1)$, $b = \|f\|_2$, and $x = \|w\|_{H^1}$. According to the minimization formula for a quadratic polynomial,

$$\min_{x\in\mathbb{R}}(cx^2 - bx) = -\frac{b}{4c}$$

for $c > 0$. Applying this to (11.13) gives

$$\mathcal{D}_f[w] \ge -\frac{\kappa^2+1}{4}\|f\|_2^2, \tag{11.14}$$

for $w \in H_0^1(\Omega)$.

If we set

$$d_0 := \inf_{w\in H_0^1(\Omega)} \mathcal{D}_f[w],$$

then (11.14) shows that $d_0 > -\infty$. By Lemma 2.1, there exists a sequence of $w_k \subset H_0^1(\Omega)$ so that

$$\lim_{k\to\infty} \mathcal{D}_f[w_k] = d_0. \tag{11.15}$$

Our strategy is to argue that the sequence $\{w_k\}$ is Cauchy in $H_0^1(\Omega)$, and therefore converges by completeness.

The quadratic structure of $\mathcal{E}[\cdot]$ implies that

$$\mathcal{E}\left[\frac{u+v}{2}\right] = \frac{1}{2}\mathcal{E}[u] + \frac{1}{2}\mathcal{E}[v] - \frac{1}{4}\mathcal{E}[u-v] \tag{11.16}$$

for all $u, v \in H_0^1(\Omega)$. This allows us to compute

$$\mathcal{D}_f\left[\frac{w_k + w_m}{2}\right] = \frac{1}{2}\mathcal{E}[w_k] + \frac{1}{2}\mathcal{E}[w_m] - \frac{1}{4}\mathcal{E}[w_k - w_m] - \frac{1}{2}\langle f, w_k + w_m\rangle$$
$$= \frac{1}{2}\mathcal{D}_f[w_k] + \frac{1}{2}\mathcal{D}_f[w_m] - \frac{1}{4}\mathcal{E}[w_k - w_m].$$

Because d_0 is the infimum of $\mathcal{D}_f[\cdot]$, this implies

$$d_0 \le \frac{1}{2}\mathcal{D}_f[w_k] + \frac{1}{2}\mathcal{D}_f[w_m] - \frac{1}{4}\mathcal{E}[w_k - w_m]. \tag{11.17}$$

Turning this inequality around gives

$$\mathcal{E}[w_k - w_m] \le 2\mathcal{D}_f[w_k] + 2\mathcal{D}_f[w_m] - 4d_0.$$

By (11.15),

$$\lim_{k,m\to\infty} \big(2\mathcal{D}_f[w_k] + 2\mathcal{D}_f[w_m] - 4d_0\big) = 0,$$

and since $\mathcal{E}[\cdot] \ge 0$ this yields

$$\lim_{k,m\to 0} \mathcal{E}[w_k - w_m] = 0. \tag{11.18}$$

Using the coercivity estimate (11.11), it follows from (11.18) that

$$\lim_{k,m\to 0} \|w_k - w_m\|_{H^1} = 0,$$

i.e., the sequence $\{w_k\}$ is Cauchy in $H_0^1(\Omega)$. Therefore, by completeness, there exists a function $u \in H_0^1(\Omega)$ such that

$$u := \lim_{k\to\infty} w_k.$$

Convergence in H^1 implies that

$$\mathcal{D}_f[u] = \lim_{k\to\infty} \mathcal{D}_f[w_k] = d_0.$$

Therefore u minimizes $\mathcal{D}_f[\cdot]$.

To see that the minimizing function is unique, suppose that both u_1 and u_2 satisfy $\mathcal{D}_f[u_j] = d_0$. By the same reasoning used to derive (11.17) we have

$$d_0 = \frac{1}{2}\mathcal{D}_f[u_1] + \frac{1}{2}\mathcal{D}_f[u_2] - \frac{1}{4}\mathcal{E}[u_1 - u_2].$$

By assumption both $\mathcal{D}_f[u_1]$ and $\mathcal{D}_f[u_2]$ are equal to d_0, so this implies

$$\mathcal{E}[u_1 - u_2] = 0.$$

Theorem 11.2 then implies $\|u_1 - u_2\|_2 = 0$, so that $u_1 \equiv u_2$. □

Note the crucial role that completeness plays in the proof of Theorem 11.3. If we had taken the domain of the Dirichlet energy to $C^2(\overline{\Omega})$, then there would be no way to deduce convergence of the sequence $\{w_k\}$ from the energy limit (11.18).

Corollary 11.4 *For $f \in L^2(\Omega)$ the weak formulation* (11.5) *of the Poisson equation admits a unique solution $u \in H^1_0(\Omega)$.*

Proof Existence of the solution follows from Theorems 11.1 and 11.3. To prove uniqueness, suppose that u_1 and u_2 both satisfy (11.5). Subtracting the equations gives

$$\int_\Omega \nabla(u_1 - u_2) \cdot \nabla\psi \, d^n\boldsymbol{x} = 0$$

for all $\psi \in C^\infty_{\text{cpt}}(\Omega)$. By the definition (10.18) of $H^1_0(\Omega)$ we can take a sequence of $\psi_k \in H^1_0(\Omega)$ such that $\psi_k \to u_1 - u_2$ in the H^1 norm. This implies in particular that $\nabla\psi_k \to \nabla(u_1 - u_2)$ in the L^2 sense, so that

$$\begin{aligned}\mathcal{E}[u_1 - u_2] &= \lim_{k\to\infty} \int_\Omega \nabla(u_1 - u_2) \cdot \nabla\psi_k \, d^n\boldsymbol{x} \\ &= 0.\end{aligned}$$

It follows from the Poincaré inequality (Theorem 11.2) that $\|u_1 - u_2\| = 0$, hence $u_1 \equiv u_2$. □

11.4 Elliptic Regularity

If $-\Delta u = f$ in the classical sense, then f has a level of differentiability 2 orders below that of u. Our goal in this section is to develop a converse to this statement that allows us to deduce the Sobolev regularity of a weak solution u from that of f. This type of regularity result holds for elliptic equations in general, but we will focus on the Laplacian for simplicity.

To avoid complications near the boundary of the domain, we introduce local versions of the Sobolev spaces. For $\Omega \subset \mathbb{R}^n$ let

$$H^m_{\text{loc}}(\Omega) := \left\{u \in L^1_{\text{loc}}(\Omega);\ u\psi \in H^m(\Omega) \text{ for all } \psi \in C^\infty_{\text{cpt}}(\Omega)\right\}.$$

Theorem 11.5 (Interior regularity) *Suppose that $u \in H^1_{\text{loc}}(\Omega)$ is a weak solution of*

$$-\Delta u = f.$$

If $f \in H^m_{\text{loc}}(\Omega)$ for $m \geq 0$ then $u \in H^{m+2}_{\text{loc}}(\Omega)$.

The Sobolev embedding theorem (Theorem 10.11) gives

$$H^m_{\text{loc}}(\Omega) \subset C^k(\Omega)$$

for $k < m - \frac{n}{2}$. Thus Theorem 11.5 shows that the weak solution of the Poisson equation obtained in Corollary 11.4 is a classical solution provided $f \in H^m_{loc}(\Omega)$ with $m > \frac{n}{2}$. Furthermore, if $f \in C^\infty(\Omega)$ then $u \in C^\infty(\Omega)$ also.

It is possible to include regularity up to the boundary, although this is technically much more difficult. For example, if $\partial\Omega$ is C^∞ and $u \in H^1_0(\Omega)$ is a weak solution of $-\Delta u = f$ for $f \in C^\infty(\Omega)$, then $u \in C^\infty(\overline{\Omega})$.

Our proof of Theorem 11.5 makes use of the Fourier analysis on $\mathbb{T}^n$ that we introduced in Sect. 10.4.

Lemma 11.6 *Suppose that $u \in H^1(\mathbb{T}^n)$ solves $-\Delta u = f$ in the weak sense. If $f \in H^m(\mathbb{T}^n)$ for $m \in \mathbb{N}_0$, then $u \in H^{m+2}(\mathbb{T}^n)$.*

Proof For $u \in H^1(\mathbb{T}^n)$ and $f \in L^2(\mathbb{T}^n)$, we assume that

$$\int_{\mathbb{T}^n} [\nabla u \cdot \nabla \psi - f\psi]\, d^n\boldsymbol{x} = 0, \tag{11.19}$$

for all $\psi \in C^\infty(\mathbb{T})$. Setting $\psi(\boldsymbol{x}) = e^{-i\boldsymbol{k}\cdot\boldsymbol{x}}$ in this equation gives

$$\int_{\mathbb{T}^n} [-i\boldsymbol{k} \cdot \nabla u(\boldsymbol{x}) - f(\boldsymbol{x})]\, e^{-i\boldsymbol{k}\cdot\boldsymbol{x}}\, d^n\boldsymbol{x} = 0.$$

Using 10.26, we can translate this into a relation between the Fourier coefficients,

$$\begin{aligned} c_{\boldsymbol{k}}[f] &= -\sum_{j=1}^{n} ik_j c_{\boldsymbol{k}}\left[\frac{\partial u}{\partial x_j}\right] \\ &= |\boldsymbol{k}|^2\, c_{\boldsymbol{k}}[u]. \end{aligned} \tag{11.20}$$

According to Theorem 10.13, if $f \in H^m(\mathbb{T}^n)$ then

$$\sum_{\boldsymbol{k}\in\mathbb{Z}^n} |\boldsymbol{k}|^{2m}\, |c_{\boldsymbol{k}}[f]|^2 < \infty.$$

By (11.20) this implies that

$$\sum_{k \in \mathbb{Z}^n} |k|^{2m+4} |c_k[u]|^2 < \infty,$$

which gives $u \in H^{m+2}(\mathbb{T}^n)$ by Theorem 10.13. □

We can now deduce the interior elliptic regularity result from Lemma 11.6 by localizing.

Proof of Theorem 11.5. Suppose that $-\Delta u = f$ in the weak sense of (11.5), with $u \in H_0^1(\Omega)$ and $f \in L^2_{\text{loc}}(\Omega)$. By rescaling and translating, if necessary, we can assume that $\Omega \subset [0, 2\pi]^n$, which allows us to identify Ω with a subset of $\mathbb{T}^n$. For $\chi \in C^\infty_{\text{cpt}}(\Omega)$ we can extend $u\chi$ by zero to a function on $[0, 2\pi]^n$. This allows us to consider $u\chi$ as a function in $H^1(\mathbb{T}^n)$. Our goal is to apply Lemma 11.6 to the localized function $u\chi$.

We must first show that $u\chi$ satisfies a weak Poisson equation on $\mathbb{T}^n$. Given $\psi \in C^\infty(\mathbb{T}^n)$ we can use the test function $\chi\psi \in C^\infty_{\text{cpt}}(\Omega)$ in the weak solution condition (11.5) to obtain

$$\int_{\mathbb{T}^n} [\nabla u \cdot \nabla(\chi\psi) - f\chi\psi] \, d^n\boldsymbol{x} = 0.$$

Using the product rule for the gradient, we can rewrite this as

$$\begin{aligned} 0 &= \int_{\mathbb{T}^n} [\chi \nabla u \cdot \nabla\psi + \psi \nabla u \cdot \nabla\chi - f\chi\psi] \, d^n\boldsymbol{x} \\ &= \int_{\mathbb{T}^n} [\nabla(u\chi) \cdot \nabla\psi - u\nabla\chi \cdot \nabla\psi + \psi\nabla u \cdot \nabla\chi - f\chi\psi] \, d^n\boldsymbol{x}. \end{aligned} \tag{11.21}$$

In order to interpret this as a weak equation for $u\chi$, we need to rewrite the second term as an integral involving ψ rather than $\nabla\psi$.

Since the components of $u\nabla\chi$ are in $C^\infty_{\text{cpt}}(\Omega)$, the definition of the weak derivative ∇u gives

$$\begin{aligned} \int_\Omega \nabla u \cdot (\psi\nabla\chi) \, d^n\boldsymbol{x} &= -\int_\Omega u\nabla \cdot (\psi\nabla\chi) \, d^n\boldsymbol{x} \\ &= -\int_\Omega [u\nabla\psi \cdot \nabla\chi + u\psi\Delta\chi] \, d^n\boldsymbol{x}. \end{aligned}$$

Using this formula on the term $u\nabla\chi \cdot \nabla\psi$ in (11.21) gives

$$0 = \int_{\mathbb{T}^n} \Big[\nabla(u\chi) \cdot \nabla\psi + \big(2\nabla u \cdot \nabla\chi + u\Delta\chi - f\chi\big)\psi\Big] \, d^n\boldsymbol{x}. \tag{11.22}$$

This holds for all $\psi \in C^\infty(\mathbb{T}^n)$, implying that $u\chi$ is a weak solution, in the sense of (11.19), of the equation

$$-\Delta(u\chi) = -2\nabla u \cdot \nabla\chi - u\Delta\chi + f\chi. \tag{11.23}$$

Because $\chi \in C^\infty_{\text{cpt}}(\Omega)$, the right-hand side of (11.23) lies in $L^2(\mathbb{T}^n)$ by the assumptions on u and f. Hence $u\chi \in H^2(\mathbb{T}^n)$ by Lemma 11.6. This holds for all $\chi \in C^\infty_{\text{cpt}}(\Omega)$, implying that $u \in H^2_{\text{loc}}(\mathbb{T}^n)$.

We can now apply the same argument inductively. If $u \in H^q_{\text{loc}}(\Omega)$, and $f \in H^{q-1}_{\text{loc}}(\Omega)$, then the right-hand side of (11.23) lies in $H^{q-1}(\mathbb{T}^n)$ for $\chi \in C^\infty_{\text{cpt}}(\Omega)$. Lemma 11.6 then gives $u \in H^{q+1}_{\text{loc}}(\Omega)$. □

11.5 Eigenvalues by Minimization

In Sect. 7.6, we mentioned that the spectral theorem for finite-dimensional matrices has an extension to certain differential operators. In this section, we will prove this result in the classical setting of the Laplacian on a bounded domain with Dirichlet boundary conditions.

Theorem 11.7 (Spectral theorem for the Dirichlet Laplacian) *Let $\Omega \subset \mathbb{R}^n$ be a bounded domain. There exists an orthonormal basis $\{\phi_k\}_{k\in\mathbb{N}}$ for $L^2(\Omega)$ such that*

$$-\Delta\phi_k = \lambda_k\phi_k$$

with $\phi_k \in H^1_0(\Omega;\mathbb{R}) \cap C^\infty(\Omega)$. Furthermore, $\lambda_k > 0$ for all k and

$$\lim_{k\to\infty} \lambda_k = \infty.$$

In the case of $L^2[0,\pi]$, the sequence of Dirichlet eigenfunctions is given by $\phi_k(x) = \sin(kx)$ for $k \in \mathbb{N}$, with eigenvalues $\lambda_k = k^2$. If follows from Theorem 8.6 that this sequence yields a basis, as shown in Exercise 8.4.

We can solve the eigenvalue problem in the general case by adapting the variational method used for the Poisson equation earlier in this chapter. This method gives more than just existence of the basis; it also suggests a natural strategy for numerical approximation of eigenvalues and eigenfunctions, which we will explore in Sect. 11.7.

The weak formulation of the eigenvalue equation $-\Delta\phi = \lambda\phi$ with Dirichlet boundary conditions is a special case of (10.34). For $\phi \in H^1_0(\Omega)$, the condition is that

$$\int_\Omega [\nabla\phi\cdot\nabla\psi - \lambda\phi\psi]\, d^n\boldsymbol{x} = 0 \tag{11.24}$$

for all $\psi \in C^\infty_{\text{cpt}}(\Omega)$.

If we substitute $\bar\psi$ in place of ψ then the second term in (11.24) is λ times the L^2 inner product $\langle\phi,\psi\rangle$. By the same token we could interpret the first term as an "inner product" form of the Dirichlet energy (11.1). We will denote this by

$$\mathcal{E}[u, v] := \int_\Omega \nabla u \cdot \overline{\nabla v}\, d^n \boldsymbol{x},$$

so that $\mathcal{E}[u] = \mathcal{E}[u, u]$. The L^2 and H^1 inner products are related by

$$\langle u, v\rangle_{H^1} = \langle u, v\rangle + \mathcal{E}[u, v].$$

With this convention we can write the weak eigenvalue equation (11.24) in an equivalent form as

$$\mathcal{E}[\phi, \psi] = \lambda\, \langle \phi, \psi\rangle \tag{11.25}$$

for all $\psi \in C^\infty_{\text{cpt}}(\Omega)$.

For the minimization argument, it will prove convenient to enlarge the space of test functions from $C^\infty_{\text{cpt}}(\Omega)$ to $H^1_0(\Omega)$. To justify this, note that by definition a function $v \in H^1_0(\Omega)$ can be approximated by a sequence of $\psi_k \in C^\infty_{\text{cpt}}(\Omega)$ with respect to the H^1 norm. This implies in particular that

$$\lim_{k\to\infty} \mathcal{E}[\phi, \psi_k] = \mathcal{E}[\phi, v], \qquad \lim_{k\to\infty} \langle \phi, \psi_k\rangle = \langle \phi, v\rangle\,.$$

We can thus conclude that ϕ solves (11.24) if and only if

$$\mathcal{E}[\phi, v] = \lambda\, \langle \phi, v\rangle \tag{11.26}$$

for all $v \in H^1_0(\Omega)$.

The formulation of the eigenvalue equation as a minimization problem is known as *Rayleigh's principle*, after the physicist Lord Rayleigh. For $v \in H^1_0(\Omega)$, $v \neq 0$, the ratio

$$\mathcal{R}[v] := \frac{\mathcal{E}[v]}{\|v\|_2^2}$$

is called the *Rayleigh quotient* for v. Note also that if ϕ satisfies (11.26) then

$$\mathcal{R}[\phi] = \lambda. \tag{11.27}$$

Furthermore, the Poincaré inequality (Theorem 11.2) shows that

$$\mathcal{R}[v] \geq \frac{1}{\kappa^2} > 0 \tag{11.28}$$

for $v \in H^1_0(\Omega)$, $v \neq 0$. This suggests that the smallest eigenvalue is related to the Poincaré constant, and that we can locate it by minimizing $\mathcal{R}[\cdot]$.

The argument for existence of a minimum is a little trickier than the analysis of the Dirichlet principle in Sect. 11.3. To understand why, note that

$$\mathcal{R}[cv] = \mathcal{R}[v]$$

for $c \in \mathbb{C}\backslash\{0\}$, so the minimizing function is not unique. Therefore it is quite possible to have a sequence that minimizes the Rayleigh quotient but does not converge in H^1.

The tool that allows us to resolve this issue was developed by Franz Rellich in the early 20th century.

Theorem 11.8 (Rellich's theorem) *Suppose* $\Omega \subset \mathbb{R}^n$ *is a bounded domain and* $\{v_k\}$ *is a sequence in* $H_0^1(\Omega)$ *that satisfies a uniform bound*

$$\|v_k\|_{H^1} \leq M$$

for all $k \in \mathbb{N}$. *Then* $\{v_k\}$ *has a subsequence that converges in* $L^2(\Omega)$.

It is important to note that Rellich's theorem refers to two different norms. The sequence is assumed bounded in the H^1 norm, and the subsequence is guaranteed to converge with respect to the L^2 norm. This is crucial to the result and not a mere technicality. We will defer the proof of Rellich's theorem to Sect. 11.6. The remainder of this section is devoted to its application to the Rayleigh minimization scheme.

Theorem 11.9 (First eigenvalue) *There exists* $\phi_1 \in H_0^1(\Omega; \mathbb{R}) \cap C^\infty(\Omega)$ *satisfying*

$$-\Delta\phi_1 = \lambda_1\phi_1$$

for $\lambda_1 > 0$, *such that*

$$\lambda_1 \leq \mathcal{R}[v] \tag{11.29}$$

for all $v \in H_0^1(\Omega)$, $v \neq 0$.

Proof By (11.28) the values of $\mathcal{R}[\cdot]$ are bounded below by the Poincaré constant. Therefore, the infimum

$$\lambda_1 := \inf_{v \in H_0^1(\Omega)\backslash\{0\}} \mathcal{R}[v]$$

exists and is strictly positive.

By Lemma 2.1 there exists a sequence $\{v_k\} \subset H_0^1(\Omega)\backslash\{0\}$ such that

$$\lim_{k\to\infty} \mathcal{R}[v_k] = \lambda_1. \tag{11.30}$$

After rescaling each v_k by a constant we can assume that $\|v_k\|_2 = 1$, so that

$$\mathcal{R}[v_k] = \mathcal{E}[v_k].$$

The sequence of energies $\mathcal{E}[v_k]$ is bounded by (11.30). This also implies that the sequence $\{v_k\}$ is bounded with respect to the H^1 norm, because

$$\|v_k\|_{H^1}^2 = 1 + \mathcal{E}[v_k]$$

by the relation (11.10).

According to Theorem 11.8 there exists a subsequence of $\{v_k\}$ that converges in the L^2 sense to some function $\phi_1 \in L^2$. By restrict our attention to this subsequence and relabeling if needed, we can assume that

$$\lim_{k\to\infty} \|v_k - \phi_1\|_2 = 0. \tag{11.31}$$

The next goal is to improve this to a statement of convergence in H^1. We use the same strategy as in Sect. 11.3, starting from

$$\mathcal{E}\left[\frac{v_k + v_m}{2}\right] = \frac{1}{2}\mathcal{E}[v_k] + \frac{1}{2}\mathcal{E}[v_m] - \frac{1}{4}\mathcal{E}[v_k - v_m]. \tag{11.32}$$

By the definition of λ_1,

$$\mathcal{R}\left[\frac{v_k + v_m}{2}\right] \geq \lambda_1.$$

Therefore

$$\mathcal{E}\left[\frac{v_k + v_m}{2}\right] \geq \frac{\lambda_1}{4}\|v_k + v_m\|_2^2.$$

Using this in (11.32) gives

$$\mathcal{E}[v_k - v_m] \leq 2\mathcal{E}[v_k] + 2\mathcal{E}[v_m] - \lambda_1\|v_k + v_m\|_2^2. \tag{11.33}$$

By (11.31), we have

$$\lim_{k,m\to\infty} \|v_k + v_m\|_2^2 = \|2\phi_1\|_2^2 = 4,$$

and by construction $\mathcal{E}[v_k] \to \lambda_1$. Hence (11.33) implies that

$$\lim_{k,m\to\infty} \mathcal{E}[v_k - v_m] \to 0.$$

Since we already know that $\{v_k\}$ converges in the L^2 norm, we conclude that

$$\lim_{k,m\to\infty} \|v_k - v_m\|_{H^1} \to 0.$$

That is, the sequence $\{v_k\}$ is Cauchy with respect to the H^1 norm.

By completeness $\{v_k\}$ converges with respect to the H^1 norm to some $u \in H_0^1(\Omega)$. Since the L^2 norm is bounded above by the H^1 norm, this means $v_k \to u$ in the L^2 sense also. Hence $u \equiv \phi_1$, which proves that $\phi_1 \in H_0^1(\Omega)$. It then follows from (11.30) that

$$\mathcal{E}[\phi_1] = \mathcal{R}[\phi_1] = \lambda_1.$$

To see that this implies the weak solution condition, suppose $w \in H_0^1(\Omega)$. Using the inner product form of $\mathcal{E}[\cdot]$, we can expand

$$\mathcal{E}[\phi_1 + tw] = \lambda_1 + 2t \operatorname{Re} \mathcal{E}[\phi_1, w] + t^2 \mathcal{E}[w], \tag{11.34}$$

for $t \in \mathbb{R}$. Similarly,

$$\|\phi_1 + tw\|_2^2 = 1 + 2t \operatorname{Re} \langle \phi_1, w \rangle + t^2 \|w\|_2^2. \tag{11.35}$$

By the definition of λ_1, the function $t \mapsto \mathcal{R}[\phi_1 + tw]$ has a minimum at $t = 0$, so that

$$\frac{d}{dt} \mathcal{R}[\phi_1 + tw]\bigg|_{t=0} = 0.$$

Computing this derivative using (11.34) and (11.35) gives

$$\frac{d}{dt} \mathcal{R}[\phi_1 + tw]\bigg|_{t=0} = 2 \operatorname{Re} \mathcal{E}[\phi_1, w] - 2\lambda_1 \operatorname{Re} \langle \phi_1, w \rangle .$$

We thus conclude that

$$\operatorname{Re} \mathcal{E}[\phi_1, w] = \lambda_1 \operatorname{Re} \langle \phi_1, w \rangle$$

for all $w \in H_0^1(\Omega)$. By replacing w by iw, we can deduce also that

$$\operatorname{Im} \mathcal{E}[\phi_1, w] = \lambda_1 \operatorname{Im} \langle \phi_1, w \rangle$$

for all $w \in H_0^1(\Omega)$. In combination these give (11.26), so ϕ_1 is a weak solution of

$$-\Delta \phi_1 = \lambda_1 \phi_1.$$

In principle, ϕ_1 could be complex-valued at this point, but since its real and imaginary parts each satisfy (11.26) separately, we can select one of these to specialize to the real-valued case.

To deduce the regularity of ϕ_1, we apply Theorem 11.5 with $f = \lambda\phi_1$. The fact that $\phi_1 \in H_0^1(\Omega)$ then implies that $\phi_1 \in H_{\mathrm{loc}}^3(\Omega)$. Starting from $H_{\mathrm{loc}}^3(\Omega)$ then gives $\phi_1 \in H_{\mathrm{loc}}^5(\Omega)$, and so on. This inductive argument shows that $\phi_1 \in H_{\mathrm{loc}}^q(\Omega)$ for each $q \in \mathbb{N}$. We conclude that $\phi_1 \in C^\infty(\Omega)$ by Theorem 10.11. □

It is clear that the λ_1 produced in Theorem 11.9 is the smallest eigenvalue, since all eigenvalues occur as values of the Rayleigh quotient by (11.27). An example of ϕ_1 is shown in Fig. 11.1. We will see in the exercises that the first eigenfunction cannot have zeros in Ω. This can be used to show that the first eigenfunction is unique up to a multiplicative constant, i.e., λ_1 has multiplicity one.

To find other eigenvalues, the strategy is to restrict to subspaces and then apply the same construction used for λ_1. For a subset $A \in L^2(\Omega)$ the *orthogonal complement* is

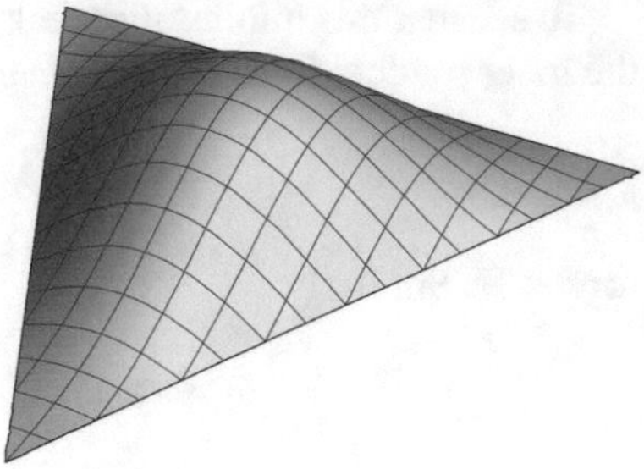

Fig. 11.1 The first eigenfunction of a equilateral triangle domain

$$A^\perp := \left\{w \in L^2(\Omega);\ \langle w, v\rangle = 0 \text{ for all } v \in A\right\}.$$

The orthogonal complement of a set is a subspace, by the linearity of the inner product.

In the argument below we will consider subspaces of $H_0^1(\Omega)$ of the form

$$W = A^\perp \cap H_0^1(\Omega),$$

where A is a finite list of eigenfunctions. We claim that W is closed as a subspace of $H_0^1(\Omega)$. To see this, suppose that $w_k \to w$ in the H^1 norm, with $w_k \in W$. Since $\|w_k - w\|_2 \le \|w_k - w\|_{H^1}$ by (11.10), this implies that $w_k \to w$ with respect to the L^2 norm also. Thus for $v \in A$,

$$\langle w, v\rangle = \lim_{k\to\infty} \langle w_k, v\rangle = 0.$$

This shows that $w \in W$. Therefore W is closed. By Lemma 7.8, this implies that W is a Hilbert space with respect to the H^1 inner product

Proof of Theorem 11.7 Let $\phi_1 \in H_0^1(\Omega;\mathbb{R})$ be the eigenvector obtained in Theorem 11.9, normalized so that $\|\phi_1\|_2 = 1$. The subspace

$$W_1 := \{\phi_1\}^\perp \cap H_0^1(\Omega)$$

is a Hilbert space with respect to the H^1 norm, by the remarks above.

Applying the minimization procedure used in Theorem 11.9 to the restriction of the Rayleigh quotient to W_1 gives $\phi_2 \in W_1$ such that $\|\phi_2\|_2 = 1$ and

$$\lambda_2 := \mathcal{R}[\phi_2] \le \mathcal{R}[w] \tag{11.36}$$

for all $w \in W_1 \backslash \{0\}$. By the same variational argument used for ϕ_1, this implies that

$$\mathcal{E}[\phi_2, w] = \lambda_2 \langle \phi_2, w\rangle \tag{11.37}$$

for all $w \in W_1$.

To extend this formula to the full weak solution condition, note that $\langle \phi_2, \phi_1 \rangle = 0$ because $\phi_2 \in W_1$. The fact that ϕ_1 satisfies (11.26) also gives

$$\mathcal{E}[\phi_1, \phi_2] = \lambda_1 \langle \phi_1, \phi_2 \rangle = 0. \tag{11.38}$$

Now consider a general $v \in H_0^1(\Omega)$. For $c := \langle v, \phi_1 \rangle$ we have

$$\langle v - c\phi_1, \phi_1 \rangle = 0,$$

so that $v - c\phi_1 \in W_1$. After setting $w := v - c\phi_1$, we can expand

$$\begin{aligned} \mathcal{E}[\phi_2, v] - \lambda_2 \langle \phi_2, v \rangle_{L^2} &= c\Big(\mathcal{E}[\phi_2, \phi_1] - \lambda_2 \langle \phi_2, \phi_1 \rangle\Big) \\ &\quad + \mathcal{E}[\phi_2, w] - \lambda_2 \langle \phi_2, w \rangle \end{aligned}$$

The first line on the right is zero by (11.38) and the second is zero by (11.37). Thus (11.26) is satisfied for all $v \in H_0^1(\Omega)$, showing that $-\Delta\phi_2 = \lambda_2\phi_2$ in the weak sense. By taking the real or imaginary part we can assume that ϕ_2 is real-valued.

Subsequent eigenvalues are obtained by repeating this process inductively. After k eigenfunctions have been found, we set

$$W_k := \{\phi_1, \ldots, \phi_k\}^\perp \cap H_0^1(\Omega), \tag{11.39}$$

and minimize the Rayleigh quotient over W_k to find λ_{k+1} and ϕ_{k+1}. Note that W_k always contains nonzero vectors, because $H_0^1(\Omega)$ is infinite-dimensional. The regularity argument from the end of the proof of Theorem 11.9 applies to any solution of (11.26), so that $\phi_k \in C^\infty(\Omega)$ for each k.

This process produces an orthonormal sequence of eigenfunctions $\{\phi_k\}$ with eigenvalues satisfying

$$\lambda_1 \le \lambda_2 \le \lambda_3 \le \ldots.$$

To see that $\lambda_k \to \infty$, note that $\|\phi_k\|_2 = 1$ and $\mathcal{R}[\phi_k] = \lambda_k$ by construction, so that

$$\|\phi_k\|_{H^1}^2 = 1 + \lambda_k. \tag{11.40}$$

Suppose the sequence $\{\lambda_k\}$ is bounded. Then (11.40) shows that the sequence $\{\phi_k\}$ is bounded with respect to the H^1 norm. Theorem 11.8 then implies that a subsequence of $\{\phi_k\}$ converges in $L^2(\Omega)$. But the ϕ_k are orthonormal with respect to the L^2 norm, so that

$$\|\phi_k - \phi_m\|_2 = \sqrt{2}$$

for all $k \neq m$. Convergence of a subsequence of $\{\phi_k\}$ is therefore impossible in $L^2(\Omega)$. This contradiction shows that $\{\lambda_k\}$ cannot be bounded. Since the sequence is increasing, this implies

$$\lim_{k\to\infty} \lambda_k = \infty.$$

The final claim is that $\{\phi_k\}$ forms an orthonormal basis of $L^2(\Omega)$. After obtaining the full sequence from the inductive procedure, let us set

$$W_\infty := \{\phi_1, \phi_2, \dots\}^\perp \cap H_0^1(\Omega).$$

Suppose that W_∞ contains a nonzero vector. Applying the Rayleigh quotient minimization as above produces yet another eigenvalue λ. Since the λ_k's were constructed by minimizing the Rayleigh quotient on subspaces $W_k \supset W_\infty$, this new eigenvalue satisfies $\lambda \geq \lambda_k$ for all $k \in \mathbb{N}$. This is impossible because $\lambda_k \to \infty$. Hence $W_\infty = \{0\}$.

In other words, the only vector in $H_0^1(\Omega)$ that is orthogonal to all of the ϕ_k is 0. Since $C^\infty_{\text{cpt}}(\Omega) \subset H_0^1(\Omega)$ and $C^\infty_{\text{cpt}}(\Omega)$ is dense in $L^2(\Omega)$ by Theorem 7.5, this implies that the only vector in $L^2(\Omega)$ that is orthogonal to all of the ϕ_k is 0. Hence $\{\phi_k\}$ is a basis by Theorem 7.10. □

11.6 Sequential Compactness

In this section we take up the proof of Rellich's theorem (Theorem 11.8). Results of this type, that force convergence of an approximating sequence, are a crucial component of variational strategies for PDE.

In a normed vector space, a subset A is said to be *sequentially compact* if every sequence within A contains a subsequence converging to a limit in A. A fundamental result in analysis called the Bolzano-Weierstrass theorem (Theorem A.1) says that in $\mathbb{R}^n$ this is equivalent to the definition of compact given in Sect. 2.3. That is, a subset of $\mathbb{R}^n$ is sequentially compact if and only if it is closed and bounded.

Rellich's theorem could be paraphrased as the statement that a closed and bounded subset of $H_0^1(\Omega)$ is sequentially compact as a subset of $L^2(\Omega)$, provided we are careful about the two different norms referenced in this statement. Our strategy will be to reduce Rellich's theorem to an application of Bolzano-Weierstrass using Fourier series. We start with the periodic case.

Theorem 11.10 *Suppose that $\{v_j\}$ is a sequence in $H^1(\mathbb{T}^n)$ that satisfies a uniform bound*

$$\|v_j\|_{H^1} \leq M \tag{11.41}$$

for all $j \in \mathbb{N}$. Then $\{v_j\}$ has a subsequence that converges in $L^2(\mathbb{T}^n)$.

Proof The argument is essentially the same in any dimension, so let us take $n = 1$ to simplify the notation. Suppose that $\{v_j\}$ is a sequence in $H^1(\mathbb{T})$ satisfying (11.41). The periodic Fourier coefficients are defined by

$$c_k[v_j] := \frac{1}{2\pi}\int_{-\pi}^{\pi} v_j(x)e^{-ikx}\,dx \tag{11.42}$$

for $k \in \mathbb{Z}$, with corresponding partial sums

$$S_m[v_j] = \sum_{k=-m}^{m} c_k[v_j]e^{ikx}.$$

Applying the Cauchy-Schwarz inequality (Theorem 7.1) to (11.42) gives

$$\left|c_k[v_j]\right| \le \frac{1}{\sqrt{2\pi}}\|v_j\|_2.$$

The assumption (11.41) implies also that the L^2 norms $\|v_j\|_2$ are bounded by M, so that

$$\left|c_k[f_j]\right| \le \frac{M}{\sqrt{2\pi}}$$

for all $j \in \mathbb{N}$ and $k \in \mathbb{Z}$.

We start the process of finding a convergent subsequence by making the first few coefficients converge. The collection of points $(c_{-1}[v_j], c_0[v_j], c_1[v_j])$ forms a bounded sequence in $\mathbb{C}^3$. Applying Bolzano-Weierstrass (Theorem A.1) to this sequence gives a point $(a_{-1}, a_0, a_1) \subset \mathbb{C}^3$ and a subsequence

$$\left\{v_j^{(1)}\right\} \subset \{v_j\},$$

such that

$$\lim_{j\to\infty} c_k\left[v_j^{(1)}\right] = a_k$$

for $k = -1, 0, 1$. Since the partial sum S_1 involves only these three coefficients, this implies the uniform convergence

$$\lim_{j\to\infty} S_1\left[v_j^{(1)}\right] = \sum_{k=-1}^{1} a_k e^{ikx},$$

which also gives convergence in the L^2 sense.

The same reasoning can be applied to the Fourier coefficients with $k = -2, \dots, 2$ to obtain a subsequence

$$\left\{v_j^{(2)}\right\} \subset \left\{v_j^{(1)}\right\},$$

such that

$$\lim_{j\to\infty} c_k\left[v_j^{(2)}\right] = a_k$$

for $k = -2, \dots, 2$. This process can be continued inductively to produce a family of subsequences $\left\{v_j^{(l)}\right\}$ such that

$$\lim_{j\to\infty} S_l\left[v_j^{(l)}\right] = \sum_{k=-l}^{l} a_k e^{ikx} \tag{11.43}$$

in $L^2(\mathbb{T})$.

To complete the proof, set

$$w_j := v_j^{(j)}.$$

Because w_j is an element of the lth subsequence for $l \le j$, we deduce from (11.43) that

$$\lim_{j\to\infty} S_m\left[w_j\right] = \sum_{k=-m}^{m} a_k e^{ikx} \tag{11.44}$$

in $L^2(\mathbb{T})$ for all $m \in \mathbb{N}$.

We now claim that the sequence w_j converges in $L^2(\mathbb{T})$. In order to deduce this from (11.44) we need to control the rate at which $S_m[w_j]$ converges to w_j as $m \to \infty$. This is where the H^1 bound (11.41) becomes crucial. In terms of Fourier coefficients, the H^1 norm could be written

$$\|f\|_{H^1}^2 = \sum_{k=-\infty}^{\infty} (1+k^2)\,|c_k[f]|^2\,.$$

Hence (11.41) implies that

$$\sum_{k=-\infty}^{\infty} (1+k^2)\left|c_k[w_j]\right|^2 \le M^2$$

for all j. This leads to an estimate:

$$\begin{aligned}\|w_j - S_m[w_j]\|_2^2 &= \sum_{|k|>m} \left|c_k[w_j]\right|^2 \\ &\le \sum_{|k|>m} \frac{1+k^2}{1+m^2}\left|c_k[w_j]\right|^2 \\ &\le \frac{M^2}{1+m^2},\end{aligned} \tag{11.45}$$

independent of j.

By the triangle inequality,

$$\|w_i - w_j\|_2 \le \|w_i - S_m[w_i]\|_2 + \|S_m[w_i] - S_m[w_j]\|_2 + \|w_j - S_m[w_j]\|_2.$$

Given $\varepsilon > 0$, fix m large enough that

$$\frac{M}{\sqrt{1+m^2}} < \varepsilon.$$

By (11.45) this reduces our triangle estimate to

$$\|w_i - w_j\|_2 < \|S_m[w_i] - S_m[w_j]\|_2 + 2\varepsilon. \tag{11.46}$$

Since $S_m[w_j]$ converges in $L^2(\mathbb{T})$ by (11.44), we can choose N sufficiently large so that

$$\|S_m[w_i] - S_m[w_j]\|_{L^2} < \varepsilon$$

for all $i, j \geq N$. By (11.46) this implies that

$$\|w_i - w_j\|_2 < 3\varepsilon$$

for $i.j \geq N$. This shows that the subsequence $\{w_j\}$ is Cauchy with respect to the L^2 norm. By completeness the subsequence converges in $L^2(\mathbb{T})$. □

Proof of Theorem 11.8 We can assume $\overline{\Omega} \subset (-\pi, \pi)^n$, after rescaling if needed. By Lemma 10.10, elements of $H_0^1(\Omega)$ can be extended by zero to $H^1([-\pi, \pi]^n)$. We can then make these functions periodic to give the inclusion

$$H_0^1(\Omega) \subset H^1(\mathbb{T}^n). \tag{11.47}$$

Given a sequence $\{v_j\} \subset H_0^1(\Omega)$ that is uniformly bounded in the H^1 norm, applying Theorem 11.10 to the extension given by (11.47) gives a subsequence $\{w_j\}$ that converges in $L^2(\mathbb{T}^n)$. By construction the restriction of w_j to $[-\pi, \pi]^n$ vanishes outside Ω, so this also gives a convergent subsequence in $L^2(\Omega)$. □

11.7 Estimation of Eigenvalues

As we saw in Sect. 11.5, the Rayleigh principle for the eigenvalue problem is very useful as a theoretical tool. It also leads to some very practical applications in terms of estimating eigenvalues or calculating them numerically.

The basic strategy is to exploit the formula for eigenvalues that appeared in the proof of Theorem 11.7. Assuming the Dirichlet eigenvalues $\{\lambda_k\}$ of Ω are written in increasing order and repeated according to multiplicity,

$$\lambda_k = \min_{w \in W_{k-1} \setminus \{0\}} \mathcal{R}[w], \tag{11.48}$$

where

$$W_{k-1} := \{\phi_1, \ldots, \phi_{k-1}\}^\perp \cap H_0^1(\Omega),$$

with ϕ_k the eigenfunction corresponding to λ_k The only problem with this formula is that determining the kth eigenfunction requires knowledge of the first $k-1$ eigenfunctions. This issue is resolved by the following:

Theorem 11.11 (Minimax principle) *For a bounded domain $\Omega \subset \mathbb{R}^n$, let Λ_k denote the set of all k-dimensional subspaces of $H_0^1(\Omega)$. Let $\{\lambda_k\}$ denote the sequence of Dirichlet eigenvalues of Ω in increasing order. Then*

$$\lambda_k = \min_{V \in \Lambda_k} \left\{ \max_{u \in V \setminus \{0\}} \mathcal{R}[u] \right\} \tag{11.49}$$

for each $k \in \mathbb{N}$.

Proof Let $\{\phi_k\} \subset H_0^1(\Omega)$ denote the eigenfunction basis. By the weak solution condition (11.25), orthonomality in $L^2(\Omega)$ implies also that

$$\mathcal{E}[\phi_i, \phi_j] = \begin{cases} \lambda_i, & i = j, \\ 0, & i \neq j. \end{cases} \tag{11.50}$$

Let us set

$$V_{\text{eig}} = [\phi_1, \ldots, \phi_k] \subset H_0^1(\Omega),$$

where $[\ldots]$ denotes the linear span of a collection of vectors. For $u \in V_{\text{eig}}$, expanded as

$$u = \sum_{j=1}^{k} c_j \phi_k,$$

we see from (11.50) that

$$\mathcal{E}[u] = \sum_{j=1}^{k} \lambda_j \left|c_j\right|^2.$$

The Rayleigh quotient is thus

$$\mathcal{R}[u] = \frac{\sum_{j=1}^{k} \lambda_j \left|c_j\right|^2}{\sum_{j=1}^{k} \left|c_j\right|^2}.$$

Since $\lambda_1 \leq \cdots \leq \lambda_k$ by assumption, it is clear that

$$\mathcal{R}[u] \leq \lambda_k$$

for $u \in V_{\text{eig}} \setminus \{0\}$. Moreover, $\mathcal{R}[\phi_k] = \lambda_k$, so that

$$\max_{u \in V_{\text{eig}} \setminus \{0\}} \mathcal{R}[u] = \lambda_k. \tag{11.51}$$

Now consider a general subspace $V \in \Lambda_k$. Since $\dim[\phi_1, \ldots, \phi_{k-1}] = k - 1$, there exists a nonzero vector

$$w \in V \cap [\phi_1, \ldots, \phi_{k-1}]^\perp.$$

Since $w \in [\phi_1, \ldots, \phi_{k-1}]^\perp$, it follows from (11.48) that

$$\lambda_k \leq \mathcal{R}[w].$$

Hence,

$$\lambda_k \leq \max_{u \in V \setminus \{0\}} \mathcal{R}[u] \tag{11.52}$$

for $V \in \Lambda_k$.

In combination, (11.51) and (11.52) show that the minimum on the right-hand side of (11.49) exists and is equal to λ_k. □

As a sample application, we can use the minimax principle to compare eigenvalues of nested domains. This is possible because of the inclusion

$$H_0^1(\Omega) \subset H_0^1(\tilde{\Omega}), \tag{11.53}$$

provided by Lemma 10.10.

Corollary 11.12 *Consider two bounded domains in $\mathbb{R}^n$ satisfying $\Omega \subset \tilde{\Omega}$. Assuming the Dirichlet eigenvalue sequences are arranged in increasing order,*

$$\lambda_k(\tilde{\Omega}) \leq \lambda_k(\Omega)$$

for all $k \in \mathbb{N}$.

Proof Let $\{\phi_k\} \subset H_0^1(\Omega)$ be the sequence of eigenfunctions of Ω. By (11.53), we can consider $[\phi_1, \ldots, \phi_k]$ to be a subspace of $H_0^1(\tilde{\Omega})$. If $\tilde{\Lambda}_k$ denotes the set of k-dimensional subspaces of $H_0^1(\tilde{\Omega})$, then this implies

$$\min_{W \in \tilde{\Lambda}_k} \left\{ \max_{u \in W \setminus \{0\}} \mathcal{R}[u] \right\} \leq \max_{u \in [\phi_1, \ldots, \phi_k] \setminus \{0\}} \mathcal{R}[u]$$

By Theorem 11.11, the left-hand side is $\lambda_k(\tilde{\Omega})$, while the right-hand side equals $\lambda_k(\Omega)$ by (11.51). □

Another way to make use the Rayleigh and minimax principles is to approximate eigenvalues by restricting subspaces of computationally simple functions within $H_0^1(\Omega)$. This approach can give surprisingly good estimates even when the subspaces are small.

Example 11.13 On the unit disk $\mathbb{D} = \{r < 1\} \subset \mathbb{R}^2$, consider the family of functions

$$w_\alpha(r) := 1 - r^\alpha$$

for $\alpha > 0$. Because w_α is radial, the energy can be computed by

$$\begin{aligned}\mathcal{E}[w_\alpha] &= \int_0^1 \left(\frac{\partial w_\alpha}{\partial r}\right)^2 2\pi r\, dr \\ &= 2\pi\alpha^2 \int_0^1 r^{2\alpha-1}\, dr \\ &= \pi\alpha.\end{aligned}$$

The L^2 norm is

$$\begin{aligned}\|w_\alpha\|_2^2 &= \int_0^1 (1 - r^\alpha)^2\, 2\pi r\, dr \\ &= \frac{\pi\alpha^2}{2 + 3\alpha + \alpha^2}.\end{aligned}$$

Hence the Rayleigh quotient gives the bound

$$\lambda_1 \le \mathcal{R}[w_\alpha] = \frac{2}{\alpha} + 3 + \alpha$$

for $\alpha > 0$. The optimal choice is $\alpha = \sqrt{2}$, which gives

$$\lambda_1 \le 3 + 2\sqrt{2} \doteq 5.828. \tag{11.54}$$

Compare this to the exact value computed in Example 5.5 in terms of zeros of the Bessel J-function,

$$\lambda_1 = j_{0,1}^2 \doteq 5.783. \tag{11.55}$$

The optimal choice of w_α gives a reasonable approximation to the true eigenfunction, as Fig. 11.2 demonstrates. ◊

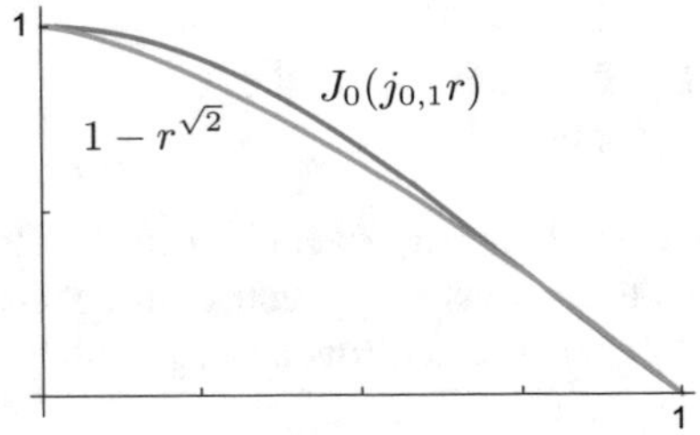

Fig. 11.2 The Bessel function for the first eigenfunction and $w_{\sqrt{2}}$

We can estimate higher eigenvalues and improve the accuracy by using a larger subspace. This computational strategy was introduced by Walter Ritz in 1909, and is referred to as the *Rayleigh-Ritz method*. Given a finite dimensional subspace $A \subset H_0^1(\Omega)$, we let $\Lambda_k(A)$ denote the k-dimensional subspaces of A. The approximate eigenvalues associated to A are then given by

$$\eta_k := \min_{W \in \Lambda_k(A)} \left\{ \max_{u \in W \setminus \{0\}} \mathcal{R}[u] \right\}, \tag{11.56}$$

for $k = 1, \ldots, \dim A$.

Since A is finite dimensional, the calculation of (11.56) can be recast as a matrix eigenvalue problem. By the same arguments used in the proof of Theorem 11.7, the values η_k are associated to vectors $v_k \in A$ satisfying the approximate weak eigenvalue equation,

$$\mathcal{E}[v_k, w] = \eta_k \langle v_k, w \rangle \tag{11.57}$$

for all $w \in A$.

To interpret (11.57) as a matrix eigenvalue equation we fix a basis $\{w_j\}_{j=1}^m$ for A. In terms of this basis, the energy functional and L^2 inner product define matrices

$$E_{ij} := \mathcal{E}[w_i, w_j], \qquad F_{ij} := \langle w_i, w_j \rangle. \tag{11.58}$$

If v_k is expanded as

$$v_k = \sum_{j=1}^m c_j w_j,$$

then (11.57) is equivalent to

$$\sum_{i=1}^m c_i (E_{ij} - \eta_k F_{ij}) = 0 \tag{11.59}$$

for $j = 1, \ldots, m$. This equation has a nontrivial solution only if the rows of the matrix $E - \eta_k F$ are linearly dependent, which is equivalent to the vanishing of the determinant. The values η_k can thus be calculated as the roots of a polynomial

$$\{\eta_1, \ldots, \eta_k\} = \{\eta : \det(E - \eta F) = 0\}. \tag{11.60}$$

In other words, the η_j are the eigenvalues of EF^{-1}.

Example 11.14 Consider $\mathbb{D}$ as in Example 11.13, but now take the subspace $A = [w_1, w_2, w_3]$, where

$$w_j(r) := 1 - r^{2j}.$$

Straightforward computations give the matrices (11.58) as

$$E = \begin{pmatrix} 2\pi & \frac{8\pi}{3} & 3\pi \\ \frac{8\pi}{3} & 4\pi & \frac{24\pi}{5} \\ 3\pi & \frac{24\pi}{5} & 6\pi \end{pmatrix}, \qquad F = \begin{pmatrix} \frac{\pi}{3} & \frac{5\pi}{12} & \frac{9\pi}{20} \\ \frac{5\pi}{12} & \frac{8\pi}{15} & \frac{7\pi}{12} \\ \frac{9\pi}{20} & \frac{7\pi}{12} & \frac{9\pi}{4} \end{pmatrix}.$$

The roots of $\det(E - \eta F)$ are

$$\eta_1 \doteq 5.783, \quad \eta_2 \doteq 30.712, \quad \eta_3 \doteq 113.505.$$

The estimate η_1 matches the exact value (11.55) very closely; in fact,

$$|\lambda_1 - \eta_1| \approx 10^{-6}.$$

The second value η_2 is a reasonable approximation to

$$\lambda_3 = j_{0,2}^2 \doteq 30.471.$$

However, we missed the second eigenvalue

$$\lambda_2 = j_{1,1}^2 \doteq 14.682.$$

The problem is that the space A consists entirely of radial functions, so that the second eigenfunction,

$$\phi_2(r, \theta) = J_1(\sqrt{\lambda_2} r) e^{i\theta},$$

is orthogonal to A. ◊

The missing eigenvalue in Example 11.14 illustrates a potential flaw in the Rayleigh-Ritz scheme. We need to make sure the subspace A covers $H_0^1(\Omega)$ sufficiently well in order to catch all low-lying eigenvalues. At the same time, we also need a means of producing this subspace efficiently. The *finite element method* is an approach that addresses both of these concerns.

To set up the finite element method we subdivide Ω into small polygonal domains, producing a *mesh* (or *triangulation*). Figure 11.3 illustrates a mesh for a two-dimensional domain. To each interior vertex is associated a piecewise linear function called an "element" that is positive at the vertex and decays linearly to zero on the neighboring faces. The span of these elements defines a subspace A with dimension equal to the number of interior vertices.

Example 11.15 Let us consider the domain $(0, \pi)$, for which the exact Dirichlet eigenfunctions are $\phi_k(x) = \sin kx$ for $k \in \mathbb{N}$, with $\lambda_k = k^2$. Define a mesh by subdividing $(0, \pi)$ into $m + 1$ intervals of length $\pi/(m + 1)$. To the jth vertex we associate the element

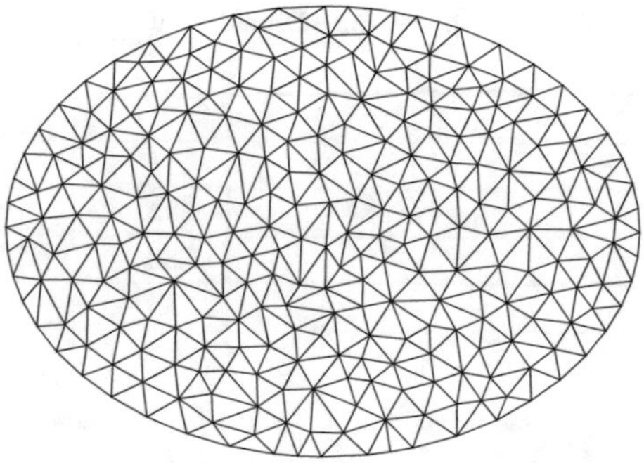

Fig. 11.3 Discrete mesh for an oval domain in $\mathbb{R}^2$

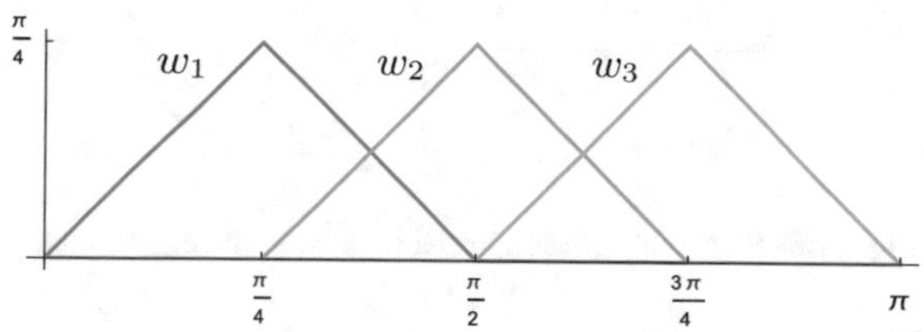

Fig. 11.4 Piecewise linear elements for $(0, \pi)$ with $m = 3$

$$w_j(x) = \begin{cases} x - \frac{j-1}{m+1}\pi, & \frac{j-1}{m+1}\pi \le x \le \frac{j}{m+1}\pi, \\ -x + \frac{j+1}{m+1}\pi, & \frac{j}{m+1}\pi \le x \le \frac{j+1}{m+1}\pi, \\ 0, & \text{otherwise}, \end{cases}$$

for $j = 1, \ldots m$. These elements are illustrated in Fig. 11.4.

The energy and inner product matrices can be computed by straightforward integrals,

$$E_{ij} = \begin{cases} \frac{2\pi}{m+1} & i = j, \\ -\frac{\pi}{m+1} & |i - j| = 1, \\ 0 & \text{otherwise}, \end{cases} \qquad F_{ij} = \begin{cases} \frac{2\pi^3}{3(m+1)^3} & i = j, \\ \frac{\pi^3}{6(m+1)^3} & |i - j| = 1, \\ 0 & \text{otherwise}. \end{cases}$$

Table 11.1 shows the resulting approximate eigenvalues as a function of the number of elements m. ◊

The finite element method produces approximate eigenfunctions as well as eigenvalues. Once the eigenvalue η_k is determined, the coefficients c_j can be computed from (11.59). These coefficients represent the eigenvectors of $(EF^{-1})^T$ in the basis $\{w_j\}$. The function $v_k = \sum_{j=1}^{m} c_j w_j$ is best approximation within the subspace A to the true eigenfunction ϕ_k. Figure 11.5 shows some approximate eigenfunctions determined by the calculations in Example 11.15.

The Rayleigh-Ritz strategy proves to be quite adaptable to more complicated problems. The procedure consists of: (1) generating a mesh for a given domain, (2) computing the matrix entries E_{ij} and F_{ij} corresponding to the elements of the

Table 11.1 Approximate Dirichlet eigenvalues for $(0, \pi)$ computed using m elements

m	3	10	25	50	exact
η_1	1.05	1.01	1.00	1.00	1
η_2	4.86	4.11	4.02	4.01	4
η_3	12. 84	9.56	9.10	9.03	9
η_4	–	17.80	16.31	16.08	16
η_5	–	29.45	25.77	25.20	25

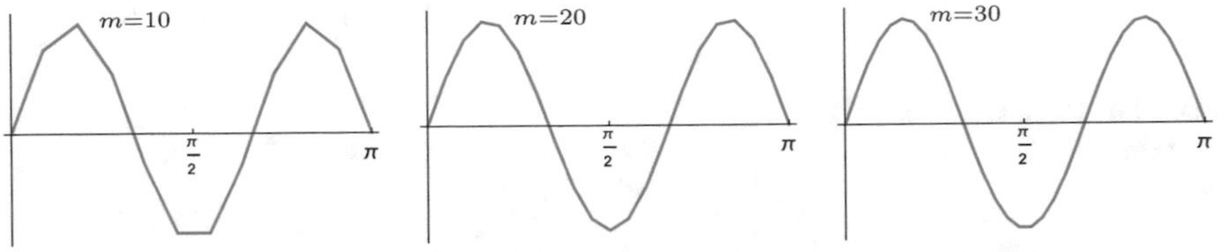

Fig. 11.5 Piecewise linear approximations to the eigenfunction $\phi_3(x) = \cos 3x$

Fig. 11.6 A approximation of the eigenfunction ϕ_6 for a star-shaped domain

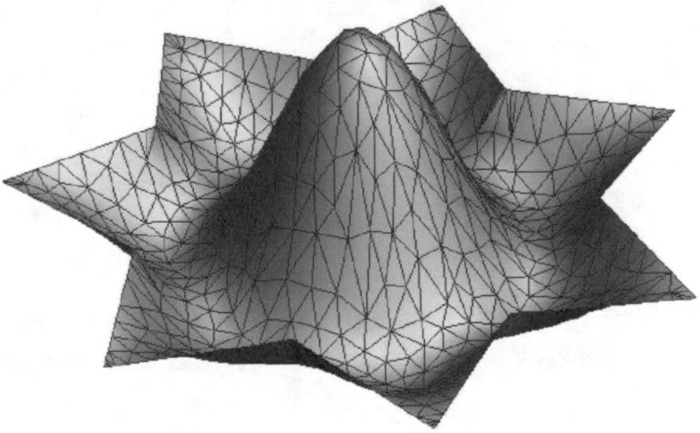

mesh, and (3) solving a finite-dimensional eigenvalue problem. A two-dimensional example is shown in Fig. 11.6.

11.8 Euler-Lagrange Equations

In the mid-18th century, Lagrange and Euler jointly developed a framework for expressing problems in classical mechanics in terms of the minimization of an *action functional*. Euler coined the term *calculus of variations* to describe this approach, which proved adaptable to a great variety of problems.

In the original classical mechanics setting, the action functional was the integral of a *Lagrangian function*, defined as kinetic energy minus potential energy. These might be energies of a single particle or a system.

In a typical PDE application on a bounded domain $\Omega \subset \mathbb{R}^n$, we take the Lagrangian $L(\boldsymbol{p}, w, \boldsymbol{x})$ to be a smooth function

$$L : \mathbb{R}^n \times \mathbb{R} \times \overline{\Omega} \to \mathbb{R}.$$

The action functional is defined by

$$S[w] := \int_\Omega L(\nabla w(\boldsymbol{x}), w(\boldsymbol{x}), \boldsymbol{x})\, d^n\boldsymbol{x}$$

for $w \in C^1(\overline{\Omega}; \mathbb{R})$. Suppose that u is a critical point of S, in the sense that

$$\frac{d}{dt} S[u + t\psi]\Big|_{t=0} = 0$$

for every $\psi \in C^\infty_{\mathrm{cpt}}(\Omega)$. This implies a (possibly nonlinear) PDE for u, called the *Euler-Lagrange equation* of L.

The Dirichlet principle gives the most basic example of this setup. For

$$L(\boldsymbol{p}, w, \boldsymbol{x}) := \frac{|\boldsymbol{p}|^2}{2} \tag{11.61}$$

the action functional is the Dirichlet energy $\mathcal{E}[w]$, and the Euler-Lagrange equation is the Laplace equation. To formulate the Poisson equation, we modify the Lagrangian to include the forcing term $f \in L^2(\Omega; \mathbb{R})$,

$$L(\boldsymbol{p}, w, \boldsymbol{x}) := \frac{|\boldsymbol{p}|^2}{2} - fw. \tag{11.62}$$

In this case the action is the functional $\mathcal{D}_f[w]$ defined in (11.6).

A classic nonlinear example is the surface area minimization problem. For $\Omega \in \mathbb{R}^2$, the graph of a function $w : \Omega \to \mathbb{R}$ defines a surface in $\mathbb{R}^3$. According to (2.8), the surface area of this patch of surface is given by

$$A[w] := \int_\Omega \sqrt{1 + |\nabla w|^2}\, d^2\boldsymbol{x}.$$

We can interpret this as an action functional corresponding to the Lagrangian

$$L(\boldsymbol{p}, w, \boldsymbol{x}) := \sqrt{1 + |\boldsymbol{p}|^2}.$$

For $f : \partial\Omega \to \mathbb{R}$, the problem of minimizing $A[w]$ under the constraint $w|_{\partial\Omega} = f$ was first studied by Lagrange. But historically this is called the *Plateau problem* after the 19th century physicist Joseph Plateau who conduct experiments on minimal surfaces using soap films.

Let us work out the Euler-Lagrange equation for the surface area functional. For $\psi \in C^\infty_{\text{cpt}}(\Omega)$, we have

$$\begin{aligned}\frac{d}{dt}A[u+t\psi]\Big|_{t=0} &= \frac{d}{dt}\int_\Omega \sqrt{1+|\nabla u + t\nabla\psi|^2}\, d^2\boldsymbol{x}\Big|_{t=0} \\ &= \int_\Omega \frac{\nabla u\cdot\nabla\psi}{\sqrt{1+|\nabla u|^2}}\, d^2\boldsymbol{x}.\end{aligned}$$

By Green's first identity (Theorem 2.10), and the fact that ψ vanishes near the boundary,

$$\frac{d}{dt}A[u+t\psi]\Big|_{t=0} = \int_\Omega \psi\nabla\cdot\left(\frac{\nabla u}{\sqrt{1+|\nabla u|^2}}\right) d^2\boldsymbol{x}.$$

Setting this equal to zero for all ψ gives the Euler-Lagrange equation

$$\nabla\cdot\left(\frac{\nabla u}{\sqrt{1+|\nabla u|^2}}\right) = 0. \qquad (11.63)$$

This nonlinear PDE is called the *minimal surface equation*.

There is a well-developed existence and regularity theory for general Euler-Lagrange equations, but this is too technical for us to go into here. In many cases, the finite element method can be used to effectively reduce the numerical approximation of solutions to linear algebra. Figure 11.7 shows a solution of the minimal surface equation calculated using finite elements.

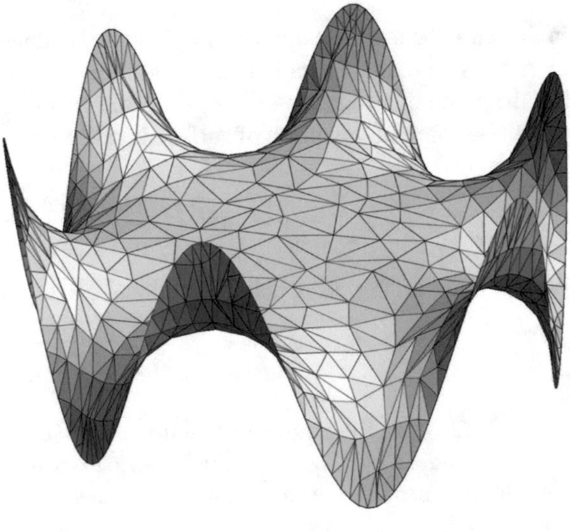

Fig. 11.7 The minimal surface over the unit disk associated to the boundary function $f(\theta) = \cos 6\theta$

11.9 Exercises

11.1 For a finite-dimensional subspace $A \subset H_0^1(\Omega)$, suppose we approximate the solution of the Poisson equation (11.4) for $f \in L^2(\Omega)$ by setting

$$u = \min_{w \in A} \mathcal{D}_f[w].$$

Given a basis $w_1, \dots w_m$ for A (not necessarily orthonormal), set $u = \sum_{i=1}^m c_i w_i$. Find equation an equation for $(c_1, \dots, c_m)$ in terms of f and the matrices E and F defined in (11.58).

11.2 To demonstrate the role that ellipticity plays in Theorem 11.5, consider the operator $L = r^2\Delta$ on the unit ball $\mathbb{B} = \{r < 1\} \subset \mathbb{R}^3$, where $r := |\boldsymbol{x}|$. For $f \in L^2(\mathbb{B})$, a weak solution of the equation $Lu = f$ with Dirichlet boundary conditions is defined as a function $u \in H_0^1(\mathbb{B})$ satisfying

$$\int_{\mathbb{B}} \left[\nabla u \cdot \nabla(r^2\psi) + f\psi\right] d^2\boldsymbol{x} = 0 \tag{11.64}$$

for all $\psi \in C^\infty_{\mathrm{cpt}}(\mathbb{B})$.

(a) Compute the weak partial derivatives of the function $\log r$ and show that $\log r \in H_0^1(\mathbb{B})$.
(b) Show that $u(\boldsymbol{x}) = \log r$ satisfies (11.64) with $f = 1$. Note that even though $\partial\mathbb{B}$ and f are C^∞, the solution u is not even H^2. (It is not a coincidence that the singularity of u occurs at the point where ellipticity of the operator fails.)

11.3 Let $\Omega \subset \mathbb{R}^2$ be the equilateral triangle with vertices $(0, 0)$, $(2, 0)$, and $(1, \sqrt{3})$. Define $w \in H_0^1(\Omega)$ to be the piecewise linear function whose graph forms a tetrahedron over Ω, with the top vertex at $(1, 1/\sqrt{3}, 1/\sqrt{3})$. Approximate the first eigenvalue by computing the Rayleigh quotient $\mathcal{R}[w]$. (For comparison, the exact value is $\lambda_1 = \frac{4\pi^2}{3}$; this corresponds to the eigenfunction shown in Fig. 11.1.)

11.4 Let $\mathbb{B}^3$ denote the unit ball $\{r < 1\} \subset \mathbb{R}^3$. Find an upper bound on the first eigenvalue by computing the Rayleigh quotient of the radial function $w(r) = 1 - r$. (Compare your answer to the exact value $\lambda_1 = \pi^2$ found in Exercise 5.8.)

11.5 For a bounded domain $\Omega \subset \mathbb{R}^n$, let $\phi_1 \in H_0^1(\Omega; \mathbb{R}) \cap C^\infty(\Omega)$ be the first eigenfunction as obtained in Theorem 11.9, normalized so that $\|\phi_1\|_2 = 1$. In this problem we will show that ϕ_1 has no zeros in Ω and that this eigenfunction is unique up to a multiplicative constant.

(a) Set

$$\phi_\pm(\boldsymbol{x}) := \max\{\pm\phi_1(\boldsymbol{x}), 0\}$$

for $\boldsymbol{x} \in \Omega$, and show that

$$\phi_1 = \phi_+ - \phi_-$$

with $\phi_\pm \geq 0$ and $\phi_+\phi_- \equiv 0$.

(b) Note Lemma 10.10 implies that $\phi_\pm \in H_0^1(\Omega; \mathbb{R})$. Assuming that ϕ_1 is normalized $\|\phi_1\|_2 = 1$, show that

$$\|\phi_+\|^2 + \|\phi_-\|^2 = 1,$$
$$\mathcal{E}[\phi_+] + \mathcal{E}[\phi_-] = \lambda_1.$$

(c) Since λ_1 minimizes the Rayleigh quotient,

$$\mathcal{E}[\phi_\pm] \geq \lambda_1 \|\phi_\pm\|_2.$$

Use this, together with the fact that $\mathcal{E}[\phi_1] = \lambda_1$, to deduce that

$$\mathcal{E}[\phi_\pm] = \lambda_1 \|\phi_\pm\|_2.$$

Hence, by the proof of Theorem 11.9, $\phi_\pm \in C^\infty(\Omega)$ and

$$-\Delta\phi_\pm = \lambda_1\phi_\pm. \tag{11.65}$$

(d) Use the strong maximum principle of Theorem 9.5 to deduce from (11.65) that if $\phi_\pm$ has a zero within Ω then $\phi_\pm \equiv 0$. Conclude that ϕ_1 has no zeros in Ω.

(e) If $u \in H_0^1(\Omega; \mathbb{R}) \cap C^\infty(\Omega)$ is some other eigenfunction with eigenvalue λ_1, then $u - c\phi_1$ is also an eigenfunction for each $c \in \mathbb{R}$. Show that c can be chosen so that $u - c\phi_1$ has a zero in Ω. Conclude that $u \equiv c\phi_1$.

11.6 Determine the Euler-Lagrange equations on $\Omega \subset \mathbb{R}^n$ corresponding to the following Lagrangians:

(a) $L(\boldsymbol{p}, w, \boldsymbol{x}) = \frac{1}{2}|\boldsymbol{p}|^2 + \frac{1}{2}a(\boldsymbol{x})w^2$ for $a \in C^0(\Omega)$.

(b) $L(\boldsymbol{p}, w, \boldsymbol{x}) = \frac{1}{2}\sum_{i,j=1}^{n} a_{ij}(\boldsymbol{x})p_i p_j$ with $a_{ij} \in C^1(\Omega)$.

(c) $L(\boldsymbol{p}, w, \boldsymbol{x}) = |\boldsymbol{p}|$.

(d) $L(\boldsymbol{p}, w, \boldsymbol{x}) = \frac{1}{2}|\boldsymbol{p}|^2 + F(w)$ for $F \in C^1(\mathbb{R})$.

Chapter 12
Distributions

To define weak derivatives in Chap. 10, we measured the values of a function $f \in L^1_{\text{loc}}(\Omega)$ by integrating against test functions. One way to interpret this process is that f defines a functional $C^\infty_{\text{cpt}}(\Omega) \to \mathbb{C}$ given by

$$\psi \mapsto \int_\Omega f\psi \, d^n x.$$

A *distribution* on $\Omega \subset \mathbb{R}^n$ is a more general functional $C^\infty_{\text{cpt}}(\Omega) \to \mathbb{C}$, not necessarily expressible as an integral. To qualify as a distribution, a functional is required to satisfy conditions that insure that weak derivatives and other basic operations are well defined.

As with weak derivatives, the concept of a distribution was inspired by idealized situations in physics. Indeed, the term "distribution" was inspired by charge distributions in electrostatics, an example that we will discuss in Sect. 12.1. Distributions generalize the notion of weak solutions, in the sense that every function in $L^1_{\text{loc}}(\Omega)$ also defines a distribution. The trade-off for the increased generality is that some basic operations for functions cannot be applied to distributions. The product of two distributions is not generally well defined, for example.

There are some technicalities in the mathematical theory of distributions that require more background on the topology of function spaces than we assume for this text. We will treat these technicalities rather lightly; our focus will be on exploring the PDE applications.

12.1 Model Problem: Coulomb's Law

Coulomb's law of electrostatics is an empirical observation developed by 18th century physicist Charles-Augustin de Coulomb. It says that a particle with electric charge q_0, located at the origin, generates an electric field given by

D. Borthwick, *Introduction to Partial Differential Equations*, Universitext, DOI 10.1007/978-3-319-48936-0_12

$$E(x) = \frac{kq_0 x}{|x|^3}, \tag{12.1}$$

where k (Coulomb's constant) depends on the properties of the medium surrounding the charges.

In Sect. 11.1 we discussed another important empirical law of electrostatics, Gauss's law. With the same convention for physical constants as in (12.1), the differential form of the law says that

$$\nabla \cdot E = 4\pi k\rho, \tag{12.2}$$

where ρ is the charge per unit volume as a function of position.

These two empirical laws present something of a mathematical conundrum, in that the field specified by Coulomb is not differentiable at $x = 0$, not even weakly. On the other hand, for $x \neq 0$,

$$\begin{aligned} \nabla \cdot \frac{x}{r^3} &= \frac{\nabla \cdot x}{r^3} - \frac{3x}{r^4} \cdot \nabla r \\ &= \frac{3}{r^3} - \frac{3x}{r^4} \cdot \frac{x}{r} \\ &= 0 \end{aligned} \tag{12.3}$$

This is consistent with (12.2), in that Coulomb assumes the charge density is zero for $x \neq 0$. However, if a function in L^1_{loc} vanishes except at a single point, then that function is zero by the equivalence (7.6). Thus a point charge density has no meaningful interpretation as a locally integrable function.

To reconcile (12.1) with Gauss's law, let us consider the weak form of (12.2),

$$\int_{\mathbb{R}^3} E \cdot \nabla \psi \, d^3x = -4\pi k \int_{\mathbb{R}^3} \rho\psi \, d^3x \tag{12.4}$$

for all $\psi \in C^\infty_{\text{cpt}}(\mathbb{R}^3)$. The left side of (12.4) is well defined because the components of E are locally integrable.

Since the Coulomb field is smooth away from the origin, we can integrate by parts as long as we exclude the origin from the region of integration by writing the integral as a limit,

$$\int_{\mathbb{R}^3} E \cdot \nabla \psi \, d^3x = \lim_{\varepsilon \to 0} \int_{\{r \geq \varepsilon\}} E \cdot \nabla \psi \, d^3x. \tag{12.5}$$

The region $\{r \geq \varepsilon\}$ has boundary given by the sphere $\{r = \varepsilon\}$. In this case the "outward" unit normal is a radial unit vector pointing towards the origin,

$$\nu = -\frac{x}{r}\bigg|_{r=\varepsilon}. \tag{12.6}$$

By the divergence theorem (Theorem 2.6), and the fact that $\nabla \cdot E = 0$ for $r > 0$ by (12.3),

$$\int_{\{r\geq\varepsilon\}} \boldsymbol{E}\cdot\nabla\psi\, d^3\boldsymbol{x} = \int_{\{r=\varepsilon\}} \boldsymbol{\nu}\cdot\boldsymbol{E}\,\psi\, dS.$$

Hence by (12.1) and (12.6),

$$\int_{\{r\geq\varepsilon\}} \boldsymbol{E}\cdot\nabla\psi\, d^3\boldsymbol{x} = -\int_{\{r=\varepsilon\}} \frac{kq_0}{\varepsilon^2}\psi\, dS$$

Taking $\varepsilon \to 0$ now gives

$$\int_{\mathbb{R}^3} \boldsymbol{E}\cdot\nabla\psi\, d^3\boldsymbol{x} = -\lim_{\varepsilon\to 0}\int_{\{r=\varepsilon\}} \frac{kq_0}{\varepsilon^2}\psi\, dS. \tag{12.7}$$

Because ψ is continuous, the average of ψ over the sphere $\{r = \varepsilon\}$ approaches $\psi(0)$ as $\varepsilon \to 0$, i.e.,

$$\lim_{\varepsilon\to 0}\frac{1}{4\pi\varepsilon^2}\int_{\{r=\varepsilon\}} \psi\, dS = \psi(0).$$

Applying this to (12.7) gives

$$\int_{\mathbb{R}^3} \frac{kq\boldsymbol{x}}{r^3}\cdot\nabla\psi\, d^3\boldsymbol{x} = -4\pi kq_0\psi(0). \tag{12.8}$$

The weak condition (12.4) thus requires that

$$\int_{\mathbb{R}^3} \rho\psi\, d^3\boldsymbol{x} = q_0\psi(0),$$

for every $\psi \in C^\infty_{\mathrm{cpt}}(\mathbb{R}^3)$. This is consistent with the physical interpretation of ρ as a charge located exactly at the origin.

The concept of a "point density" was widely used in physics applications in the 18th and 19th centuries. In a 1930 book on quantum mechanics, the physicist Paul Dirac described such densities in terms of a *delta function* $\delta(\boldsymbol{x})$, whose defining property is that

$$\int_{\mathbb{R}^n} f(\boldsymbol{x})\delta(\boldsymbol{x})\, d^n\boldsymbol{x} := f(0), \tag{12.9}$$

for a continuous function f. This terminology and notation are potentially misleading, because δ is not a function and (12.9) is not actually an integral. However, Dirac's formulation hints at the proper mathematical interpretation, which is that δ should be understood as a functional $f \mapsto f(0)$.

If we accept the intuitive definition of the delta function for the moment, then we can interpret the calculation (12.8) as showing that

$$\nabla\cdot\frac{\boldsymbol{x}}{r^3} = 4\pi\delta. \tag{12.10}$$

12.2 The Space of Distributions

A *distribution* on a domain $\Omega \subset \mathbb{R}^n$ is a continuous linear functional $C^\infty_{\text{cpt}}(\Omega) \to \mathbb{C}$. The map defined by a distribution u is usually written as a pairing of u with a test function, i.e.,

$$\psi \mapsto (u, \psi) \in \mathbb{C} \tag{12.11}$$

for $\psi \in C^\infty_{\text{cpt}}(\Omega)$. Linearity means that

$$(u, c_1\psi_1 + c_2\psi_2) = c_1(f, \psi_1) + c_2(f, \psi_2),$$

for all $c_1, c_2 \in \mathbb{C}$ and $\psi_1, \psi_2 \in C^\infty_{\text{cpt}}(\Omega)$.

The definition of distribution also includes the word "continuous". To define continuity for functionals we must first specify what convergence means in $C^\infty_{\text{cpt}}(\Omega)$. The standard definition is that for a sequence $\{\psi_k\}$ to converge to ψ in $C^\infty_{\text{cpt}}(\Omega)$ means that all ψ_k have support in some fixed compact set $K \subset \Omega$, and the sequence of functions and all sequences of partial derivatives converge uniformly on K. Continuity of the functional (12.11) is then defined by the condition that convergence of a sequence $\psi_k \to \psi$ in $C^\infty_{\text{cpt}}(\Omega)$ implies that

$$\lim_{k\to\infty} (u, \psi_k) = (u, \psi). \tag{12.12}$$

In finite dimensions continuity is implied by linearity. That is not the case here, but in practice it is quite difficult to come up with a functional that is linear but not continuous.

The set of distributions on Ω forms a vector space denoted by $\mathcal{D}'(\Omega)$. Linear combinations of distributions are defined in the obvious way by

$$(c_1u_1 + c_2u_2, \psi) := c_1(u_1, \psi) + c_2(u_2, \psi),$$

for $u_1, u_2 \in \mathcal{D}'(\Omega)$ and $c_1, c_2 \in \mathbb{C}$. The mathematical theory of distributions was developed independently in the mid-20th century by Sergei Sobolev and Laurent Schwartz. Schwartz used $\mathcal{D}$ as a notation for C^∞_{cpt}, and the prime accent on $\mathcal{D}'$ comes from the notation for the dual of a vector space in linear algebra.

A locally integrable function $f \in L^1_{\text{loc}}(\Omega)$ defines a distribution through the integral pairing

$$(f, \psi) := \int_\Omega f\psi \, d^n\boldsymbol{x}. \tag{12.13}$$

Under this convention there is an inclusion

$$L^1_{\text{loc}}(\Omega) \subset \mathcal{D}'(\Omega).$$

In particular, all L^p functions can be interpreted as distributions.

As we saw with the point charge density in Sect. 12.1, not all distributions are given by functions. We use the notation $\delta_{\boldsymbol{x}}$ for the delta function centered at $\boldsymbol{x} \in \Omega$, defined by

$$(\delta_{\boldsymbol{x}}, \psi) := \psi(\boldsymbol{x}). \tag{12.14}$$

By convention the subscript is dropped for $\boldsymbol{x} = 0$, i.e., $\delta := \delta_0$.

Multiplication by smooth functions preserves the space $C^\infty_{\text{cpt}}(\Omega)$. Therefore it makes sense to multiply a distribution $u \in \mathcal{D}'(\Omega)$ by a function $f \in C^\infty(\Omega)$. The product distribution is defined by

$$(fu, \psi) := (u, f\phi).$$

It does not make sense, however, to multiply two distributions together. This fact was intuitively clear in early applications: the product of two charge densities makes no physical sense.

Convergence of a sequence of distributions is defined in a very straightforward way. We say that $u_k \to u$ in $\mathcal{D}'(\Omega)$ if

$$\lim_{k\to\infty} (u_k, \psi) = (u, \psi)$$

for all $\psi \in C^\infty_{\text{cpt}}(\Omega)$. All distributions can in fact be approximated by smooth functions by such a limit, although we are not equipped to prove that here. We will present one useful special case, a construction of the delta function as a limit of integrable functions.

Lemma 12.1 *Given $f \in L^1(\mathbb{R}^n)$ satisfying*

$$\int_{\mathbb{R}^n} f \, d^n\boldsymbol{x} = 1, \tag{12.15}$$

define the rescaled function,

$$f_a(\boldsymbol{x}) := a^n f(a\boldsymbol{x})$$

for $a > 0$. Then

$$\lim_{a\to\infty} f_a = \delta, \tag{12.16}$$

as a distributional limit.

Proof For $\psi \in C^\infty_{\text{cpt}}(\mathbb{R}^n)$ we can evaluate the pairing with f_a using a change variables,

$$\begin{aligned}(f_a, \psi) &= \int_{\mathbb{R}^n} a^n f(a\boldsymbol{x})\psi(\boldsymbol{x}) \, d^n\boldsymbol{x} \\ &= \int_{\mathbb{R}^n} f(\boldsymbol{x})\psi(\boldsymbol{x}/a) \, d^n\boldsymbol{x}.\end{aligned}$$

By the assumption (12.15), we can also write

$$\psi(0) = \int_{\mathbb{R}^n} f(\boldsymbol{x})\psi(0)\, d^n\boldsymbol{x},$$

which gives the estimate

$$\left|(f_a, \psi) - \psi(0)\right| \leq \int_{\mathbb{R}^n} |f(\boldsymbol{x})|\left|\psi(\boldsymbol{x}/a) - \psi(0)\right| d^n\boldsymbol{x}. \tag{12.17}$$

Given $\varepsilon > 0$, the fact that f is integrable implies that exists R sufficiently large so that

$$\int_{|\boldsymbol{x}|\geq R} |f|\, d^n\boldsymbol{x} < \varepsilon. \tag{12.18}$$

By the continuity of ψ we can also choose $\delta > 0$ so that

$$|\psi(\boldsymbol{x}) - \psi(0)| < \varepsilon$$

for $|\boldsymbol{x}| < \delta$. For $a \geq R/\delta$ this implies that

$$\left|\psi(\boldsymbol{x}/a) - \psi(0)\right| < \varepsilon \tag{12.19}$$

for all $|\boldsymbol{x}| \leq R$. Using (12.18) and (12.19) to estimate the difference (12.17) gives

$$\begin{aligned}\left|(f_a, \psi) - \psi(0)\right| &\leq 2\|\psi\|_\infty \int_{|\boldsymbol{x}|\geq R} |f(\boldsymbol{x})|\, d^n\boldsymbol{x} + \varepsilon \int_{|\boldsymbol{x}|\leq R} |f(\boldsymbol{x})|\, d^n\boldsymbol{x} \\ &\leq \big(2\|\psi\|_\infty + \|f\|_1\big)\varepsilon,\end{aligned}$$

for $a \geq R/\delta$. Since ε was arbitrary, this shows that

$$\lim_{a\to\infty} (f_a, \psi) = \psi(0).$$

□

The rescaling used in Lemma 12.1 is illustrated in Fig. 12.1. Note that this looks very similar to Fig. 9.1, and in fact the proof of Lemma 12.1 uses essentially the same argument as that of Theorem 9.1. We saw another case of this construction in the proof of Theorem 6.2. Indeed, we can now interpret the result of Theorem 6.2 as a distributional limit of the heat kernel,

$$\lim_{t\to 0} H_t = \delta,$$

where H_t was defined in (6.16).

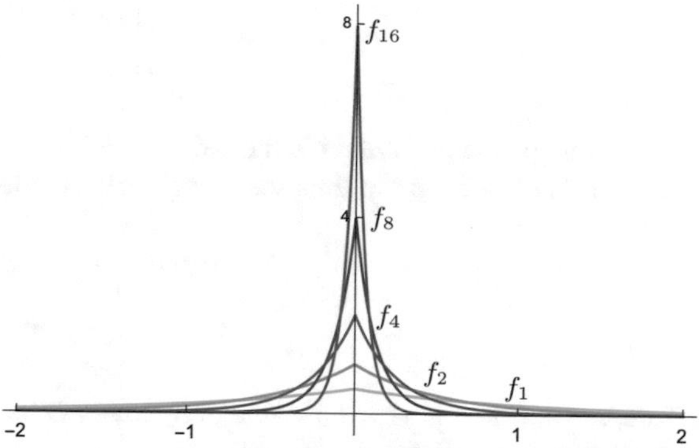

Fig. 12.1 Rescaled functions f_a for $f(x) = \frac{1}{2}e^{-|x|}$

12.3 Distributional Derivatives

The *distributional derivative* extends the concept of the weak derivative introduced in Sect. 10.1. By analogy with (10.7), for $u \in \mathcal{D}'(\Omega)$ and we define the distribution $D^\alpha u$ by

$$(D^\alpha u, \psi) := (-1)^{|\alpha|}(u, D^\alpha \psi), \tag{12.20}$$

with

$$D^\alpha := \frac{\partial^{\alpha_1}}{\partial x_1^{\alpha_1}} \cdots \frac{\partial^{\alpha_n}}{\partial x_n^{\alpha_n}}$$

and $|\alpha| := \alpha_1 + \cdots + \alpha_n$, as before. The pairing (12.20) is well defined as a distribution because D^α is both linear and continuous as a map $C^\infty_{\mathrm{cpt}}(\Omega) \to C^\infty_{\mathrm{cpt}}(\Omega)$.

The terms "distributional" and "weak" are frequently used interchangeably to describe derivatives, since the definitions overlap to a considerable extent. The only difference is that a weak derivative is representable as a locally integrable function. Weak derivatives may not exist, whereas all distributions are infinitely differentiable.

Example 12.2 Let us reconsider Example 10.3, where we considered the derivative of $w \in L^1_{\mathrm{loc}}(\Omega)$ defined by

$$w(t) = \begin{cases} w_-(t), & t < 0, \\ w_+(t), & t \ge 0, \end{cases}$$

where $w_\pm \in C^1(\mathbb{R})$. As part of that calculation we showed that

$$-\int_{-\infty}^{\infty} w\psi'\,dt = \big[w_+(0) - w_-(0)\big]\psi(0) + \int_{-\infty}^{\infty} h\psi\,dt, \tag{12.21}$$

where h is the piecewise derivative

$$h(t) := \begin{cases} w'_-(t), & t < 0, \\ w'_+(t), & t > 0. \end{cases}$$

The left-hand side of (12.21) is the pairing (w', ψ) by the definition (12.20). From the right-hand side we can thus see that the distributional derivative is

$$w' = h + \big[w_+(0) - w_-(0)\big]\delta.$$

◊

Example 12.3 For $\delta_{\boldsymbol{x}} \in \mathcal{D}'(\mathbb{R}^n)$, the derivatives $D^\alpha \delta_{\boldsymbol{x}}$ are easily computed from the definition (12.20). For $\psi \in C^\infty_{\mathrm{cpt}}(\mathbb{R}^n)$,

$$\begin{aligned}(D^\alpha \delta_{\boldsymbol{x}}, \psi) &= (-1)^{|\alpha|}(\delta_{\boldsymbol{x}}, D^\alpha \psi) \\ &= (-1)^{|\alpha|} D^\alpha \psi(\boldsymbol{x}).\end{aligned}$$

In other words, the distribution $D^\alpha \delta_{\boldsymbol{x}}$ evaluates the derivative of the test function at the point $\boldsymbol{x}$, up to a sign. ◊

Example 12.4 The function $\ln |x|$ is locally integrable on $\mathbb{R}$ and so defines a distribution in $\mathcal{D}'(\mathbb{R})$. Therefore $(\ln |x|)'$ exists in the distribution sense. This is puzzling because

$$\frac{d}{dx} \ln |x| = \frac{1}{x}$$

for $x \neq 0$, and x^{-1} is not locally integrable.

To understand what is happening here, we must return to the distributional definition,

$$\begin{aligned}\big((\ln |x|)', \psi\big) &:= -(\ln |x|, \psi') \\ &= -\int_{-\infty}^{\infty} \psi'(x) \ln |x| \, dx\end{aligned}$$

for $\psi \in C^\infty_{\mathrm{cpt}}(\mathbb{R})$. To compute this we avoid the singularity at 0 by writing

$$\big((\ln |x|)', \psi\big) = -\lim_{\varepsilon \to 0} \int_{|x| \geq \varepsilon} \psi'(x) \ln |x| \, dx. \tag{12.22}$$

Integration by parts gives

$$\begin{aligned}-\int_{-\infty}^{-\varepsilon} \psi'(x) \ln |x| \, dx &= -\psi(x) \ln |x| \Big|_{-\infty}^{-\varepsilon} + \int_{-\infty}^{-\varepsilon} \frac{\psi(x)}{x} \, dx \\ &= -\psi(-\varepsilon) \ln \varepsilon + \int_{-\infty}^{-\varepsilon} \frac{\psi(x)}{x} \, dx,\end{aligned}$$

and similarly

$$-\int_{\varepsilon}^{\infty} \psi'(x) \ln |x| \, dx = \psi(\varepsilon) \ln \varepsilon + \int_{\varepsilon}^{\infty} \frac{\psi(x)}{x} \, dx.$$

After combining these two halves, we obtain

$$\int_{|x| \geq \varepsilon} \psi'(x) \ln |x| \, dx = \big[\psi(\varepsilon) - \psi(-\varepsilon)\big] \ln \varepsilon + \int_{|x| \geq \varepsilon} \frac{\psi(x)}{x} \, dx.$$

By the definition of the derivative,

$$\lim_{\varepsilon \to 0} \frac{\psi(\varepsilon) - \psi(-\varepsilon)}{2\varepsilon} = \psi'(0).$$

Therefore

$$\begin{aligned} \lim_{\varepsilon \to 0} \big[\psi(\varepsilon) - \psi(-\varepsilon)\big] \ln \varepsilon &= 2\psi'(0) \lim_{\varepsilon \to 0} \varepsilon \ln \varepsilon \\ &= 0. \end{aligned}$$

Hence (12.22) reduces to

$$\big((\ln |x|)', \psi\big) = \lim_{\varepsilon \to 0} \int_{|x| \geq \varepsilon} \frac{\psi(x)}{x} \, dx. \tag{12.23}$$

The limit on the right exists for $\psi \in C^{\infty}_{\text{cpt}}(\mathbb{R})$, even though x^{-1} is not integrable, because the limit is taken symmetrically. This limiting procedure defines a distribution called the *principal value* of x^{-1}, written as $\text{PV}[x^{-1}]$. We could rephrase (12.23) as

$$\frac{d}{dx} \ln |x| = \text{PV}\left[x^{-1}\right].$$

◊

Example 12.5 Let us reinterpret the discussion from Sect. 12.1 in terms of distributional derivatives. We already noted that the components of $\boldsymbol{x}/r^3$ are locally integrable, so we can consider the Coulomb formula (12.1) for $\boldsymbol{E}$ as the definition of a vector-valued distribution. The distributional divergence of $\boldsymbol{x}/r^3$ is defined by the condition that

$$\left(\nabla \cdot \frac{\boldsymbol{x}}{r^3}, \psi\right) := -\int_{\mathbb{R}^3} \frac{\boldsymbol{x}}{r^3} \cdot \nabla \psi \, d^3x,$$

for $\psi \in C^{\infty}_{\text{cpt}}(\mathbb{R}^3)$. The derivation of (12.8) thus shows that

$$\nabla \cdot \frac{\boldsymbol{x}}{r^3} = 4\pi\delta. \tag{12.24}$$

We can also consider the corresponding result for the Coulomb electric potential

$$\phi(\boldsymbol{x}) = \frac{1}{r}.$$

(ignoring the physical constants). The gradient of ϕ exists in the weak sense and is given by

$$\nabla\left(\frac{1}{r}\right) = -\frac{\boldsymbol{x}}{r^3}.$$

Since $\Delta = \nabla \cdot \nabla$, we deduce from (12.24) that

$$-\Delta\left(\frac{1}{4\pi r}\right) = \delta. \tag{12.25}$$

◊

12.4 Fundamental Solutions

Because the Poisson equation is linear, it makes sense to construct a solution with a continuous density by superimposing a field of point sources. With a change of variables, we can see from (12.25) that the potential function corresponding to a point source at $\boldsymbol{y} \in \mathbb{R}^3$ is

$$\phi_{\boldsymbol{y}}(\boldsymbol{x}) := \frac{1}{4\pi|\boldsymbol{x} - \boldsymbol{y}|}.$$

Weighting the point sources by the density ρ and summing them with an integral gives

$$u(\boldsymbol{x}) = \frac{1}{4\pi}\int_{\mathbb{R}^3} \frac{\rho(\boldsymbol{y})}{|\boldsymbol{x} - \boldsymbol{y}|}\, d^3\boldsymbol{y}. \tag{12.26}$$

This formula, which is often stated as the integral form of Coulomb's law, does indeed yield a solution of the Poisson equation on $\mathbb{R}^3$ under certain conditions. For example if $\rho \in C^1_{\text{cpt}}(\mathbb{R}^3)$ then one can confirm that $-\Delta u = \rho$ by direct computation. The C^1 condition is stronger than necessary here, but continuity alone would not be sufficient. (The precise notion of regularity needed for this problem is something called *Hölder continuity*.)

This idea of constructing of general solutions by superposition of point sources is the inspiration for the concept of a *fundamental solution*. For a constant-coefficient differential operator L acting on $\mathbb{R}^n$, of the form

$$L = \sum_{|\alpha| \le m} a_\alpha D^\alpha,$$

with $a^\alpha \in \mathbb{C}$, a fundamental solution is a distribution $\Phi \in \mathcal{D}'(\mathbb{R}^n)$ such that

$$L\Phi = \delta. \tag{12.27}$$

For example, in the Coulomb case the calculation (12.25) gives the fundamental solution of $-\Delta$ on $\mathbb{R}^3$. Fundamental solutions are especially important for classical problems involving the Laplacian.

The solution formula (12.26) resembles the convolution used to solve the heat equation in Sect. 6.3. For $f, g \in L^1(\mathbb{R}^n)$ the convolution is defined as

$$f * g(\boldsymbol{x}) := \int_{\mathbb{R}^n} f(\boldsymbol{y})g(\boldsymbol{x}-\boldsymbol{y})\, d^n\boldsymbol{y}.$$

A simple change of variables shows that this product is symmetric,

$$f * g = g * f.$$

In order to produce solution formulas from fundamental solutions, we need to understand how to take convolutions with distributions.

For $f, g \in L^1(\mathbb{R}^n)$, the distributional pairing of $f * g$ with $\psi \in C^\infty_{\text{cpt}}(\Omega)$ gives

$$(f * g, \psi) = \int_{\mathbb{R}^n}\int_{\mathbb{R}^n} f(\boldsymbol{y})g(\boldsymbol{x}-\boldsymbol{y})\psi(\boldsymbol{x})\, d^n\boldsymbol{y}\, d^n\boldsymbol{x}. \tag{12.28}$$

The $\boldsymbol{x}$ integration looks almost like the convolution of ψ with g, except with the argument switched from $\boldsymbol{y}-\boldsymbol{x}$ to $\boldsymbol{x}-\boldsymbol{y}$. With the reflection defined by

$$g^-(\boldsymbol{x}) := g(-\boldsymbol{x}),$$

we have

$$g^- * \psi(\boldsymbol{y}) = \int_{\mathbb{R}^n} g(\boldsymbol{x}-\boldsymbol{y})\psi(\boldsymbol{x})\, d^n\boldsymbol{x}.$$

Thus (12.28) reduces to

$$(f * g, \psi) := (f, g^- * \psi).$$

If $\phi, \psi \in C^\infty_{\text{cpt}}(\mathbb{R}^n)$, then it is easy to check that $\phi^- * \psi \in C^\infty_{\text{cpt}}(\mathbb{R}^n)$ also. Moreover, the map $\psi \mapsto \phi^- * \psi$ is linear and continuous. We can thus define $u * \phi$ for $u \in \mathcal{D}'(\mathbb{R}^n)$ and $\phi \in C^\infty_{\text{cpt}}(\mathbb{R}^n)$ by

$$(u * \phi, \psi) := (u, \phi^- * \psi) \tag{12.29}$$

for $\psi \in C^\infty_{\text{cpt}}(\mathbb{R}^n)$.

The distribution δ plays a special role with regard to convolutions. By the definition (12.29),

$$\begin{aligned}(\delta * \phi, \psi) &:= (\delta, \phi^- * \psi)\\ &= \phi^- * \psi(0)\\ &= \int_\Omega \phi(\boldsymbol{x})\psi(\boldsymbol{x})\, d^n\boldsymbol{x}\\ &= (\phi, \psi).\end{aligned}$$

This shows that

$$\delta * \phi = \phi. \tag{12.30}$$

In other words, convolution by δ is the identity map.

Let Φ be the fundamental solution for the constant coefficient operator L. Our goal is to show that the equation $Lu = f$ is solved by the convolution $u = \Phi * f$, at least for $f \in C^\infty_{\text{cpt}}(\mathbb{R}^n)$. To check this, we need to know how to evaluate derivatives of the convolution.

Lemma 12.6 *For $w \in \mathcal{D}'(\mathbb{R}^n)$ and $\psi \in C^\infty_{\text{cpt}}(\mathbb{R}^n)$,*

$$D^\alpha(w * f) = (D^\alpha w) * \phi = w * (D^\alpha f).$$

Proof For $\phi, \psi \in C^\infty_{\text{cpt}}(\Omega)$, we compute directly that

$$\begin{aligned}D^\alpha(\phi * \psi)(\boldsymbol{x}) &= \int_\Omega \phi(\boldsymbol{x} - \boldsymbol{y})\psi(\boldsymbol{y})\, d^n\boldsymbol{y}\\ &= \int_\Omega D^\alpha\phi(\boldsymbol{x} - \boldsymbol{y})\psi(\boldsymbol{y})\, d^n\boldsymbol{y}\\ &= (D^\alpha\phi) * \psi(\boldsymbol{x}).\end{aligned}$$

Since the convolution is symmetric, the same formula holds with ψ and ϕ switched. Thus the formula

$$D^\alpha(\psi * \phi) = (D^\alpha\psi) * \phi = \psi * (D^\alpha\phi) \tag{12.31}$$

holds for test functions.

For $w \in \mathcal{D}'(\mathbb{R})$ and $\psi \in C^\infty_{\text{cpt}}(\mathbb{R}^n)$, it follows from the definitions that

$$(D^\alpha(f * \phi), \psi) = (-1)^{|\alpha|}(f, \phi^- * (D^\alpha\psi)).$$

By (12.31) this gives

$$\begin{aligned}(D^\alpha(f * \phi), \psi) &= (-1)^{|\alpha|}(f, D^\alpha(\phi^- * \psi))\\ &= (D^\alpha f, \phi^- * \psi)\\ &= ((D^\alpha f) * \phi, \psi),\end{aligned}$$

and also

$$\begin{aligned}(D^\alpha(f * \phi), \psi) &= (-1)^{|\alpha|}(f, (D^\alpha\phi)^- * \psi)) \\ &= (f * (D^\alpha\phi), \psi).\end{aligned}$$

□

Theorem 12.7 *If L is a constant coefficient operator on $\mathbb{R}^n$ with fundamental solution Φ, then for $f \in C^\infty_{\text{cpt}}(\mathbb{R}^n)$ the equation*

$$Lu = f$$

is solved by

$$u = \Phi * f.$$

Proof By Lemma 12.6,

$$\begin{aligned}L(\Phi * f) &= \sum_{|\alpha|\le m} a_\alpha D^\alpha(\Phi * f) \\ &= \sum_{|\alpha|\le m} a_\alpha (D^\alpha \Phi) * f \\ &= (L\Phi) * f.\end{aligned}$$

Note that the second step only works because the coefficients a_α are assumed to be constant. Since $L\Phi = \delta$, we see from (12.30) that

$$L(\Phi * f) = f.$$

□

A result called the Malgrange-Ehrenpreis theorem, proven in the 1950s, says that every constant coefficient differential operator on $\mathbb{R}^n$ admits a fundamental solution. The fundamental solution of the Laplacian, which we will now work out for any dimension, is the most important case.

Theorem 12.8 *On $\mathbb{R}^n$ the operator $-\Delta$ has the fundamental solution*

$$\Phi(\mathbf{x}) = \begin{cases} -\frac{1}{2\pi}\ln r, & n = 2, \\ \frac{1}{(n-2)A_n r^{n-2}}, & n \ge 3, \end{cases} \tag{12.32}$$

where A_n denotes the volume of the unit sphere in dimension n.

Proof We start from the distributional derivative,

$$(-\Delta\Phi, \psi) = -(\Phi, \Delta\psi)$$

for $\psi \in C^\infty_{\mathrm{cpt}}(\mathbb{R}^n)$. To evaluate this, it is useful to first compute the gradient,

$$\nabla \Phi(\boldsymbol{x}) = -\frac{\boldsymbol{x}}{A_n r^n}.$$

The function $\boldsymbol{x}/r^n$ is locally integrable in $\mathbb{R}^n$ and ψ has compactly support. Therefore we can deduce from Green's first identity (Theorem 2.10) that

$$\begin{aligned}\int_{\mathbb{R}^n} \Phi \Delta\psi \, d^n\boldsymbol{x} &= -\int_{\mathbb{R}^n} \nabla\Phi \cdot \nabla\psi \, d^n\boldsymbol{x} \\ &= \frac{1}{A_n}\int_{\mathbb{R}^n} \frac{\boldsymbol{x}}{r^n} \cdot \nabla\psi \, d^n\boldsymbol{x} \\ &= \frac{1}{A_n}\int_{\mathbb{R}^n} \frac{1}{r^{n-1}} \frac{\partial\psi}{\partial r} \, d^n\boldsymbol{x}.\end{aligned}$$

The integral can be evaluated using radial coordinates as in (2.10):

$$\begin{aligned}\int_{\mathbb{R}^n} \Phi \Delta\psi \, d^n\boldsymbol{x} &= \frac{1}{A_n}\int_{\mathbb{S}^{n-1}} \int_0^\infty \frac{\partial\psi}{\partial r} \, dr \, dS \\ &= -\frac{1}{A_n}\int_{\mathbb{S}^{n-1}} \psi(0) \, dS \\ &= -\psi(0).\end{aligned}$$

This shows that

$$(-\Delta\Phi, \psi) = \psi(0),$$

hence $-\Delta\Phi = \delta$. □

12.5 Green's Functions

Although fundamental solutions are defined only for the domain $\mathbb{R}^n$, one of their principle applications is to boundary value problems on a bounded domain $\Omega \subset \mathbb{R}^n$. The connection comes from a integral formula introduced in 1828 by George Green.

For this section, let Φ denote the fundamental solution of the Laplacian on $\mathbb{R}^n$, as given by (12.32). For $\boldsymbol{y} \in \mathbb{R}^n$ we set

$$\Phi_{\boldsymbol{y}}(\boldsymbol{x}) := \Phi(\boldsymbol{x} - \boldsymbol{y}). \tag{12.33}$$

Theorem 12.9 (Green's representation formula) *Suppose that $\Omega \subset \mathbb{R}^n$ is a bounded domain with piecewise C^1 boundary. For $u \in C^2(\overline{\Omega})$,*

$$u(y) = -\int_{\Omega} \Phi_y \Delta u \, d^n x + \int_{\partial\Omega} \left[\Phi_y \frac{\partial u}{\partial \nu} - u \frac{\partial \Phi_y}{\partial \nu} \right] dS$$

for $y \in \Omega$.

Proof Because the point $y \in \Omega$ is fixed, for notational convenience we can change variables to assume $y = 0$. For $\varepsilon > 0$ set

$$B_\varepsilon := B(0; \varepsilon),$$

and assume that ε is small enough that $\overline{B}_\varepsilon \subset \Omega$.

On $\overline{\Omega} - B_\varepsilon$, Φ is smooth and satisfies $\Delta\Phi = 0$. Therefore, applying Green's second identity (Theorem 2.11) on this domain with $v = \Phi$ gives

$$\int_{\Omega - \overline{B}_\varepsilon} \Phi \, \Delta u \, d^n x = \int_{\partial\Omega} \left(\Phi \frac{\partial u}{\partial \nu} - u \frac{\partial \Phi}{\partial \nu} \right) dS + \int_{\partial B_\varepsilon} \left(\Phi \frac{\partial u}{\partial \nu} - u \frac{\partial \Phi}{\partial \nu} \right) dS. \tag{12.34}$$

Because Δu is continuous and Φ is locally integrable,

$$\lim_{\varepsilon \to 0} \int_{\Omega - \overline{B}_\varepsilon} \Phi \, \Delta u \, d^n x = \int_{\Omega} \Phi \, \Delta u \, d^n x,$$

To prove the representation formula we must therefore show that

$$\lim_{\varepsilon \to 0} \int_{\partial B_\varepsilon} \left(\Phi \frac{\partial u}{\partial r} - u \frac{\partial \Phi}{\partial r} \right) dS = u(0). \tag{12.35}$$

To handle the first term in (12.35), note that

$$\left| \frac{\partial u}{\partial r} \right| \le |\nabla u|,$$

for $r > 0$. Therefore, since ∇u is continuous by assumption, we have a bound

$$\max_{\partial B_\varepsilon} \left| \frac{\partial u}{\partial r} \right| \le M$$

for $\varepsilon > 0$, with M independent of ε. Using the fact that $\operatorname{vol}(\partial B_\varepsilon) = A_n \varepsilon^{n-1}$ and the formula (12.32) for Φ, we can estimate

$$\left| \int_{\partial B_\varepsilon} \Phi \frac{\partial u}{\partial r} \, dS \right| \le M \begin{cases} \varepsilon \ln \varepsilon, & n = 2, \\ \frac{\varepsilon}{n-2}, & n \ge 3. \end{cases}$$

This shows that

$$\lim_{\varepsilon \to 0} \int_{\partial B_\varepsilon} \Phi \frac{\partial u}{\partial r} \, dS = 0.$$

For the second term in (12.35), we use the fact that

$$\frac{\partial \Phi}{\partial r} = -\frac{1}{A_n r^{n-1}},$$

for $r > 0$, to compute

$$-\int_{\partial B_\varepsilon} u \frac{\partial \Phi}{\partial r}\, dS = \frac{1}{A_n \varepsilon^{n-1}} \int_{\partial B_\varepsilon} u\, dS.$$

The right-hand side is the average value of u over the sphere ∂B_ε. By continuity,

$$\lim_{\varepsilon \to 0} \frac{1}{\mathrm{vol}(\partial B_\varepsilon)} \int_{\partial B_\varepsilon} u\, dS = u(0).$$

This proves (12.35), and thus establishes the representation formula. □

The representation formula of Theorem 12.9 has many applications. The original goal that Green had in mind was a solution formula for the Poisson problem with inhomogeneous Dirichlet boundary conditions, which we will now describe.

Suppose there exists a family of functions $H_y \in C^2(\overline{\Omega})$, for $\boldsymbol{y} \in \Omega$, satisfying

$$\Delta H_y = 0, \qquad H_y\big|_{\partial\Omega} = \Phi_y\big|_{\partial\Omega}. \tag{12.36}$$

Then the *Green's function* of Ω is

$$G_y := \Phi_y - H_y. \tag{12.37}$$

It is possible to show that H_y exists under general regularity conditions on $\partial\Omega$, but this is too technical for us to get into here. We will focus on cases where H_y can be computed explicitly, which requires the geometry of Ω to be very simple.

Theorem 12.10 *Suppose $\Omega \subset \mathbb{R}^n$ is a bounded domain with piecewise C^1 boundary that admits a Green's function G_y, Then the Poisson problem on Ω,*

$$-\Delta u = f, \qquad u\big|_{\partial\Omega} = g,$$

for $f \in C^0(\Omega)$, $g \in C^0(\partial\Omega)$, is solved by the function

$$u(\boldsymbol{y}) = -\int_\Omega f G_y\, d^n\boldsymbol{x} - \int_{\partial\Omega} g \frac{\partial G_y}{\partial \nu}\, dS.$$

Proof Setting $v = H_y$ in Green's second identity (Theorem 2.11) gives

$$\int_\Omega \Big(H_y \Delta u - u \Delta H_y\Big) d^n\boldsymbol{y} = \int_{\partial\Omega} H_y \frac{\partial u}{\partial \nu}\, dS. \tag{12.38}$$

The result then follows by subtracting (12.38) from the representation formula of Theorem 12.9. □

Example 12.11 The Green's function for the unit disk $\mathbb{D} \subset \mathbb{R}^2$ can be derived using a trick from electrostatics called the *method of images*. This involves placing charges outside the domain in order to solve the boundary value problem. For the unit disk, in order to find H_y we consider a charge placed at the point $\tilde{\boldsymbol{y}}$ given by "reflecting" $\boldsymbol{y} \in \mathbb{C}\backslash\{0\}$ across the unit circle, i.e.,

$$\tilde{\boldsymbol{y}} := \frac{\boldsymbol{y}}{|\boldsymbol{y}|^2}. \tag{12.39}$$

Note that $\Phi_{\tilde{\boldsymbol{y}}}$ is harmonic on $\mathbb{D}$ because $\tilde{\boldsymbol{y}} \notin \mathbb{D}$.

For $\boldsymbol{x} \in \partial\mathbb{D}$, let φ denote the angle from $\boldsymbol{y}$ to $\boldsymbol{x}$, as shown in Fig. 12.2. By the law of cosines on the triangle made by 0, $\boldsymbol{y}$ and $\boldsymbol{x}$,

$$|\boldsymbol{x} - \boldsymbol{y}|^2 = 1 + |\boldsymbol{y}|^2 - 2|\boldsymbol{y}| \cos\varphi.$$

If $\boldsymbol{y}$ is replaced by $\tilde{\boldsymbol{y}}$, the corresponding formula is

$$\begin{aligned} |\boldsymbol{x} - \tilde{\boldsymbol{y}}|^2 &= 1 + |\tilde{\boldsymbol{y}}|^2 - 2|\tilde{\boldsymbol{y}}| \cos\varphi \\ &= 1 + |\boldsymbol{y}|^{-2} - 2|\boldsymbol{y}|^{-1} \cos\varphi. \end{aligned}$$

Solving for $\cos\varphi$ in these expressions gives the relation

$$\frac{|\boldsymbol{x} - \boldsymbol{y}|^2 - 1 - |\boldsymbol{y}|^2}{2|\boldsymbol{y}|} = \frac{|\boldsymbol{x} - \tilde{\boldsymbol{y}}|^2 - 1 - |\boldsymbol{y}|^{-2}}{2|\boldsymbol{y}|^{-1}},$$

which simplifies to

$$|\boldsymbol{x} - \tilde{\boldsymbol{y}}| = \frac{|\boldsymbol{x} - \boldsymbol{y}|}{|\boldsymbol{y}|} \tag{12.40}$$

for $\boldsymbol{x} \in \partial\mathbb{D}$.

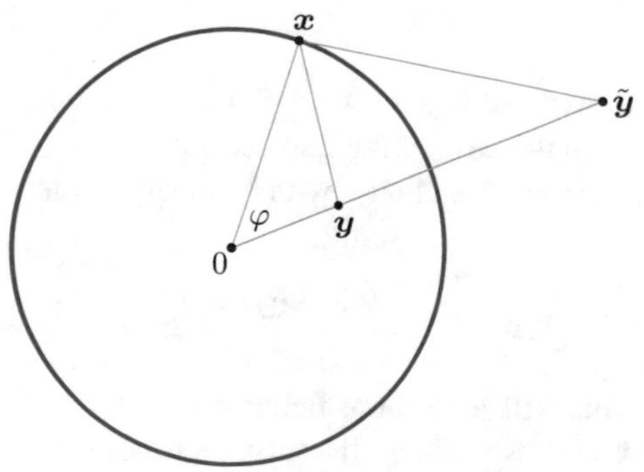

Fig. 12.2 Geometry for the method of images on the unit disk

Since $\Phi(x) = -\frac{1}{2\pi} \ln |x|$ in $\mathbb{R}^2$, taking the logarithm of (12.40) gives

$$\Phi_{\tilde{y}}(x) = \Phi_y(x) + \frac{1}{2\pi} \ln |y|$$

for $y \neq 0$ or x. Thus we can solve (12.36) for $y \neq 0$ by setting

$$H_y := \Phi_{\tilde{y}} - \frac{1}{2\pi} \ln |y|.$$

For $y = 0$ the obvious solution is $H_0 := 0$ because $\Phi|_{\partial\mathbb{D}} = 0$.

The Green's function is thus

$$G_y(x) = \begin{cases} -\frac{1}{2\pi} \ln \left(\frac{|x-y|}{|x-\tilde{y}|\,|y|} \right), & y \neq 0 \\ -\frac{1}{2\pi} \ln |x|, & y = 0. \end{cases}$$

To apply this in the solution formula, we need the radial derivative of G. For y fixed and $r := |x|$ we compute

$$\begin{aligned} \frac{\partial}{\partial r} \ln |x - y| &= x \cdot \nabla \ln |x - y| \\ &= x \cdot \frac{(x - y)}{|x - y|^2} \\ &= \frac{1 - x \cdot y}{|x - y|^2}. \end{aligned}$$

Applying the corresponding result for $|x - \tilde{y}|$ and subtracting gives

$$\frac{\partial G_y}{\partial r}(x) = -\frac{1}{2\pi} \left(\frac{1 - x \cdot y}{|x - y|^2} - \frac{1 - x \cdot \tilde{y}}{|x - \tilde{y}|^2} \right),$$

which by (12.40) simplifies to

$$\frac{\partial G_y}{\partial r}(x) = -\frac{1 - |y|^2}{2\pi |x - y|^2}. \tag{12.41}$$

Conveniently, the calculation for $y = 0$ leads to the same expression.

In the case of the Laplace equation on $\mathbb{D}$, Theorem 12.10 gives the formula for a harmonic function u with boundary value g as

$$u(y) = \frac{1}{2\pi} \int_{\partial\mathbb{D}} \frac{1 - |y|^2}{|x - y|^2} g(x)\, dS(x).$$

This will look more familiar in polar coordinates. With $y = (r \cos\theta, r \sin\theta)$ and $x = (\cos\eta, \sin\eta)$, the formula becomes

$$u(r, \theta) = \frac{1}{2\pi} \int_0^{2\pi} \frac{1 - r^2}{1 + r^2 - 2r\cos(\eta - \theta)} g(\cos\eta, \sin\eta)\, d\eta.$$

This is the classical Poisson formula (9.4) that we derived from Fourier series. ◇

12.6 Time-Dependent Fundamental Solutions

To adapt the concept of a fundamental solution to evolution equations, we need to consider time-dependent distributions on $\mathbb{R}^n$. We will use a subscript to denote the time dependence, to avoid confusion with the spatial variables. Thus a map $\mathbb{R} \to \mathcal{D}'(\mathbb{R}^n)$ will be written

$$t \mapsto w_t.$$

For $\psi \in C^\infty_{\text{cpt}}(\mathbb{R}^n)$ the pairing (w_t, ψ) is a complex-valued function of t.

The function $t \mapsto w_t$ is differentiable with respect to time if there exists a family of distributions $\frac{\partial w_t}{\partial t} \in \mathcal{D}'(\mathbb{R}^n)$ such that

$$\frac{d}{dt}(w_t, \psi) = \left(\frac{\partial w_t}{\partial t}, \psi \right), \tag{12.42}$$

for all $\psi \in C^\infty_{\text{cpt}}(\Omega)$. Higher derivatives are defined in the same way.

Example 12.12 In $\mathbb{R}$, consider the derivatives of δ_t, the delta function supported at t. By definition,

$$(\delta_t, \psi) := \psi(t),$$

so that

$$\left(\frac{\partial^n}{\partial t^n} \delta_t, \psi \right) = \psi^{(n)}(t).$$

Compare this to the spatial derivatives, defined according to (12.20),

$$\begin{aligned} \left(\frac{\partial^n}{\partial x^n} \delta_t, \psi \right) &:= (-1)^n \left(\delta_t, \psi^{(n)} \right) \\ &= (-1)^n \psi^{(n)}(t). \end{aligned}$$

We conclude that

$$\frac{\partial^n}{\partial t^n} \delta_t = (-1)^n \frac{\partial^n}{\partial x^n} \delta_t. \tag{12.43}$$

◇

Let us try to deduce the fundamental solution for the one-dimensional wave equation from d'Alembert's formula (4.8),

$$u(t,x) = \frac{1}{2}\big[g(x+t) + g(x-t)\big] + \frac{1}{2}\int_{x-t}^{x+t} h(\tau)\,d\tau, \tag{12.44}$$

The second term in (12.44) could be interpreted as a convolution

$$\begin{aligned}\int_{x-t}^{x+t} h(\tau)\,d\tau &= \int_{-\infty}^{\infty} \chi_{[-t,t]}(x-\tau)h(\tau)\,d\tau \\ &= \chi_{[-t,t]} * h(x),\end{aligned}$$

where χ_I denotes a characteristic function as in (7.5). Therefore it makes sense to define this component of the fundamental solution as

$$W_t := \frac{1}{2}\chi_{(-t,t)}. \tag{12.45}$$

The time derivatives of W_t are computed from the pairing

$$(W_t, \psi) = \frac{1}{2}\int_{-t}^{t} \psi\,dx$$

for $\psi \in C^\infty_{\text{cpt}}(\mathbb{R})$. By the fundamental theorem of calculus,

$$\frac{d}{dt}(W_t, \psi) = \frac{1}{2}\big[\psi(t) + \psi(-t)\big], \tag{12.46}$$

which shows that

$$\frac{\partial W_t}{\partial t} = \frac{1}{2}\big(\delta_t + \delta_{-t}\big). \tag{12.47}$$

Differentiating again using (12.43) gives

$$\frac{\partial^2 W_t}{\partial t^2} = \frac{1}{2}\left(-\delta_t' + \delta_{-t}'\right). \tag{12.48}$$

On the other hand, x-derivatives of W_t are defined by (12.20). In particular,

$$\left(\frac{\partial^2 W_t}{\partial x^2}, \psi\right) := \left(W_t, \psi''\right)$$

This can be evaluated by direct integration,

$$\begin{aligned}\left(\frac{\partial^2 W_t}{\partial x^2}, \psi\right) &:= \frac{1}{2}\int_{-t}^{t} \psi''\,dx \\ &= \frac{1}{2}\big[\psi'(t) - \psi'(-t)\big].\end{aligned}$$

This shows that

$$\frac{\partial^2 W_t}{\partial x^2} = \frac{1}{2}\left(-\delta_t' + \delta_{-t}'\right). \tag{12.49}$$

By (12.48) and (12.49), W_t is a distributional solution of the wave equation,

$$\left(\frac{\partial^2}{\partial t^2} - \frac{\partial^2}{\partial x^2}\right) W_t = 0.$$

In contrast to the definition (12.27) of a fundamental solution in the spatial case, W_t satisfies a homogeneous equation. The delta function appears only in the boundary conditions,

$$W_0 = 0, \quad \left.\frac{\partial W_t}{\partial t}\right|_{t=0} = \delta.$$

The distribution W_t, which is analogous to a fundamental solution, is called the *wave kernel*. By (12.47), the g component of the d'Alembert solution formula (12.44) could be written in terms of W_t as

$$\frac{1}{2}\big[g(x+t) + g(x-t)\big] = \frac{\partial W_t}{\partial t} * g(x).$$

Thus the full convolution formula for the solution reads

$$u(t, \cdot) = \frac{\partial W_t}{\partial t} * g + W_t * h.$$

12.7 Exercises

12.1 Define the distribution $u \in \mathcal{D}'(\mathbb{R})$ by

$$(u, \psi) := \int_{-1}^{1} \frac{\psi(x) - \psi(0)}{x}\, dx + \int_{|x|\geq 1} \frac{\psi(x)}{x}\, dx.$$

Show that $u = \mathrm{PV}[x^{-1}]$.

12.2 Let $f \in L^1_{\mathrm{loc}}(\mathbb{R})$ be the function

$$f(x) = \begin{cases} \log x, & x > 0, \\ -\log(-x), & x < 0. \end{cases}$$

For $x \neq 0$, $f'(x) = |x|^{-1}$, but this is not locally integrable. Show that the distributional derivative is

$$(f', \psi) = \int_{-1}^{1} \frac{\psi(x) - \psi(0)}{|x|} + \int_{|x|\geq 1} \frac{\psi(x)}{|x|}\, dx.$$

12.3 Let $\mathbb{H}$ denote the upper half-plane $\{x_2 > 0\} \subset \mathbb{R}^2$. The goal of this problem is to show that the Laplace equation on $\mathbb{H}$,

$$\Delta u = 0, \qquad u(\cdot, 0) = g,$$

has the solution

$$u(\boldsymbol{y}) = \frac{1}{\pi} \int_{-\infty}^{\infty} \frac{y_2}{(x - y_1)^2 + y_2^2}\, g(x)\, dx$$

for $g \in C^{\infty}_{\mathrm{cpt}}(\mathbb{R})$.

(a) Derive this formula from Theorem 12.10 using the method of images as in Example 12.11. In this case the reflection of $\boldsymbol{y} \in \mathbb{H}$ is given by $\overline{(y_1, y_2)} = (y_1, -y_2)$ (the complex conjugate).
(b) Show that the fact that $u(\cdot, 0) = g$ could also be derived by using Lemma 12.1 to deduce that

$$\lim_{x \to 0} \frac{y}{\pi(x^2 + y^2)} = \delta(y).$$

12.4 In $\mathbb{R}^3$ show that

$$(-\Delta - k^2) \frac{e^{ikr}}{4\pi r} = \delta$$

for all $k \in \mathbb{R}$.

12.5 For $n \geq 3$ let $\mathbb{B}^n$ denote the unit ball $\{r < 1\} \subset \mathbb{R}^n$.

(a) Apply the method of images as in Example 12.11 to derive the solution $H_{\boldsymbol{y}}$ of (12.36) for $\boldsymbol{y} \in \mathbb{B}^n$. and compute the Green's function. (Note that the formulas (12.39) and (12.40) remain valid in any dimension.)
(b) Show that the radial derivative of the Green's function satisfies

$$\frac{\partial G_{\boldsymbol{y}}}{\partial r}(\boldsymbol{x}) = -\frac{1 - |\boldsymbol{y}^2|}{A_n |\boldsymbol{x} - \boldsymbol{y}|^{n-1}}.$$

(c) Find the resulting solution formula from Theorem 12.10, and show that this generalizes the mean value formula for harmonic functions obtained in Theorem 9.3.

Chapter 13
The Fourier Transform

For a bounded domain $\Omega \subset \mathbb{R}^n$, Theorem 11.7 shows that we can effectively "diagonalize" the Laplacian by choosing an orthonormal basis for $L^2(\Omega)$ consisting of eigenfunctions. Such a result is not possible on $\mathbb{R}^n$ itself; the Laplacian has no eigenfunctions in $L^2(\mathbb{R}^n)$.

The closest analog to eigenfunctions on $\mathbb{R}^n$ are the spatial components of the plane wave solutions introduced in Exercise 4.8,

$$\phi_{\boldsymbol{\xi}}(\boldsymbol{x}) := e^{i\boldsymbol{\xi}\cdot\boldsymbol{x}},$$

associated to a frequency vector $\boldsymbol{\xi} \in \mathbb{R}^n$. These functions satisfy a convenient differentiation formula,

$$D^\alpha \phi_{\boldsymbol{\xi}} = (i\boldsymbol{\xi})^\alpha \phi_{\boldsymbol{\xi}},$$

and in particular

$$-\Delta e^{i\boldsymbol{\xi}\cdot\boldsymbol{x}} = |\boldsymbol{\xi}|^2 e^{i\boldsymbol{\xi}\cdot\boldsymbol{x}}.$$

The appropriate generalization of the Fourier series to $L^2(\mathbb{R}^n)$ is an integral transform based on these plane waves. Although the technical details are quite different from Fourier series, the transform serves a similar purpose in that it exchanges the roles of differentiation and multiplication.

13.1 Fourier Transform

The *Fourier transform* of a function in $f \in L^1(\mathbb{R}^n)$ is a function of the frequency $\boldsymbol{\xi} \in \mathbb{R}^n$ defined by

$$\hat{f}(\boldsymbol{\xi}) := \int_{\mathbb{R}^n} e^{-i\boldsymbol{\xi}\cdot\boldsymbol{x}} f(\boldsymbol{x})\, d^n\boldsymbol{x}. \tag{13.1}$$

D. Borthwick, *Introduction to Partial Differential Equations*,
Universitext, DOI 10.1007/978-3-319-48936-0_13

Note that the integral is well defined by the integrability of f, and in fact

$$\left|\hat{f}(\boldsymbol{\xi})\right| \leq \|f\|_{L^1} \tag{13.2}$$

for all $\boldsymbol{\xi} \in \mathbb{R}^n$. As a map the transform is denoted by

$$\mathcal{F} : f \mapsto \hat{f}.$$

To develop the properties of the Fourier transform, it proves convenient to introduce a particular class of test functions, called *Schwartz functions*. The space $\mathcal{S}$ consists of smooth functions which, along with all derivatives, decay *rapidly* at infinity. The precise meaning of "rapid" is "faster than any power of r." An alternate form of this definition is

$$\mathcal{S}(\mathbb{R}^n) := \left\{ f \in C^\infty(\mathbb{R}^n);\ \left\|\boldsymbol{x}^\alpha D^\beta f\right\|_\infty < \infty \text{ for all } \alpha, \beta \right\}, \tag{13.3}$$

where $\|\cdot\|_\infty$ is the sup norm introduced in Sect. 7.3.

A basic example of a Schwartz function is a *Gaussian function* of the form

$$f(\boldsymbol{x}) = e^{-a|\boldsymbol{x}|^2},$$

with $a > 0$. We also have

$$C^\infty_{\mathrm{cpt}}(\mathbb{R}^n) \subset \mathcal{S}(\mathbb{R}^n),$$

because compactly supported functions obviously satisfy the decay requirement.

As in the discrete case, the Fourier transform interchanges the operations of differentiation and multiplication in a convenient way. For this statement, we let $D^\alpha_{\boldsymbol{x}}$ and $D^\alpha_{\boldsymbol{\xi}}$ denote partial derivatives with respect to $\boldsymbol{x}$ or $\boldsymbol{\xi}$, respectively.

Lemma 13.1 *For $\psi \in \mathcal{S}(\mathbb{R}^n)$,*

$$\mathcal{F}[D^\alpha_{\boldsymbol{x}} \psi](\boldsymbol{\xi}) = (i\boldsymbol{\xi})^\alpha \hat{\psi}(\boldsymbol{\xi}), \tag{13.4}$$

and

$$\mathcal{F}[\boldsymbol{x}^\alpha \psi](\boldsymbol{\xi}) = (i D_{\boldsymbol{\xi}})^\alpha \hat{\psi}(\boldsymbol{\xi}). \tag{13.5}$$

Proof The first identity follows from integration by parts,

$$\begin{aligned}
\mathcal{F}[D^\alpha \psi](\boldsymbol{\xi}) &= \int_{\mathbb{R}^n} e^{-i\boldsymbol{\xi}\cdot\boldsymbol{x}} D^\alpha_{\boldsymbol{x}} \psi(\boldsymbol{x})\, d^n\boldsymbol{x} \\
&= \int_{\mathbb{R}^n} \psi(\boldsymbol{x}) (i D_{\boldsymbol{\xi}})^\alpha (e^{-i\boldsymbol{\xi}\cdot\boldsymbol{x}})\, d^n\boldsymbol{x} \\
&= \int_{\mathbb{R}^n} \psi(\boldsymbol{x}) (i\boldsymbol{\xi})^\alpha e^{-i\boldsymbol{\xi}\cdot\boldsymbol{x}}\, d^n\boldsymbol{x} \\
&= (i\boldsymbol{\xi})^\alpha \hat{\psi}(\boldsymbol{\xi}).
\end{aligned}$$

The second is also a direct computation,

$$\begin{aligned}\mathcal{F}[\boldsymbol{x}^\alpha f](\boldsymbol{\xi}) &= \int_{\mathbb{R}^n} e^{-i\boldsymbol{\xi}\cdot\boldsymbol{x}}\boldsymbol{x}^\alpha f(\boldsymbol{x})\, d^n\boldsymbol{x} \\ &= \int_{\mathbb{R}^n} (iD_{\boldsymbol{\xi}})^\alpha (e^{-i\boldsymbol{\xi}\cdot\boldsymbol{x}}) f(\boldsymbol{x})\, d^n\boldsymbol{x} \\ &= (iD_{\boldsymbol{\xi}})^\alpha \hat{f}(\boldsymbol{\xi}).\end{aligned}$$

Pulling the differentiation outside the integral in the final step is justified by the smoothness and decay assumptions on ψ. □

In Lemma 13.1 we can see Schwartz's motivation for the definition of $\mathcal{S}$. Under the Fourier transform, smoothness translates to rapid decay, and vice versa. These properties are balanced in the definition of $\mathcal{S}$, which leads to the following result.

Lemma 13.2 *The Fourier transform $\mathcal{F}$ maps $\mathcal{S}(\mathbb{R}^n) \to \mathcal{S}(\mathbb{R}^n)$.*

Proof Suppose that $f \in \mathcal{S}$. In order to show that $\hat{f}$ is Schwartz, we need to produce a bound on the function $\boldsymbol{\xi}^\beta D^\alpha \hat{f}$ for each α, β. By (13.4) and (13.5),

$$\boldsymbol{\xi}^\beta D_{\boldsymbol{\xi}}^\alpha \hat{f}(\boldsymbol{\xi}) = i^{|\alpha|+|\beta|} \int_{\mathbb{R}^n} e^{-i\boldsymbol{\xi}\cdot\boldsymbol{x}}\boldsymbol{x}^\alpha D_{\boldsymbol{x}}^\beta f(\boldsymbol{x})\, d^n\boldsymbol{x}. \tag{13.6}$$

To estimate, we set

$$M_{N,\alpha,\beta} := \left\| (1+|\boldsymbol{x}|^2)^N \boldsymbol{x}^\alpha D_{\boldsymbol{x}}^\beta f \right\|_\infty,$$

which is finite by the definition (13.3). Because $(1+|x|^2)^{-N}$ is integrable for N sufficiently large, we can estimate (13.6) by

$$\left|\boldsymbol{\xi}^\beta D^\alpha \hat{f}(\boldsymbol{\xi})\right| \le M_{N,\alpha,\beta} \int_{\mathbb{R}^n} \frac{1}{(1+|x|^2)^N}\, d^n\boldsymbol{x}.$$

The right-hand side is independent of $\boldsymbol{\xi}$, so this yields the required estimate. □

Example 13.3 Consider the one-dimensional Gaussian function

$$\varphi(x) := e^{-ax^2}$$

for $a > 0$. Note that φ satisfies the ODE

$$\frac{d\varphi}{dx} = -2ax\varphi.$$

Taking the Fourier transform of both sides and applying Lemma 13.1 gives

$$i\xi\hat{\varphi} = -2ai\frac{d\hat{\varphi}}{d\xi},$$

which reduces to

$$\frac{d\hat{\varphi}}{d\xi} = -\frac{\xi}{2a}\hat{\varphi}.$$

Separating variables and integrating yields the solution

$$\hat{\varphi}(\xi) = \hat{\varphi}(0)e^{-\xi^2/4a}.$$

To fix the constant, we can use (2.19) with $n = 1$ to compute

$$\begin{aligned}\hat{\varphi}(0) &= \int_{-\infty}^{\infty} e^{-ax^2}\,dx \\ &= \frac{1}{\sqrt{a}}\int_{-\infty}^{\infty} e^{-x^2}\,dx \\ &= \sqrt{\frac{\pi}{a}}.\end{aligned} \tag{13.7}$$

Thus,

$$\hat{\varphi}(\xi) = \sqrt{\frac{\pi}{a}}e^{-\xi^2/4a}.$$

◊

The computation from Example 13.3 can be generalized to $\mathbb{R}^n$ by factoring the integrals,

$$\begin{aligned}\mathcal{F}\left[e^{-a|\boldsymbol{x}|^2}\right](\boldsymbol{\xi}) &= \int_{\mathbb{R}^n} e^{-i\boldsymbol{x}\cdot\boldsymbol{\xi}-a|\boldsymbol{x}|^2}\,d^n\boldsymbol{x} \\ &= \prod_{j=1}^{n}\left(\int_{-\infty}^{\infty} e^{-ix_j\xi_j-ax_j^2}\,dx_j\right) \\ &= \prod_{j=1}^{n}\sqrt{\frac{\pi}{a}}e^{-\xi_j^2/4a}.\end{aligned}$$

Thus

$$\mathcal{F}\left[e^{-a|\boldsymbol{x}|^2}\right](\boldsymbol{\xi}) = \left(\frac{\pi}{a}\right)^{\frac{n}{2}} e^{-|\boldsymbol{\xi}|^2/4a} \tag{13.8}$$

for $a > 0$.

For $f, g \in \mathcal{S}(\mathbb{R}^n)$, consider the integral

$$\int_{\mathbb{R}^n}\int_{\mathbb{R}^n} f(\boldsymbol{x})e^{-i\boldsymbol{x}\cdot\boldsymbol{y}}g(\boldsymbol{y})\,d^n\boldsymbol{x}\,d^n\boldsymbol{y}. \tag{13.9}$$

The integrals over $\boldsymbol{x}$ and $\boldsymbol{y}$ can be taken in either order, yielding the useful identity:

$$\int_{\mathbb{R}^n} f\hat{g}\, d^n\boldsymbol{x} = \int_{\mathbb{R}^n} \hat{f} g\, d^n\boldsymbol{y} \tag{13.10}$$

for $f, g \in \mathcal{S}(\mathbb{R}^n)$,

Theorem 13.4 *The Fourier transform on $\mathcal{S}(\mathbb{R}^n)$ has an inverse $\mathcal{F}^{-1}$ given by*

$$f(\boldsymbol{x}) = (2\pi)^{-n} \int_{\mathbb{R}^n} e^{i\boldsymbol{\xi}\cdot\boldsymbol{x}} \hat{f}(\boldsymbol{\xi})\, d^n\boldsymbol{\xi}. \tag{13.11}$$

Proof In (13.10) let us set $g = e^{-a\boldsymbol{x}^2}$ for $a > 0$, By (13.8), this implies

$$\left(\frac{\pi}{a}\right)^{\frac{n}{2}} \int_{\mathbb{R}^n} f(\boldsymbol{x}) e^{-\boldsymbol{x}^2/4a}\, d^n\boldsymbol{x} = \int_{\mathbb{R}^n} \hat{f}(\boldsymbol{y}) e^{-a\boldsymbol{y}^2}\, d^n\boldsymbol{y}. \tag{13.12}$$

On the left-hand side we can use the same argument as in the proof of Lemma 12.1 to show that

$$\begin{aligned}\lim_{a\to 0}\left[\left(\frac{\pi}{a}\right)^{\frac{n}{2}} \int_{\mathbb{R}^n} f(\boldsymbol{x}) e^{-\boldsymbol{x}^2/4a}\, d^n\boldsymbol{x}\right] &= \pi^{\frac{n}{2}} \lim_{a\to 0} \int_{\mathbb{R}^n} f(\sqrt{a}\boldsymbol{x}) e^{-\boldsymbol{x}^2/4}\, d^n\boldsymbol{x} \\ &= (2\pi)^n f(0).\end{aligned}$$

We claim that the corresponding limit on the right-hand side of (13.12) is

$$\lim_{a\to 0} \int_{\mathbb{R}^n} \hat{f}(\boldsymbol{y}) e^{-a\boldsymbol{y}^2}\, d^n\boldsymbol{y} = \int_{\mathbb{R}^n} \hat{f}(\boldsymbol{y})\, d^n\boldsymbol{y}. \tag{13.13}$$

Because the convergence is not uniform, we will check this carefully. The difference of the two sides can be estimated by

$$\left|\int_{\mathbb{R}^n} \hat{f}(\boldsymbol{y}) e^{-a\boldsymbol{y}^2}\, d^n\boldsymbol{y} - \int_{\mathbb{R}^n} \hat{f}(\boldsymbol{y})\, d^n\boldsymbol{y}\right| \le \int_{\mathbb{R}^n} \left|\hat{f}(\boldsymbol{y})\right| \left(1 - e^{-a\boldsymbol{y}^2}\right) d^n\boldsymbol{y}.$$

Given $\varepsilon > 0$ we can choose R large enough that

$$\int_{|\boldsymbol{x}|>R} \left|\hat{f}(\boldsymbol{y})\right| d^n\boldsymbol{y} < \varepsilon,$$

since $\hat{f}$ is integrable. Splitting the integral at $|\boldsymbol{x}| = R$ gives the estimate

$$\begin{aligned}\int_{\mathbb{R}^n} \left|\hat{f}(\boldsymbol{y})\right| \left(1 - e^{-a\boldsymbol{y}^2}\right) d^n\boldsymbol{y} &\le \varepsilon + \int_{B(0;R)} \left|\hat{f}(\boldsymbol{y})\right| \left(1 - e^{-a\boldsymbol{y}^2}\right) d^n\boldsymbol{y} \\ &\le \varepsilon + \left(1 - e^{-aR^2}\right) \|\hat{f}\|_1.\end{aligned}$$

The second term approaches zero as $a \to 0$, so that

$$\int_{\mathbb{R}^n} \left|\hat{f}(\boldsymbol{y})\right| \left(1 - e^{-ay^2}\right) d^n\boldsymbol{y} < 2\varepsilon,$$

for a sufficiently small. This establishes (13.13).

By these calculations, the limit of (13.12) as $a \to 0$ yields

$$(2\pi)^n f(0) = \int_{\mathbb{R}^n} \hat{f}(\boldsymbol{x})\, d^n\boldsymbol{x}. \tag{13.14}$$

This is a special case of the desired formula.

The general inverse formula can be deduced from (13.14) by a simple translation argument. For $\boldsymbol{w} \in \mathbb{R}^n$, define the translation operator $T_{\boldsymbol{w}}$ on $\mathcal{S}(\mathbb{R}^n)$ by

$$T_{\boldsymbol{w}} f(\boldsymbol{y}) := f(\boldsymbol{y} + \boldsymbol{w}).$$

A change of variables shows that

$$\begin{aligned}\widehat{T_{\boldsymbol{w}} f}(\boldsymbol{x}) &= \int_{\mathbb{R}^n} e^{-i\boldsymbol{x}\cdot\boldsymbol{y}} f(\boldsymbol{y} + \boldsymbol{w})\, d^n\boldsymbol{y} \\ &= \int_{\mathbb{R}^n} e^{-i\boldsymbol{x}\cdot(\boldsymbol{y}-\boldsymbol{w})} f(\boldsymbol{y})\, d^n\boldsymbol{y} \\ &= e^{i\boldsymbol{x}\cdot\boldsymbol{w}} \hat{f}(\boldsymbol{x}).\end{aligned}$$

Since $T_{\boldsymbol{w}} f(0) = f(\boldsymbol{w})$, plugging $T_{\boldsymbol{w}} f$ into (13.14) gives

$$(2\pi)^n f(\boldsymbol{w}) = \int_{\mathbb{R}^n} e^{i\boldsymbol{x}\cdot\boldsymbol{w}} \hat{f}(\boldsymbol{x})\, d^n\boldsymbol{x}.$$

□

The pairing formula (13.10) suggests that the L^2 inner product will behave naturally under the Fourier transform. Indeed, by Theorem 13.4 we can compute

$$\begin{aligned}\int_{\mathbb{R}^n} \hat{f}(\boldsymbol{\xi})\overline{\hat{g}(\boldsymbol{\xi})}\, d^n\boldsymbol{\xi} &= \int_{\mathbb{R}^n} \left(\int_{\mathbb{R}^n} f(\boldsymbol{x}) e^{-i\boldsymbol{x}\cdot\boldsymbol{\xi}}\, d^n\boldsymbol{x}\right) \overline{\hat{g}(\boldsymbol{\xi})}\, d^n\boldsymbol{\xi} \\ &= \int_{\mathbb{R}^n} \overline{\left(\int_{\mathbb{R}^n} e^{i\boldsymbol{x}\cdot\boldsymbol{\xi}} \hat{g}(\boldsymbol{\xi})\, d^n\boldsymbol{\xi}\right)} f(\boldsymbol{x})\, d^n\boldsymbol{x} \\ &= (2\pi)^n \int_{\mathbb{R}^n} f(\boldsymbol{x})\overline{g(\boldsymbol{x})}\, d^n\boldsymbol{x},\end{aligned}$$

for $f, g \in \mathcal{S}(\mathbb{R}^n)$. In other words,

$$\langle \hat{f}, \hat{g} \rangle = (2\pi)^n \langle f, g \rangle. \tag{13.15}$$

The integral (13.1) defining the Fourier transform does not necessarily converge for $f \in L^2(\mathbb{R}^n)$, but the identity (13.15) makes it possible to define transforms on L^2 by taking limits.

Theorem 13.5 (Plancherel's theorem) *The Fourier transform extends from $\mathcal{S}(\mathbb{R}^n)$ to an invertible map on $L^2(\mathbb{R}^n)$, such that* (13.15) *holds for all $f, g \in L^2(\mathbb{R}^n)$.*

Proof First note that Theorem 7.5 implies that $\mathcal{S}(\mathbb{R}^n)$ is dense in $L^2(\mathbb{R}^n)$ because it includes the compactly supported smooth functions. Hence for $f \in L^2(\mathbb{R}^n)$ there exists a sequence of Schwartz functions $\phi_k \to f$ in L^2. As a convergent sequence, $\{\phi_k\}$ is automatically Cauchy, i.e.,

$$\lim_{k,m\to\infty} \|\phi_k - \phi_m\|_2 = 0.$$

By (13.15),

$$\left\|\hat{\phi}_k - \hat{\phi}_m\right\|_2 = (2\pi)^{n/2} \|\phi_k - \phi_m\|_2,$$

implying that $\{\hat{\phi}_k\}$ is also Cauchy in $L^2(\mathbb{R}^n)$. Since $L^2(\mathbb{R}^n)$ is complete by Theorem 7.7, this implies convergence, and we can then define

$$\hat{f} := \lim_{k\to\infty} \hat{\phi}_k,$$

with the limit taken in the L^2 sense.

To show that (13.15) extends to L^2, suppose for $f, g \in L^2$ that $\phi_k \to f$ and $\psi_m \to g$ are approximating sequences of Schwartz functions. By the property (13.15),

$$\langle \hat{\phi}_k, \hat{\psi}_m \rangle = (2\pi)^n \langle \phi_k, \psi_m \rangle.$$

Taking the limit $k, m \to \infty$ then gives

$$\langle \hat{f}, \hat{g} \rangle = (2\pi)^n \langle f, g \rangle.$$

The same argument can be used to show that $\hat{f}$ is independent of the choice of approximating sequence. □

13.2 Tempered Distributions

Since $\mathcal{F}$ maps $\mathcal{S}(\mathbb{R}^n)$ to itself, to extend the Fourier transform to distributions it is natural replace $C^\infty_{\text{cpt}}(\mathbb{R}^n)$ by $\mathcal{S}(\mathbb{R}^n)$ as the space of test functions. The result is the space of *tempered distributions*

$$\mathcal{S}'(\mathbb{R}^n) := \left\{\text{continuous linear functionals } \mathcal{S}(\mathbb{R}^n) \to \mathbb{C}\right\}.$$

Here the word "tempered" refers to a restriction on the growth at infinity. Because the Schwartz functions decay rapidly, a locally integrable function is essentially required to have a polynomial growth rate at infinity in order to define an element of $\mathcal{S}'(\mathbb{R}^n)$.

The definition of continuity of a functional on $\mathcal{S}(\mathbb{R}^n)$ depends on a notion of convergence for Schwartz functions. A sequence $\{\psi_k\} \subset \mathcal{S}(\mathbb{R}^n)$ converges if the sequences $\{x^\alpha D^\beta \psi_k\}$ converge uniformly for each α, β. To say that $u \in \mathcal{S}'(\mathbb{R}^n)$ is continuous means that $(u, \psi_k) \to (u, \psi)$ whenever $\psi_k \to \psi$ in $\mathcal{S}(\mathbb{R}^n)$.

The delta function δ_x and its derivatives are clearly tempered distributions. We claim also that

$$L^p(\mathbb{R}^n) \subset \mathcal{S}'(\mathbb{R}^n),$$

for $p \in [1, \infty]$. This follows fairly directly from the fact that $\mathcal{S}(\mathbb{R}^n) \subset L^p(\mathbb{R}^n)$ for $p \in [1, \infty]$.

The pairing formula (13.10) gives the prescription for extending $\mathcal{F}$ to the tempered distributions. For $u \in \mathcal{S}'(\mathbb{R}^n)$, we define $\hat{u}$ by

$$(\hat{u}, \phi) := (u, \hat{\phi}) \tag{13.16}$$

for $\phi \in \mathcal{S}(\mathbb{R}^n)$. To justify this definition one needs to check that the Fourier transform is continuous as a map $\mathcal{S}(\mathbb{R}^n) \to \mathcal{S}(\mathbb{R}^n)$. This essentially follows from the calculations in the proof of Lemma 13.2.

As an example, consider the function $u = 1$ as an element of $\mathcal{S}'(\mathbb{R}^n)$. For $\psi \in \mathcal{S}(\mathbb{R}^n)$,

$$\begin{aligned}(\hat{1}, \psi) &:= (1, \hat{\psi}) \\ &= \int_{\mathbb{R}^n} \hat{\psi}(\mathbf{x})\, d^n\mathbf{x}.\end{aligned}$$

According to the inverse Fourier transform formula (13.11),

$$\int_{\mathbb{R}^n} \hat{\psi}(\mathbf{x})\, d^n\mathbf{x} = (2\pi)^n \psi(0).$$

Therefore

$$\hat{1} = (2\pi)^n \delta. \tag{13.17}$$

Physicists often express this fact by writing

$$\delta(\mathbf{x}) = (2\pi)^{-n} \int_{\mathbb{R}^n} e^{-i\mathbf{x}\cdot\boldsymbol{\xi}}\, d^n\boldsymbol{\xi},$$

with the understanding that the integral on the right is not to be taken literally.

The Fourier transform of δ is a similar calculation. For $\psi \in \mathcal{S}$,

$$\begin{aligned}(\hat{\delta}, \psi) &:= (\delta, \hat{\psi}) \\ &= \hat{\psi}(0) \\ &= \int_{\mathbb{R}^n} \psi(\boldsymbol{x})\, d^n\boldsymbol{x}.\end{aligned} \tag{13.18}$$

Therefore

$$\hat{\delta} = 1. \tag{13.19}$$

Because differentiation and multiplication by polynomials are continuous operations on $\mathcal{S}(\mathbb{R}^n)$, they extend to tempered distributions. From Lemma 13.1 we immediately derive the following:

Lemma 13.6 *For* $u \in \mathcal{S}'(\mathbb{R}^n)$,

$$\mathcal{F}[D_{\boldsymbol{x}}^\alpha u] = (i\boldsymbol{\xi})^\alpha \hat{u},$$

and

$$\mathcal{F}[\boldsymbol{x}^\alpha u] = (iD)_{\boldsymbol{\xi}}^\alpha \hat{u}.$$

The Fourier transform on $\mathcal{S}'(\mathbb{R}^n)$ is particularly useful in the construction of fundamental solutions. Consider the constant coefficient operator

$$L = \sum_{|\alpha| \le m} a_\alpha D^\alpha$$

with $a_\alpha \in \mathbb{C}$. According to Lemma 13.6 and (13.19), the Fourier transform of the equation

$$L\Phi = \delta$$

is

$$P(\boldsymbol{\xi})\hat{\Phi} = 1,$$

where

$$P(\boldsymbol{\xi}) := \sum_{|\alpha| \le m} a_\alpha (i\boldsymbol{\xi})^\alpha.$$

If the reciprocal of $P(\boldsymbol{\xi})$ makes sense as a tempered distribution then we can set $\hat{\Phi}(\boldsymbol{\xi}) = 1/P(\boldsymbol{\xi})$ and take the inverse Fourier transform to construct a fundamental solution Φ as an element of $\mathcal{S}'(\mathbb{R}^n)$.

Example 13.7 The polynomial corresponding to $-\Delta$ on $\mathbb{R}^n$ is $P(\boldsymbol{\xi}) = |\boldsymbol{\xi}|^2$. For $n \ge 3$ the function $|\boldsymbol{\xi}|^{-2}$ is locally integrable and decays at infinity, so this defines a tempered distribution. Hence we should be able to compute Φ as the inverse Fourier transform of $|\boldsymbol{\xi}|^{-2}$.

Because $|\boldsymbol{\xi}|^{-2}$ is not globally integrable, we cannot apply the formula (13.11) directly. A trick to get around this is based on the fact that

$$\int_0^\infty e^{-\alpha t}\,dt = \alpha^{-1}$$

for $\alpha > 0$. Setting $\alpha = |\boldsymbol{\xi}|^2$ gives

$$|\boldsymbol{\xi}|^{-2} = \int_0^\infty e^{-t|\xi|^2}\,dt$$

for $\boldsymbol{\xi} \neq 0$. We can pair both sides with a Schwartz function $\psi(\boldsymbol{\xi})$ and integrate to show that

$$\mathcal{F}^{-1}\left[|\boldsymbol{\xi}|^{-2}\right] = \int_0^\infty \mathcal{F}^{-1}\left[e^{-t|\xi|^2}\right] dt. \tag{13.20}$$

Setting $a = 1/(4t)$ in (13.8) gives

$$\mathcal{F}^{-1}\left[e^{-t|\xi|^2}\right](\boldsymbol{x}) = (4\pi t)^{-\frac{n}{2}} e^{-|\boldsymbol{x}|^2/4t},$$

so that (13.20) reduces to

$$\mathcal{F}^{-1}\left[|\boldsymbol{\xi}|^{-2}\right] = \int_0^\infty (4\pi t)^{-\frac{n}{2}} e^{-|\boldsymbol{x}|^2/4t}\,dt.$$

To evaluate the integral we substitute $s = |\boldsymbol{x}|^2/4t$ to obtain

$$\begin{aligned}\mathcal{F}^{-1}\left[|\boldsymbol{\xi}|^{-2}\right] &= \int_0^\infty \left(\frac{\pi\,|\boldsymbol{x}|^2}{s}\right)^{-\frac{n}{2}} e^{-s}\,\frac{|\boldsymbol{x}|^2}{4s^2}\,ds\\ &= \frac{1}{4}\pi^{-\frac{n}{2}}\,|\boldsymbol{x}|^{2-n}\int_0^\infty s^{\frac{n}{2}-2}e^{-s}\,ds.\end{aligned}$$

In terms of the gamma function (2.17) this calculation gives the fundamental solution

$$\Phi(\boldsymbol{x}) = \frac{1}{4}\pi^{-\frac{n}{2}}\Gamma(\tfrac{n}{2}-1)\,|\boldsymbol{x}|^{2-n}. \tag{13.21}$$

This agrees with the formula for Φ from Theorem 12.8, because

$$A_n = \frac{2\pi^{\frac{n}{2}}}{\Gamma(\frac{n}{2})}$$

and

$$\Gamma(\tfrac{n}{2}-1) = \frac{\Gamma(\frac{n}{2})}{\frac{n}{2}-1}.$$

◊

13.3 The Wave Kernel

Following the discussion in Sect. 12.6, we can try to define the wave kernel W_t on $\mathbb{R}^n$ by solving the distributional equations

$$\left(\frac{\partial^2}{\partial t^2} - \Delta\right) W_t = 0, \quad W_0 = 0, \quad \left.\frac{\partial W_t}{\partial t}\right|_{t=0} = \delta. \tag{13.22}$$

If we assume that $W_t \in \mathcal{S}'(\mathbb{R}^n)$, then the spatial Fourier transform allows us to analyze this equation by turning it into a simple ODE.

For each t define $\hat{W}_t \in \mathcal{S}'(\mathbb{R}^n)$ by the (spatial) distributional transform (13.16). By Lemma 13.6 and (13.19), (13.22) transforms to

$$\left(\frac{\partial^2}{\partial t^2} + |\xi|^2\right) \hat{W}_t = 0, \quad \hat{W}_0 = 0, \quad \left.\frac{\partial \hat{W}_t}{\partial t}\right|_{t=0} = 1.$$

The unique solution to this ODE is

$$\hat{W}_t(\xi) = \begin{cases} \frac{\sin(t|\xi|)}{|\xi|}, & \xi \neq 0, \\ t, & \xi = 0. \end{cases} \tag{13.23}$$

The function $\hat{W}_t$ is smooth and bounded, and therefore defines a tempered distribution on $\mathbb{R}^n$. The inverse Fourier transform $W_t \in \mathcal{S}'(\mathbb{R}^n)$ thus yields a general solution formula for the wave equation on $\mathbb{R}^n$. For initial conditions $g, h \in \mathcal{S}(\mathbb{R}^n)$,

$$u(t, \cdot) = \frac{\partial W_t}{\partial t} * g + W_t * h. \tag{13.24}$$

The direct computation of the inverse Fourier transform of (13.23) is rather tricky, but we can check this formula against the results we already know. For $n = 1$ we have $W_t = \frac{1}{2}\chi_{[-t,t]}$ from the d'Alembert formula. Since this is integrable the Fourier transform can be computed directly:

$$\begin{aligned} \hat{W}_t(\xi) &= \int_{-\infty}^{\infty} \frac{1}{2}\chi_{[-t,t]}(x)\, e^{-ix\xi}\, dx \\ &= \frac{1}{2}\int_{-t}^{t} e^{-ix\xi}\, dx \\ &= \begin{cases} \frac{\sin(t\xi)}{\xi}, & \xi \neq 0, \\ t, & \xi = 0. \end{cases} \end{aligned}$$

For $n = 3$, the Kirchhoff formula from Theorem 4.10 shows that the wave kernel is the distribution defined by

$$(W_t, \psi) := \frac{1}{4\pi t}\int_{\partial B(0;t)} \psi \, dS,$$

for $\psi \in \mathcal{S}(\mathbb{R}^3)$. By definition, the Fourier transform is given by

$$\begin{aligned}\left(\hat{W}_t, \psi\right) &:= \frac{1}{4\pi t}\int_{\partial B(0;t)} \hat{\psi}(\boldsymbol{x})\, dS(\boldsymbol{x}) \\ &= \frac{1}{4\pi t}\int_{\partial B(0;t)} \left(\int_{\mathbb{R}^3} e^{-i\boldsymbol{x}\cdot\boldsymbol{\xi}}\psi(\boldsymbol{\xi})\, d^3\boldsymbol{\xi}\right) dS(\boldsymbol{x}).\end{aligned}$$

Since $\psi(\boldsymbol{y})$ has rapid decay as $\boldsymbol{y} \to \infty$ and the $\boldsymbol{x}$ integral is restricted to a sphere, we can switch the order of integration and conclude that

$$\hat{W}_t(\boldsymbol{\xi}) = \frac{1}{4\pi t}\int_{\partial B(0;t)} e^{-i\boldsymbol{x}\cdot\boldsymbol{\xi}}\, dS(\boldsymbol{x}).$$

To compute this surface integral, note that we could rotate the $\boldsymbol{x}$ coordinate without changing the result of the integration. It therefore suffices to consider the case where $\boldsymbol{\xi}$ is parallel to the x_3 axis. If we then use the spherical coordinates (r, θ, ϕ) for the $\boldsymbol{x}$ variables, this gives

$$\boldsymbol{x}\cdot\boldsymbol{\xi} = |\boldsymbol{\xi}|\, r\cos\phi.$$

For the surface integral at radius $r = t$,

$$dS(\boldsymbol{x}) = t^2 \sin\phi \, d\phi \, d\theta.$$

The Fourier transform is thus

$$\begin{aligned}\hat{W}_t(\boldsymbol{\xi}) &= \frac{1}{4\pi t}\int_0^{2\pi}\int_0^{\pi} e^{-it|\boldsymbol{\xi}|\cos\theta}\, t^2 \sin\phi\, d\phi\, d\theta \\ &= \frac{t}{2}\int_0^{\pi} e^{-it|\boldsymbol{\xi}|\cos\theta} \sin\phi\, d\phi.\end{aligned}$$

With the substitution $u = \cos\phi$ this becomes

$$\begin{aligned}\hat{W}_t(\boldsymbol{\xi}) &:= \frac{t}{2}\int_{-1}^{1} e^{-it|\boldsymbol{\xi}|u}\, du \\ &= \begin{cases}\frac{\sin(t|\boldsymbol{\xi}|)}{|\boldsymbol{\xi}|}, & \boldsymbol{\xi} \neq 0, \\ t, & \boldsymbol{\xi} = 0.\end{cases}\end{aligned}$$

Hence the Kirchhoff formula agrees with the transform solution (13.23).

13.4 The Heat Kernel

By analogy with (13.22), the *heat kernel* H_t is defined as the solution of the distributional equation

$$\left(\frac{\partial}{\partial t} - \Delta\right) H_t = 0, \qquad H_0 = \delta. \tag{13.25}$$

Assuming $H_t \in \mathcal{S}'(\mathbb{R}^n)$, let $\hat{H}_t$ denote the spatial Fourier transform of H_t. By Lemma 13.6 and (13.19), (13.25) transforms to

$$\left(\frac{\partial}{\partial t} + |\boldsymbol{\xi}|^2\right) \hat{H}_t = 0, \qquad \hat{H}_0 = 1.$$

This simple ODE has the unique solution

$$\hat{H}_t(\boldsymbol{\xi}) = e^{-t|\boldsymbol{\xi}|^2}. \tag{13.26}$$

Because $\hat{H}_t$ is a Schwartz function for $t > 0$, we can compute the inverse Fourier transform by the direct integral formula (13.11), which gives

$$H_t(\boldsymbol{x}) = (2\pi)^{-n} \int_{\mathbb{R}^n} e^{i\boldsymbol{\xi}\cdot\boldsymbol{x}} e^{-t|\boldsymbol{\xi}|^2}\, d^n\boldsymbol{\xi}.$$

According to (13.8), this inverse transform is

$$H_t(\boldsymbol{x}) = (4\pi t)^{-\frac{n}{2}} e^{-|\boldsymbol{x}|^2/4t}. \tag{13.27}$$

In Sect. 6.3 we guessed this formula from a calculation in the one-dimensional case. The Fourier transform allows for a systematic derivation.

13.5 Exercises

13.1 Let $\mathbb{H} \subset \mathbb{R}^2$ denote the upper half space $\{x_2 > 0\}$. The Poisson kernel on $\mathbb{H}$ is the distributional solution of the equation

$$\Delta P = 0, \qquad P|_{x_2=0} = \delta.$$

(a) Let $\hat{P}(\xi, x_2)$ denote the distributional Fourier transform of P with respect to the x_1 variable. Find the corresponding equation for $\hat{P}$.
(b) Show that the unique solution of the ODE with $\hat{P}(\cdot, x_2) \in \mathcal{S}'(\mathbb{R})$ is the function

$$\hat{P}(\xi, x_2) = e^{-x_2|\xi|}.$$

(c) Compute the inverse transform to show that

$$P(x) = \frac{x_2}{\pi(x_1^2 + x_2^2)}.$$

(d) For $f \in \mathcal{S}(\mathbb{R})$, use P to write an integral formula for the solution of the Laplace problem on $\mathbb{H}$:

$$\Delta u = 0, \qquad u|_{x_2=0} = f.$$

13.2 For $\psi \in \mathcal{S}(\mathbb{R})$, the *Poisson summation formula* says that

$$\sum_{k=-\infty}^{\infty} \psi(k) = \sum_{m=-\infty}^{\infty} \hat{\psi}(2\pi m).$$

Derive this formula using the steps below.

(a) Define a periodic function $f \in C^\infty(\mathbb{T})$ (where $\mathbb{T} := \mathbb{R}/2\pi\mathbb{Z}$ as in Sect. 8.2) by averaging $\hat{\psi}$,

$$f(x) := \sum_{m=-\infty}^{\infty} \hat{\psi}(x + 2\pi m).$$

Show that

$$c_k[f] = \psi(-k).$$

(b) Obtain the summation formula by comparing f to its Fourier series expansion at $x = 0$.

13.3 Recall that the heat equation on $\mathbb{T}$ was solved by Fourier series in Theorem 8.13.

(a) Use the solution formula (8.44) to show that the heat kernel on $\mathbb{T}$ is given by the series

$$h_t(x) := \frac{1}{2\pi} \sum_{k=-\infty}^{\infty} e^{-k^2 t + ikx}.$$

(b) Use the Poisson summation formula from Exercise 13.2 to show that the periodic heat kernel h_t and the heat kernel H_t on $\mathbb{R}$ are related by averaging

$$h_t(x) = \sum_{m=-\infty}^{\infty} H_t(x + 2\pi m)$$

for $t > 0$. (Note that this shows $h_t(x) > 0$ for all $x \in \mathbb{T}$, $t > 0$, which is not clear in the formula from (a).)

13.4 The Schrödinger equation on $\mathbb{R}^n$,

$$-i\frac{\partial u}{\partial t} - \Delta u = 0, \qquad u|_{t=0} = g,$$

was introduced in Exercise 4.7.

(a) Assuming that $g \in \mathcal{S}(\mathbb{R}^n)$, find a formula for the spatial Fourier transform $\hat{u}(t, \xi)$.
(b) Show that the result from Exercise 4.7,

$$\int_{\mathbb{R}^n} |u(t, \boldsymbol{x})|^2 \, d^n\boldsymbol{x} = \int_{\mathbb{R}^n} |g|^2 \, d^n\boldsymbol{x}$$

for all $t \geq 0$, follows from the Plancherel theorem (Theorem 13.5).

Erratum to: Introduction to Partial Differential Equations

Erratum to:
D. Borthwick, *Introduction to Partial Differential Equations*, Universitext, http://doi.org/10.1007/978-3-319-48936-0

In the original version of the book, the belated corrections from author for Chaps. 2, 3, 4, 6 and 11 have been incorporated.

The updated online version of these chapters can be found at
http://doi.org/10.1007/978-3-319-48936-0
http://doi.org/10.1007/978-3-319-48936-0_2
http://doi.org/10.1007/978-3-319-48936-0_3
http://doi.org/10.1007/978-3-319-48936-0_4
http://doi.org/10.1007/978-3-319-48936-0_6
http://doi.org/10.1007/978-3-319-48936-0_11

D. Borthwick, *Introduction to Partial Differential Equations*,
Universitext, http://doi.org/10.1007/978-3-319-48936-0_14

Appendix A
Analysis Foundations

In this section we will develop some implications of the completeness axiom for $\mathbb{R}$ which are referenced in the text.

The fundamental result from which the others follow is the equivalence of compactness and sequential compactness for subsets of $\mathbb{R}^n$. Recall from Sect. 11.6 that a set A is sequentially compact if every sequence within A contains a subsequence converging to a limit in A. The equivalence was first proven by Bernard Bolzano in the early 19th century, and later rediscovered by Karl Weierstrass.

Theorem A.1 (Bolzano-Weierstrass) *In $\mathbb{R}^n$ a subset is sequentially compact if and only if it is closed and bounded.*

Proof If $A \subset \mathbb{R}^n$ is unbounded, then there exists a sequence of points $\{\boldsymbol{x}_j\} \subset A$ with $|\boldsymbol{x}_j| \to \infty$. Any subsequence has the same property, so $\{\boldsymbol{x}_j\}$ has no convergent subsequence. If A is not closed, then there is some $\boldsymbol{w} \notin A$ which is a boundary point of A. Every neighborhood of $\boldsymbol{w}$ thus includes points of A, so there exists a sequence $\{\boldsymbol{x}_j\} \subset A$ converging to $\boldsymbol{w}$. All subsequences of $\{\boldsymbol{x}_j\}$ also converge to $\boldsymbol{w}$, and therefore no subsequence converges in A. We conclude that a sequentially compact subset of $\mathbb{R}^n$ is closed and bounded.

For the converse argument, let us first consider the one-dimensional case. Let $\{x_j\}$ be a sequence in a bounded set $A \subset \mathbb{R}$. For each n the real number

$$b_n := \sup \{x_k;\ k \geq n\} \tag{A.1}$$

exists by the completeness axiom. The sequence $\{b_n\}$ is decreasing, because the supremum is taken over successively smaller sets, and also bounded by the hypothesis on A. Therefore the number

$$\alpha := \inf_{n \in \mathbb{N}} b_n$$

is well-defined in $\mathbb{R}$. The fact that $\{b_n\}$ is decreasing implies $b_n \geq \alpha$ for all n.

We claim that a subsequence of x_k converges to α. For this purpose it suffices to show that the interval $(\alpha - \varepsilon, \alpha + \varepsilon)$ contains infinitely many x_k for each $\varepsilon > 0$.

D. Borthwick, *Introduction to Partial Differential Equations*,
Universitext, DOI 10.1007/978-3-319-48936-0

If this were not the case, then for some n we would have $x_k \notin (\alpha - \varepsilon, \alpha + \varepsilon)$ for all $k \geq n$. This would imply either $b_n \leq \alpha - \varepsilon$ or $b_n \geq \alpha + \varepsilon$, both of which are impossible by the definition of α.

This proves the existence of a subsequence converging to α. The fact that A is closed implies $\alpha \in A$, so this completes the argument that a closed bounded subset of $\mathbb{R}$ is sequentially compact.

To extend this argument to higher dimensions, consider a sequence $\{\boldsymbol{x}_k\}$ in a compact subset $A \subset \mathbb{R}^n$. The sequence of first coordinates of the $\boldsymbol{x}_k$ is a bounded sequence in $\mathbb{R}$, so the above argument yields a subsequence such that the first coordinates converge. We can then restrict our attention to this subsequence and apply the same reasoning to the second coordinate, and so on. After n steps this procedure produces a subsequence which converges to an element of A. □

Bolzano used sequential compactness to prove the following result, which serves as the foundation for applications of calculus to optimization problems.

Theorem A.2 (Extreme value theorem) *For a compact set $K \subset \mathbb{R}^n$, a continuous function $K \to \mathbb{R}$ achieves a maximum and minimum value on K.*

Proof Assume that $f : K \to \mathbb{R}$ is continuous. We will show first that f is bounded. Suppose there is a sequence $\boldsymbol{x}_j \in K$ such that $|f(\boldsymbol{x}_j)| \to \infty$. By Theorem A.1, after restricting to a subsequence if necessary, we can assume that $\boldsymbol{x}_j \to \boldsymbol{w} \in K$. Continuity implies $f(\boldsymbol{x}_j) \to f(\boldsymbol{w})$, but this is impossible if $|f(\boldsymbol{x}_j)| \to \infty$. Therefore a continuous function on K is bounded.

Since $f(K)$ is a bounded subset of $\mathbb{R}$, $b := \sup f(K)$ exists in $\mathbb{R}$ by the completeness axiom. To prove that f achieves a maximum, we need to show $b \in f(K)$. If $b \notin f(K)$ then the function

$$h(\boldsymbol{x}) := \frac{1}{b - f(\boldsymbol{x})}$$

is continuous on K, and therefore bounded by the above argument. However, $h(\boldsymbol{x}) \leq M$ for $\boldsymbol{x} \in K$ would imply that $\sup f(K) \leq b - 1/M$, contradicting the definition of b. Therefore $b \in f(K)$, so f achieves a maximum. A similar argument applies to the minimum. □

The final result is the completeness of $\mathbb{R}^n$ as a normed vector space, as noted in Sect. 7.4.

Theorem A.3 *In $\mathbb{R}^n$ a sequence converges if and only if it is Cauchy.*

Proof We have already noted that a convergent sequence is Cauchy in a normed vector space. Suppose that $\{\boldsymbol{x}_k\}$ is a Cauchy sequence in $\mathbb{R}^n$. This implies in particular that the sequence is bounded. Therefore, by Theorem A.1, there exists a subsequence converging to some $\boldsymbol{w} \in \mathbb{R}^n$.

By the definition of Cauchy, for $\varepsilon > 0$ there exists N sufficiently large such that

$$\left|\boldsymbol{x}_j - \boldsymbol{x}_k\right| < \varepsilon$$

for all $j, k \geq N$. We can also choose an element $\boldsymbol{x}_l$ in the subsequence such that $l \geq N$ and

$$|\boldsymbol{x}_l - \boldsymbol{w}| < \varepsilon.$$

The triangle inequality then gives

$$\begin{aligned} \left|\boldsymbol{x}_j - \boldsymbol{w}\right| &\leq \left|\boldsymbol{x}_j - \boldsymbol{x}_l\right| + |\boldsymbol{x}_l - \boldsymbol{w}| \\ &< 2\varepsilon, \end{aligned}$$

for all $j \geq N$. Since the choice of ε was arbitrary, this shows that the full sequence converges to $\boldsymbol{w}$. □

References

For additional analysis background

Apostol, T. M. (1974). Mathematical analysis (2nd ed.). Addison-Wesley.
Edwards, R. E. (1979). Fourier series. A modern introduction (2nd ed.). Graduate Texts in Mathematics 64. Springer.
Folland, G. B. (1999). Real analysis (2nd ed.). Pure and Applied Mathematics. Wiley.
Rudin, W. (1976). Principles of mathematical analysis (3rd ed.). McGraw-Hill.
Stein, E. M., & Shakarchi, R. (2003). *Fourier analysis*. Princeton Lectures in Analysis I: Princeton.
Stein, E. M., & Shakarchi, R. (2005). *Real analysis*. Princeton Lectures in Analysis III: Princeton.
Stein, E. M., & Shakarchi, R. (2011). *Functional analysis*. Princeton Lectures in Analysis IV: Princeton.
Strichartz, R.S. (2003). *A Guide to Distribution Theory and Fourier Transforms*. World Scientific.

For further study of PDE

DiBenedetto, E. (2010). Partial differential equations (3nd ed.). Birkhuser.
Evans, L. C. (2010). Partial differential equations (2nd ed.). Graduate Studies in Mathematics 19. American Mathematical Society.
Gilbarg, D., & Trudinger, N. S. (2001). *Elliptic partial differential equations of second order*. Classics in Mathematics: Springer.
Jost, J. (2013). Partial differential equations (3rd ed.). Graduate Texts in Mathematics 214. Springer.
Strauss, W. A. (2008). Partial differential equations. An introduction (2nd ed.). Wiley.
Taylor, M. E. (2011). Partial differential equations I. Basic theory, Applied Mathematical Sciences 115. Springer.
Vasy, A. (2015). Partial differential equations, Graduate Studies in Mathematics 169. American Mathematical Society.

D. Borthwick, *Introduction to Partial Differential Equations*,
Universitext, DOI 10.1007/978-3-319-48936-0

Index

D. Borthwick, *Introduction to Partial Differential Equations*,
Universitext, DOI 10.1007/978-3-319-48936-0

Printed in the United States
By Bookmasters